Contemporary France

Hilary P. M. Winchester

Longman Scientific & Technical,
Longman Group UK Limited,
Longman House, Burnt Mill, Harlow,
Essex, CM20 2JE, England
and Associated Companies throughout the world.

Copublished in the United States with John Wiley & Sons, Inc., 605 Third Avenue, New York, NY 10158

First published 1993

ISBN 0-582-30534-9

British Library Cataloguing in Publication Data
A CIP record for this book is available from the British Library

Library of Congress Cataloging-in-Publication Data

A CIP record for this book is available from the Library of Congress

Set by 13 in 10/11pt Palatino

Printed in Hong Kong
WC/01

For my mother Mary and in
memory of my father Norman
and many happy holidays
in an ever-sunny France

And then, as our talk ran on, and it turned out that I was not a pedlar, but a literary man, who drew landscapes and was going to write a book, he changed his manner of thinking as to my reception . . . Might he say that I was a geographer?
No, I thought, in the interests of truth, he positively might not.
'Very well then' (with disappointment), 'an author.'

Stevenson (1879:84-5)

Contents

Preface

France approaches the year 2000 as a major global economic power, ranked fourth or fifth in the world as an industrial and a trading nation. It is a political and military power of immense significance. This stems partly from its widespread overseas territories, strategically located in every ocean, but also from its status as a major nuclear power, and above all from its role as a founder nation and defender of the European Community. France also plays an important role as a bastion of history, culture, intellectual life and fashion. Within France, the focus of this economic, political and cultural activity lies in Paris, in many ways the real capital of Europe.

There is a long tradition of French regional geography. Classically, this has consisted of highly detailed accounts of the *pays*, the small-scale, self-sufficient landscapes of a pre-industrial nation. Much traditional French regional geography has emphasized the differences between areas, the interaction of people with their environment in the formation of cultural landscapes, and the role of history in shaping geography. In the last twenty years, however, regional geography has taken a back seat to the systematic branches of the subject, and the real world has often been used as a laboratory to test hypotheses and relationships derived from theory. This book draws on both the old and the new approaches; it is a geography of contemporary France, a human geography profoundly influenced by both its physical background and its history. The individuality of French culture and identity is always emphasized, but the generality of trends and processes operating in an era of global capitalism is also acknowledged. This book is intended for the undergraduate wishing to gain an insight into the complexity and uniqueness of a major European nation. It is a distillation derived from numerous original French sources and from many years of research.

The first chapter, on the resources and hazards of the physical environment, is essential material for an understanding of the geography of contemporary France. A brief overview of the fundamentals of geology, soils and climate is followed by an account of some of the physical resources which fashioned early agricultural development, settlement patterns and industrial location. It is recognized, however, that physical-human actions are reciprocal, and that human activity has influenced and altered the physical environment. Accordingly, this chapter deals at some length with issues such as urban climates, water pollution, soil erosion and forest damage, all of which have recently become political issues of national and supranational importance. Many of the books on France published in English in the last ten years have either ignored or skated over any mention of the physical environment on the premise that human geography is now almost totally divorced from physical geography. Such a view is at best a partial one, and a more complete appreciation of France can only be derived from a broader understanding of its physical resources and limitations.

Chapters 2 to 4 focus in turn on population, agriculture and industry. Each of these chapters outlines both pre-war and post-war developments as a context for understanding present trends. In each case, both the distinctiveness and the generality of current patterns are emphasized. French demography has undergone a distinctive history, characterized by a very early fertility decline. This has had a profound impact on attitudes to population growth, which are much more markedly pronatalist than those of other countries in a similar stage of development. Nonetheless, recent patterns of fertility decline, retirement migration and counterurbanization are similar to those experienced in other parts of the developed world.

Agriculture and industry are both clearly related to the resources of the physical environment. These chapters may be approached as free-standing essays, although cross-references to other relevant basic information are also included. The development of agriculture and industry represents a key element in the modernization of the French economy. This modernization process has been the focus of much research; many writers have traditionally emphasized the late development of a modern industrial economy and in so doing have either implicitly or explicitly compared the evolution of France with that of Britain. This book, although inevitably drawing some of the same comparisons, attempts to place the modernization of France in the context of its resources and population rather than viewing it simply as a late comer to a process of industrialization which was unique in world history. These chapters emphasize the importance of French agricultural and industrial production for the European Community, and the recent changes which are being brought about by international forces, including European Community policy and the new international division of labour. Some topics in the economic geography of France are considered relatively briefly because of constraints on the size of this volume and because these topics have been given relatively more attention by other recent authors (see, for example, Tuppen, 1983; 1991; Flockton and Kofman, 1989). Instead, this volume adopts a broader perspective ranging from the environmental to the geopolitical.

Chapters 5 and 6, perhaps more than any others, emphasize the features of French human geography which are unique because of their location and history. Chapter 5 is concerned with the urban network of provincial France, while the whole of Chapter 6 is devoted to the French capital, Paris. Most of the French population is urban, and almost a fifth lives in Paris. The pressure of urban living has spawned numerous planning and regulatory mechanisms, for controlling or promoting growth, for new towns, for conservation of heritage areas, and for the redevelopment and improvement of problem and derelict areas. These mechanisms and their impact are considered in some detail since they affect the living environment of most French people, and because they have been instrumental in maintaining Paris as the only French Euro-city.

The final chapter stresses both the global and European significance of France, and some of the internal tensions induced by these supranational roles. It also tackles the complex problem of the maintenance of French national identity amid regional linguistic and cultural diversity. The role of the former French colonial possessions, the possibility of closer economic integration in Europe, and the control of militant separatism in the Basque Country are all issues which affect the individuality of French identity and culture. They are also issues which are replicated elsewhere, and which, in affecting the geography of France today, spread their tentacles into the geography of the contemporary world.

Acknowledgements

This book would never have been started without the interest in and affection for France which was stimulated by my home environment, by my mother's early upbringing in Paris, and by numerous long summer holidays taken as a family. I am grateful for the support offered by my family over the years, particularly from my sisters, Anne-Marie Sutcliffe and Helen Russell Johnson, my brother-in-law, Victor Sutcliffe and my brother, David Mutton. I have dedicated this book to my parents.

I was drawn to undertake my postgraduate work on France, doubtless influenced also by the strong tradition of French geographical studies at Oxford. I am grateful to the late John House who was my supervisor during most of this period, and to other members of staff, particularly Jean Gottmann and Ceri Peach, who both undertook periods of supervision.

My postgraduate colleagues in the research room at Oxford at this time were a source of continued encouragement and practical assistance. The academic relationships and friendships formed at that time have endured a number of years and substantial distances: I owe a special debt of gratitude to Paul White who encouraged me back into the academic fray after some time out, and with whom I shared a number of productive field trips to Paris. I would also like to thank Philip Ogden for his friendship and deep knowledge of France.

I thank the staff of Longman Group UK Ltd for their patience and encouragement; I hope they think the wait was worthwhile. I am greatly indebted to Hugh Clout, who read and commented on early drafts of the whole manuscript, and who made many prompt and valuable suggestions.

My colleagues at The University of Newcastle, New South Wales, where the book was finally completed, have provided invaluable support and practical help. The maps were drawn by Colin Harden, some based on the earlier work of Jenny Wyatt. Some of the chapters were retyped by Sharon Francis, and editorial assistance was provided by Barbara Whitelock and Linda Peady. Some of my postgraduate students and academic colleagues, notably Lauren Costello, Kevin Dunn and Phillip O'Neill, provided academic stimulation and gentle nagging.

Several people helped me with specific parts of the text. I should like to thank Ann Young for her comments on hazards and resources; Peter Rickard for his helpful suggestions on linguistic variations in France; Mary Mutton for assistance with French language; and Yvan Chauviré and

Daniel Noin for their insights into Paris. Financial support from the British Academy and from the ESRC/CNRS enabled me to undertake specific research topics included in Chapters 6 and 7.

In particular, my thanks go to Stephen Gale for his continued good humour, interest and practical help at all stages of the manuscript. He took most of the photographs, read and commented on every chapter in a meticulous and erudite manner, assisted with proofreading, and provided both emotional and practical support. Without him, this book would never have been completed.

Note on place names and foreign words

The French has generally been used for place names of cities and regions. Other French words are explained in the text the first time that they are used. Those foreign words and abbreviations used frequently throughout the text are compiled in the glossary (pages 246–49).

We are grateful to the following for permission to reproduce copyright material:

American Meteorological Society for Fig. 1.4 (Dettwiller, 1970); Paul Chapman Publishing Ltd for Fig. 7.3 (Flockton and Kofman, 1989); Éditions Payot & Rivages for part Table 7.1 (Dauzat, 1953); Electricité de France for Figs 4.4 & 4.5 (Electricité de France, 1991); Elsevier Science Publishers BV for part Fig. 1.3. (Arléry, 1970); Dr S.J. Gale for Plates 3.1, 3.2, 4.1, 4.2, 4.3, 5.1, 5.2, 6.1, 6.3, 6.4 & 7.1; The Geographical Association for Fig. 2.5 (Dean, 1986); Helen Dwight Reid Educational Foundation. Published by Heldref Publications for Table 1.1 (Dovland, 1987) Copyright © 1987; Hodder & Stoughton Ltd. for part Table 2.5 (Dyer, 1987) Copyright © 1978 Colin Dyer; Institut National d'Études Démographiques for part Table 2.3 (Pourcher, 1964); Longman Group UK Ltd. for part Fig. 1.3 (Beaujeu-Garnier, 1975) & Fig. 5.4 (White, 1984); Macmillan Press Ltd. for Fig. 1.6 (Waltham, 1978) & part Fig. 7.4 (Offord, 1990); S.A. Masson for Fig 6.7 (Noin *et al.*, 1984); A. Metton for part Fig. 1.7 (Dézert, 1981); the author, Dr. E.L. Naylor for Fig. 3.6 (Naylor, 1985); *Norois* for Fig. 4.7 (Jeanneau, 1989); the author, Prof. P. Pinchemel for Fig. 7.1. (Pinchemel, 1986); Presses Universitaires de France for Table 2.1 (Huber, 1931); Routledge for part Table 7.1 (Rickard, 1989); Société Languedocienne de Géographie for Fig. 5.6 (Merlin, 1986); Université Louis Pasteur (Strasbourg) for Fig. 1.2 (Pihan, 1979); the author, Dr P.E. White for part Fig. 6.3 (White), Figs 6.6 & 6.8 (White); Plate 6.2 and permission to cite unpublished paper (White, 1991); V.H. Winston & Son Inc. for Figs 5.7 & 5.8 (Pumain and St Julien, 1984).

Whilst every effort has been made to trace the owners of copyright material, in a few cases this has proved impossible and we take this opportunity to offer our apologies to any copyright holders whose rights we may have unwittingly infringed.

CHAPTER 1

The physical environment: resources and hazards

'What is that big book?' said the little prince. 'What are you doing?'
'I am a geographer,' said the old gentleman.
'What is a geographer?' asked the little prince.
'A geographer is a scholar who knows the location of all the seas, rivers, towns, mountains, and deserts.'
'That is very interesting,' said the little prince. 'Here at last is a man who has a real profession!' And he cast a look around him at the planet of the geographer. It was the most magnificent and stately planet that he had ever seen.

Saint-Exupéry (1943)

The physical and human environment

This chapter aims to provide a skeleton of the physical environment of France, necessary for the understanding of contemporary social and economic structures. Three major aspects of the physical environment are considered. First, the location and extent of primary resources such as soils, water and mineral reserves are outlined. Many manufacturing industries and settlements were established at a time when the physical environment controlled and constrained location more obviously than at present. Nonetheless, the availability of primary resources continues to influence their location and character. Much of the human geography of contemporary France can only be understood with reference to its history and hence to its physical geography; Braudel in his work *The identity of France* (1988) summarized this interplay in his first volume subtitled *history and environment*.

The second aspect of the physical environment considered in this chapter is the impact of natural and induced hazards on human activity. Natural hazards such as avalanches, floods and forest fires are important processes affecting the physical environment, many with severe local and regional effects. Some hazards, including soil erosion, air and water pollution and climatic changes, may be induced or exacerbated by human

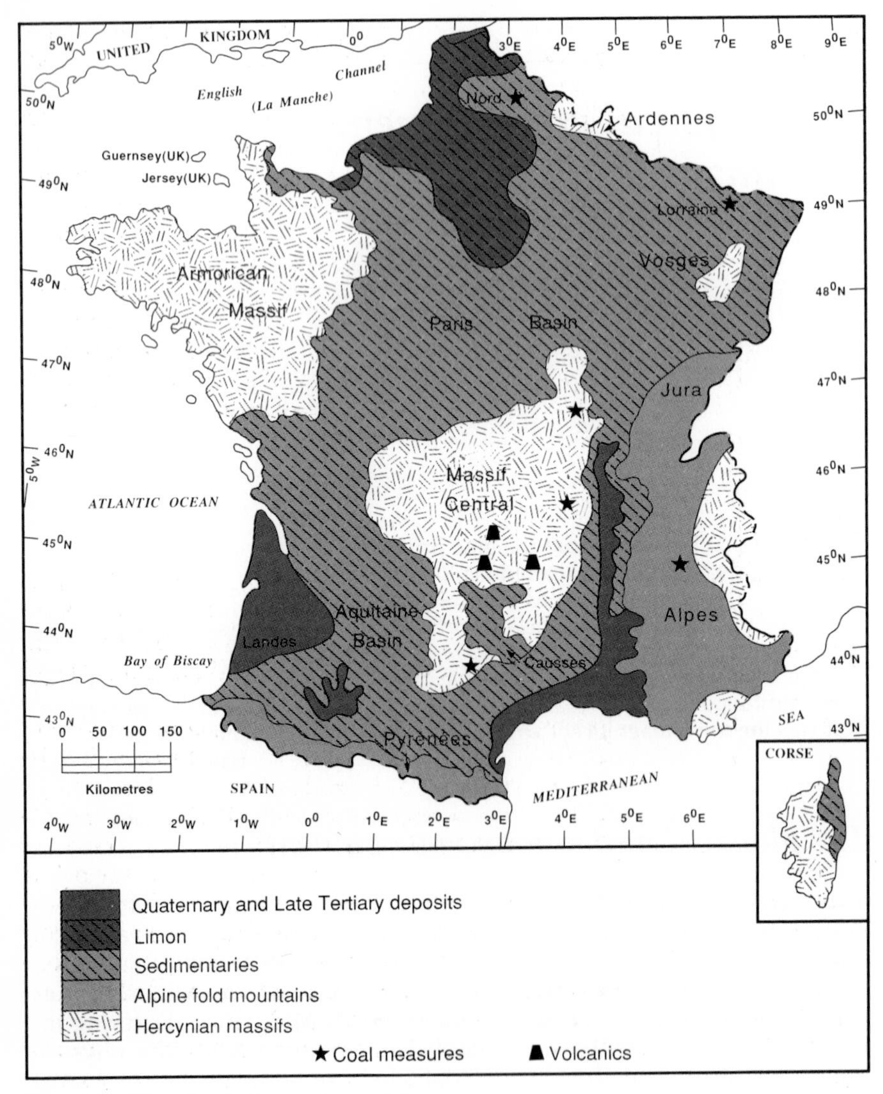

Fig. 1.1 The physical structure of France

behaviour. Many of these hazards may have effects which are felt beyond local and national boundaries.

Thirdly, the physical environment as a political arena is considered, and the politics and policies of environmental conservation briefly outlined. World-wide, the pressure for conservation has intensified since the 1970s because of concern over the greenhouse effect and the potential accompanying environmental changes such as rising sea levels and climatic deterioration. Locally, pressure for conservation of sensitive and scenic areas is evident as holiday homes and tourist resorts encroach along the coastline and into mountain areas. The Green movement has focused an increasing awareness of the sensitivity of the physical environment and of its irreplaceable nature into a political force which has helped bring about recent legislation for active conservation.

The underlying assumption of this chapter therefore is that the physical environment of a nation, its resources, its hazards and its potential, still affect the distribution and nature of population, settlement and economic activity. This is despite the apparently increasing isolation of contemporary urban and industrial society from its physical context. The relationship is not a directly deterministic one, but is tempered by history and leavened by chance. Nor is it one-sided. The challenge to understand both the constraints of the physical environment on contemporary society, and the impact of human activity on that environment, is one that has been stimulated by growing awareness of the fragility of ecological resources, the threats posed by their pollution and overuse and the necessity for conservation.

Physical structure

France consists of three main physical regions; these are the lowland basins, the older Hercynian uplands and plateaus which flank those basins, and the high Alpine mountains and valleys of the Alps and Pyrenees (Fig. 1.1). (For fuller accounts of geology and geomorphology see Pomerol, 1980; Embleton, 1984.)

The lowland basins include most of northern and western France with the exception of the Cotentin and Breton peninsulas. Structurally, the basins are continuations of the north European lowlands, and their highest point is below 600 m. The two major basins are the Paris Basin and the Aquitaine Basin, which are structural synclines separated from each other by the Poitou Sill, and from the Flemish lowland to the north by the Artois Anticline. The Paris Basin consists of an eroded syncline with a series of plains, low plateaus and shallow valleys separated by outward-facing escarpments. The basin consists of a Tertiary core underlain and surrounded by Cretaceous chalks and Jurassic sandstones, marls and limestones, with crystalline Hercynian massifs at its edge.

The Paris Basin did not experience Quaternary glacial activity, but was markedly affected by periglacial processes which have smoothed slopes by mass movement and mantled surfaces with loess, the thickest layers of which remain on the east- or southeast-facing slopes. Intense human activity since the Middle Ages, especially forest clearance, agriculture and the diversion of drainage, has profoundly affected the landscape. As Joly (1984:161) has written 'it is virtually impossible to find any part of the landscape which does not bear the human imprint' throughout France. The variation in lithology, topography and agricultural potential has given rise to a variety of cultural landscapes differentiated by local building stone and by the farming technology which could be used on different soils. These *pays*, which absorbed and delighted the regional geographers of the past (see, for example, Houston, 1953), have become more uniform as a result of ubiquitous agricultural technology and recent urbanization of the countryside, but nonetheless still form some distinctive landscapes and agricultural regions.

The Aquitaine Basin forms the most southwesterly part of the European lowland, lying between the Bay of Biscay to the west and Hercynian massifs to the east. The northern part of the Basin is similar in character to the Paris Basin, consisting predominantly of limestone giving way to younger rocks further south. However, the limestones of Périgord and Quercy give rise to karstic features; closed depressions, dry valleys, caves and gorges form a landscape quite different to that of the Paris Basin. Furthermore, the structure of the Aquitaine Basin is more faulted and fragmented than that of the Paris Basin and has fewer well-defined escarpments. The southern part of the Aquitaine Basin between the Garonne and the Pyrenees is again different in character, consisting of fluvial, marine and glaciofluvial material. This area is sheltered by the Landes, one of the largest sand-dune systems in Europe, which extends for over 200 km along the Atlantic Coast (Fig. 1.1). The sand dunes, which have built up since the mid-Holocene rise in sea level, have ponded large quantities of water to form shallow lagoons known as *étangs*. Since the 1960s, many of these *étangs* have been developed as tourist resorts. The synclinal structures of this part of southwest France provide reservoirs for natural gas, which have been exploited since the 1950s (see pages 28–9).

The Hercynian uplands which flank the lowland basins form a series of plateaus planed by erosion and punctuated by volcanic peaks. The largest and highest of these uplands is the Massif Central, consisting of an uplifted dome composed mainly of crystalline rocks, the highest point of which is formed by the volcanic structures of the Monts Dore (1886 m). The volcanoes are extinct, but the existence of geothermal energy is evident from the frequent occurrence of hot springs used as spas since the last century and more systematically exploited for heating purposes since the 1980s. A major tear fault running from Toulouse to Nevers, known as the Coal Furrow, separates the lower, less fractured western region of the Limousin from the eastern, broken volcanic region of the Auvergne-Cévennes. This tear fault also exposes mineral deposits, which have been exploited at sites such as St Etienne, Autun and Le Creusot. The southeast of this region consists of thick beds of limestones into which enormous canyons have been eroded; the gorges of the Causses and the volcanic intrusions further north combine to make the area of the Massif Central both highly individual and formerly inaccessible. A large part of this region has now been designated as National Park (see pages 31–3; Fig. 1.8).

The other Hercynian uplands are lower and have fewer volcanic features. The Vosges in eastern France and the Ardennes in the north both consist of platform blocks, with dissected plateaus and resistant ridges, those of the Vosges rising to over 1400 m. The Armorican or Breton massif in western France is much lower, rising to only 417 m; it is also relatively undissected but it is exposed to the sea on three sides. This gives rise to spectacular coastal scenery in Bretagne, with a detailed pattern complicated by differential erosion and Holocene changes in sea level. The Mediterranean island of Corse consists of an uplifted and dissected Hercynian plateau, producing a structure like a miniature Massif Central, but with a maximum altitude of only 200 m. The Hercynian massifs on the whole are areas of impressive but unrelenting landscape; throughout this century until the 1970s, permanent population was leaving and agriculture retreating from these relatively harsh environments to be replaced by temporary populations and tourism.

The Alps form a great arc of folded and overthrust mountains in southeast France, sharpened and gouged by glaciation, and rising to over 4000 m. They are separated from the Hercynian massifs by the down-faulted trough of the Rhône Valley. Pinchemel (1969: 15–17) considered this valley to be of major significance, both structurally and topographically, as a barrier to east-west communication. To the east of the Rhône, the Jura range is largely composed of limestone, forming tablelands and plateaus rising to over 1700 m, interspersed with marls and clays; it is separated from the Alps themselves by the Swiss Plateau.

The lower pre-Alps, which are predominantly calcareous and which exhibit both karstic and glacial features, rise from the Rhône Valley to the high glaciated peaks. The high Alps are complex in structure, especially in the southeast, with folding, overthrusting, and nappe formation and displacement. This is a high-energy environment with active periglacial processes producing large volumes of debris available for transfer by mass movement and by agents of erosion. Uplift has resulted in fluvial entrenchment below frost-shattered peaks. Faulting and incision have revealed exposures of coal and other minerals, while the streams have enormous capacity for the generation of hydro-electricity.

The Pyrenees, which rise to 3404 m at the Pic de Aneto, form the southern border of France with Spain (and Andorra). The eastern Pyrenees are higher and more faulted than those of the west, which are highly karstic and glaciated. The French side of the Pyrenees is narrow and relatively uniform compared with the complex Spanish side, but nonetheless contains nappes, faults and evidence of Quaternary seismic and volcanic activity. The high mountain areas of the Alps and Pyrenees are spectacular and contain extremes of landscape ranging from the wilderness of the National Park areas, through the degraded industrial settlements of the valleys, to the sophistication of modern ski resorts. It is very obvious in this region that the environment is both a resource and a hazard.

Soils

France has very few natural soils left after centuries of vegetation clearance, grazing, fertilizing and cropping. Most French soils range in type between the true Brown Forest and the Podsol. The characteristics of the soils are largely determined by climate and by parent material. The climatic zones, broadly based on latitude, give rise to zonal soils. The bulk of France, lying in cool to warm temperate climates with moderate rainfall, possesses Brown Forest soils which are only mildly leached and which contain plenty of organic matter. More podsolized soils have developed in the wetter climates of the west and in the cooler climates of the north and the mountain zones. Soils of the mountain zones tend also to be thinner and less fertile than those in lowland areas. The soils of much of France generally have good depth and texture, and contain adequate organic material. Hence, they are capable of improvement, and can support intensive farming systems, although many exhibit mild leaching. In parts of the Mediterranean, zonal Mediterranean soils have developed which are characteristically red in colour; their utility for agriculture is governed by soil-moisture conditions and by the availability of irrigation.

The soils of France are also profoundly influenced by their parent material. Agriculturally, the most significant of these azonal soils are those developed on the deposits of loess, or *limon*, which are widespread throughout northern France. Also significant are the soils developed on the limestones which cover approximately a third of the country. In lowland areas, such as in parts of the Paris Basin, the limestone-based soils are light and easy to work and easily enriched; they attracted cultivation long before technology was available to break up the heavier clays. In southern France, the elevated limestone plateaus and mountains in general have poorly-developed soils of little agricultural value.

Soil has been modified by centuries of settlement and agriculture. In the Mediterranean, scarce soil was traditionally conserved and retained by the construction of innumerable terraces on hill slopes for the cultivation of wheat, olives and vines. In Bretagne, the traditional *bocage* landscape of small fields bounded by ditches and hedges also contributed effectively to soil retention (Carnet, 1979). Similarly, in the granitic uplands of the Limousin, the processes of erosion by flow, creep and rain splash on bare soil have caused enormous banks of detritus up to 4 m in height to accumulate against hedges or dry stone walls. Charcoal from the banks has been radiocarbon dated to the twelfth century, a period of widespread forest clearances (Desbordes and Valadas, 1979).

Soil fertility was enhanced initially by fallowing and by rotation of crops. Natural fertilizers have been added for many centuries; these included night soil from urban areas, seaweed from the coast, and animal manure in pastoral districts. In the 1890s, the addition of chemical fertilizers to arable soils began on a large scale with the import of nitrates and potash; these have largely been replaced by artificial fertilizers since the Second World War. The fertility of the soil of the Mediterranean coast of Languedoc and the Atlantic coast between the Loire and the Garonne has been adversely affected by excessive salinity. This has occurred as a result of the incursion of saline marine waters into aquifers and as a result of irrigation (Szabolcs, 1974).

Soil quality is a critical factor in determining the value of land for agricultural purposes (Mori *et al.*, 1983). The best quality arable land is located in the centre and north of the Paris Basin, in parts of the Rhône Valley and the Mediterranean coast, and on the improved coastal soils of Bretagne. The poorest land, unsuitable for arable production, occurs in the Alps, Pyrenees, Massif Central, and on the sandy podsols of the Landes. Between these two extremes, there is reasonable improved land for arable purposes in much of northern and southwest France, and for pastoral purposes in Normandie and in eastern and central France

(House, 1978). Many poor soils were brought into cultivation when agriculture reached its widest spatial extent at the end of the last century, but have been subsequently abandoned.

Soil erosion

The intensive use of the soil over many centuries has frequently resulted in its significant degradation. Much of the barrenness of the limestone uplands and of the Mediterranean regions is a consequence of deforestation which began as early as Roman times. Soil erosion has been accentuated by overgrazing and, particularly in areas of annual summer drought, by unsuitable arable practices. The degradation of Mediterranean landscapes has been attributed both to climatic causes and to human exploitation over a period of at least 2000 years (Muxart *et al.*, 1985). Another interpretation, advanced by Blaikie and Brookfield (1987: 126), is that human activity initially helped to arrest erosion by means such as terracing, but that the decline of intensive agriculture about a thousand years ago has subsequently allowed the forces of wind and water to degrade the environment.

It is estimated that about 5000 km^2, approximately 10 per cent of the national territory, are seriously affected by soil erosion (Henin, 1979). Peak rates of erosion of around 15 000 tonnes per km^2 per year have been measured around the Serre-Ponçon Reservoir in the high Alps. Much lower levels of soil erosion have been measured in more moderate conditions of slope and climate, for example 17 tonnes per km^2 per year on the River Seine and 250 tonnes per km^2 per year on the River Garonne (Henin, 1979). Some recent agricultural change has apparently helped contain soil erosion; for example, in the catchment of the Seine, where pastoral land has replaced arable, the rate of erosion has decreased. Similarly, Henin (1979) considered that the increasing use of chemical herbicides instead of clearance and weeding has helped to conserve soil. On the other hand, large-scale mechanization and deforestation have contributed to erosion; mechanization of agriculture in the Val de Canche, Normandie (Roose and Masson, 1985) and land consolidation in Lorraine have increased soil erosion rates in those areas (Gras, 1979). Similarly, deforestation has been demonstrated to accelerate erosion; for example, huge areas of the Landes destroyed by forest fires in the 1940s (see page 23) have been subject to severe wind erosion, especially of the fine-grained black humus of the podsols (Juste *et al.*, 1979).

The susceptibility of the soil to present-day processes of erosion by water and wind depends largely on its coherence and on the intensity of weather events. Pihan (1979) has used an index based on the duration and the intensity of rainfall events to estimate the susceptibility of soils to water-erosion for 81 meteorological stations in France. His analysis revealed important spatial and temporal variations in potential erosion, with the highest susceptibility occurring in the Mediterranean and the extreme southwest in the months from August to November (Fig. 1.2). The values were ten times those of western Bretagne. Moderate potential erosion scores for central and northern France showed a summer peak

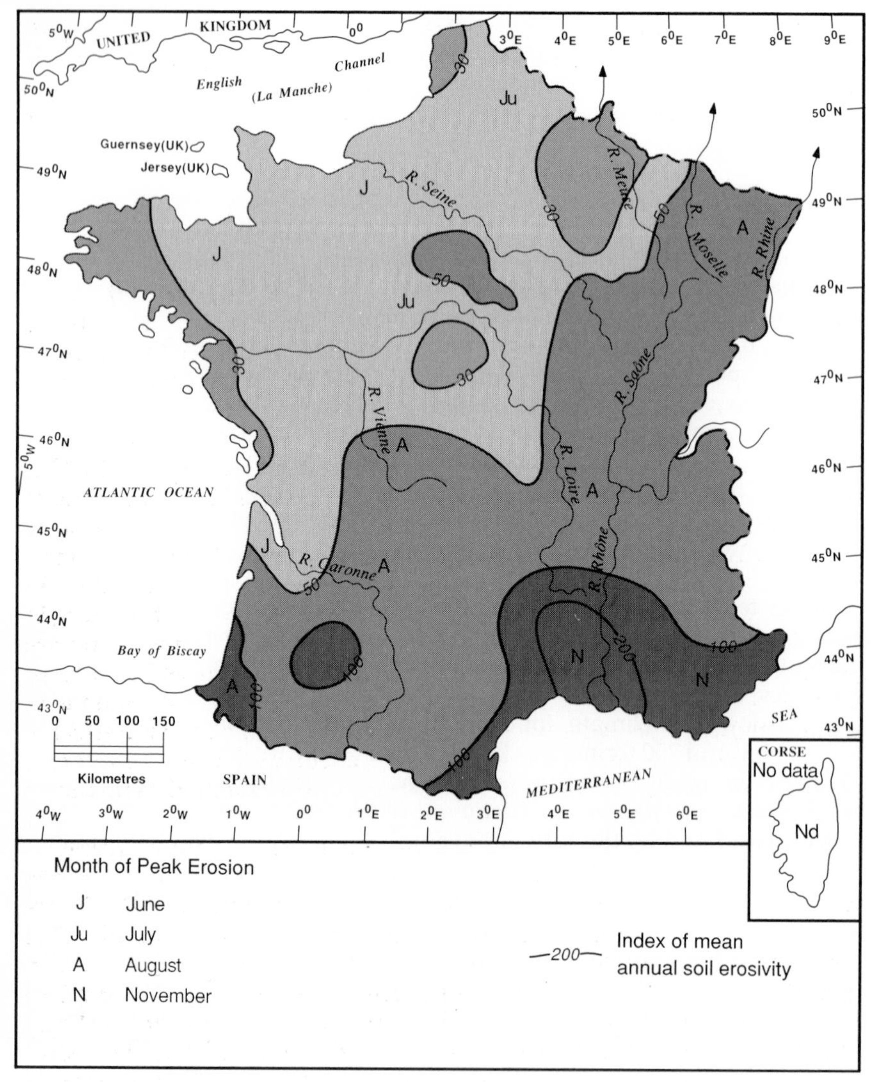

Fig. 1.2 Index of soil erosivity, France (see text for explanation of calculation of index). (Source: Pihan, 1979: 17)

from June to September. Actual erosion is heavily influenced by land use as well as by the weather; early summer rainstorms cause significant soil loss in northern maize-growing areas because of the exposed soil, but in most other areas the soil is generally better protected by growing crops (Pihan, 1979).

Collective action against soil erosion commenced early in France at the end of the nineteenth century. In areas where soil erosion was very pronounced and visible, and where it posed a threat not only to agriculture but also to other economic activities and to communications, the action taken was direct and government-supported rather than being left to individual farmers. Government schemes were developed by the Water and Forest Services (*Services des Eaux et des Forêts*) to combat water erosion in the Alps, and to combat wind erosion in the Landes and the sandy areas of the Mediterranean (Henin, 1979).

Climate

The climate of France exhibits pronounced regional variations, the most significant of which is the north-south divide between the cool temperate north and the warm temperate south. A secondary variation distinguishes the more continental east from the maritime west, while the mountain zones are affected by altitude, slope and aspect (Fig. 1.3).

The impact of latitude is shown clearly by the variation in temperature between north and south; mean annual temperatures range from 15° C in the Mediterranean to 9°C at the Belgian border, while July temperatures for the same locations range from 22°C to below 18°C. However, January temperatures, although affected by latitude, are more noticeably moderated by the warm oceanic currents of the North Atlantic, and hence Brest in Bretagne has a mild January mean temperature of 6°C, while Strasbourg in Alsace has a January mean of around freezing (Fig. 1.3). Temperatures are also significantly affected by altitude in the Alps, Pyrenees and Massif Central, where the mean summer temperatures are below 10°C and mean winter temperatures are significantly below freezing. The cooling effect of higher altitude is significantly moderated by aspect and relief; these microclimatic effects are important for the location of settlement, agriculture and ski resorts.

Annual precipitation exceeds 500 mm in most of France, rising to 2000 mm on the highest peaks. Rainfall totals are generally higher in the west than in the east, especially in Bretagne where the effects of moisture-bearing prevailing southwesterly winds are compounded by altitude. This general trend is complicated by the effects of altitude in the southeast. In general, the seasonal distribution of precipitation shows a winter maximum along the Atlantic coast, and a summer maximum in northeast France. Many parts of southeast France, in the transitional zone between a continental climate and a true Mediterranean one, exhibit complicated rainfall regimes; often the precipitation maximum occurs in or around October, when atmospheric instability results from high sea-surface temperatures (Barry and Chorley, 1992: 214–19).

The climatic regions of France result from a combination of these influences of latitude, continentality and altitude (see Arléry, 1970: 148–51, for an excellent summary). Northwest France has a mild oceanic equable temperature regime and moderate precipitation with a winter maximum, while northeast France has a more extreme continental temperature range and moderate rainfall tending towards a summer maximum. Southwest France has a transitional warm temperate climate moderated by oceanic influences and a winter precipitation maximum, while southeast France has a true Mediterranean climate with a hot dry summer and warm wet

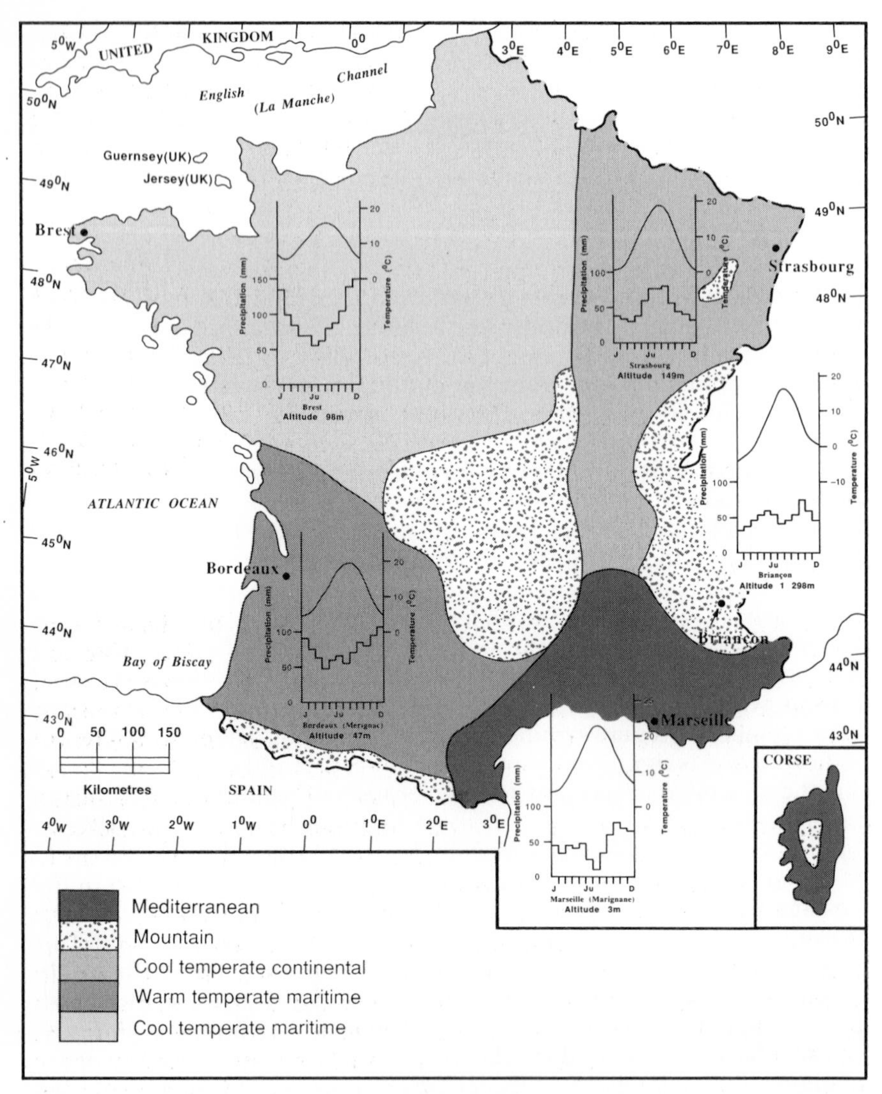

Fig. 1.3 The climatic regions of France and the mean monthly precipitation and temperature for selected meteorological stations. (Data from Arléry, 1970: 154–5; after Beaujeu-Garnier, 1975: 6)

winter. These climatic differences are illustrated by the sample climatic data in Fig. 1.3. The mountain regions have distinctively lower temperatures and higher precipitation but with marked local variations.

Most parts of France have a climate suitable for settlement and agriculture. The suitability of the climate for agriculture is determined primarily by levels of effective precipitation and by the length of the growing season. In this respect, the Mediterranean region has a long growing season but experiences minimal summer precipitation and a summer soil-moisture deficit. Irrigation is thus essential for most commercial agriculture in this region. In northwest France, especially in Bretagne, the long early growing season has been exploited for the production of early vegetables, enhanced by the widespread growth of valuable crops under cover of plastic or glass. Much of north and northeast France has a potential soil-moisture deficit in summer, which is overcome by the growth of drought-resistant cereal strains and by irrigation.

Climate obviously forms an important element in the attraction of the Mediterranean coast for tourism; the hot dry summers are much more reliable than those of the Atlantic. The climate is an even more vital part of the attraction of the Alpine areas for tourists. The sites of some new ski resorts, such as Isola 2000, have been deliberately chosen after scientific climatic study to maximize the length of snow cover (Escourrou, 1984). Climate is also perceived to be an attractive element of the environment of those mountain and coastal areas which are being increasingly developed for tourism and second homes, although the climate in these areas is frequently changeable and hazardous.

Climatic hazards

Hazards brought about by extreme fluctuations in weather have a profound effect on agriculture and other forms of economic activity. Climatic variability (in heat, sunshine and precipitation) has been shown to exert a marked impact on viticulture, causing a 13 per cent annual variation in yield, as well as changes in the alcohol content and acidity of the wines (Bourke, 1984). Climatic extremes are most likely to occur in southeast France where temperatures and rainfall are most highly seasonal. To some extent, an annual pattern of climatically-induced hazards is predictable, with winter avalanches, spring flooding and landslides, and summer droughts and forest fires (see pages 17–24).

The effect of climatic extremes may be all the more dramatic where they are unexpected. The impact of the 1976 drought, caused by a persistent blocking ridge of atmospheric high pressure, was felt most severely in northwest France. Bretagne, Normandie and the Lower Seine received less than half their normal precipitation between 1 December 1975 and 31 July 1976, while June precipitation totals were less than 10 per cent of the average (Stubbs, 1977). The drought was accompanied by a heatwave, with Paris experiencing its longest period of temperatures higher than 30°C for 120 years (Stubbs, 1977: 461). The agricultural financial loss was estimated at FF 6000 million, mainly resulting from early slaughterings of livestock because of shortage of fodder, and from poor harvests of potatoes, cereals and sugar beet (Anon, 1976). Average farm income levels fell in real terms to those of 1969–70, but the pattern of losses from poor harvests and of benefits from increased prices varied between farms and regions.

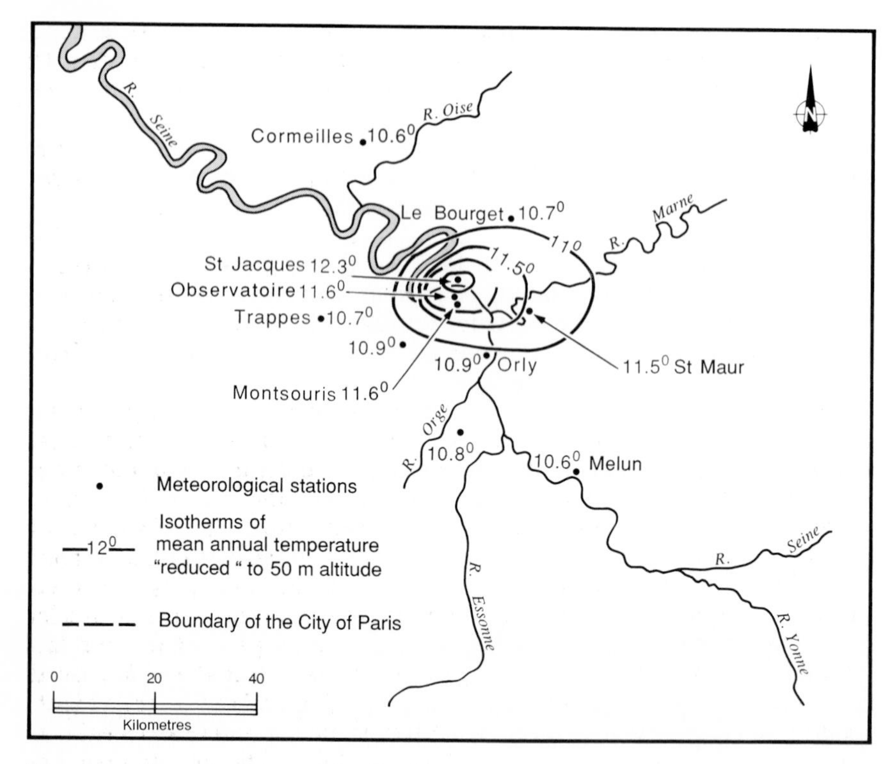

Fig. 1.4 The urban heat island of Paris, 1951–60 (temperatures in degrees Celsius). (Source: Dettwiller, 1970: 180)

Human impact on climate

Urban climate

Human modification of climate occurs at all spatial scales from the global to that of individual dwellings, slopes and fields. Some of the clearest examples of human climatic modification occur in large urban areas where changes in temperature, winds and precipitation are brought about by the concentration of heat sources and of heat-absorbing, rough and impermeable surfaces. In France, only Paris is large enough to produce measurable effects, which are comparable to those experienced in other large cities such as London.

The presence of an urban heat island around Paris (Fig. 1.4) has been demonstrated by a rise in temperatures since 1891 significantly greater than in the rest of the country (Dettwiller, 1970). Minimum temperatures rose rapidly and maximum temperatures more slowly, indicating that urbanization decreased the daily range of temperature by the storage of heat. Convincing evidence for the Parisian heat island comes from a long-term study of ground temperatures in a cellar under the Tour St Jacques in central Paris where measurements have been taken since 1670. A period of relative stability lasted from 1775 to 1890 with mean annual temperatures around 11.7°C; these have climbed steadily by about 1.5°C since the late nineteenth century (Dettwiller, 1970: 179). Comparison with adjacent weather stations over a 10-year period confirmed the urban heat island effect of about 1.7°C.

Urban landscapes not only produce effects on temperature; their configuration also leads to an increase in frictional drag, turbulence and roughness. The usual overall impact of large urban areas is to reduce wind speeds but to increase turbulence, with all its associated undesirable effects. Roughness parameters are directly related to the size of obstacles in the path of the wind (Landsberg, 1981); in Paris these range between 2 and 5 m, in suburban Nantes between 0.4 and 2.3 m, and in the open countryside less than 0.2 m (Duchêne-Marullaz, 1976, quoted in Oke, 1979).

The rise in urban temperatures around Paris has been accompanied by changes in precipitation patterns. Greater atmospheric turbulence, enhanced convection processes and more particulate matter in the atmosphere could all be expected to increase precipitation. Since 1871, warm season precipitation has increased by 38 per cent, with a 98 per cent probability that the rise is not due to chance (Dettwiller and Changnon, 1976: 519). It is thought that this has been caused mainly by increased convection processes, turbulence and the presence of condensation nuclei produced by industry; in other (non-metropolitan) parts of the mid-latitudes of the northern hemisphere intense summer storms have decreased since the 1930s. Weekday precipitation, averaging 1.93 mm over an eight-year period, has also been found to be significantly greater than weekend levels, which average 1.47 mm (Dettwiller, 1970). The higher weekday precipitation cannot be attributed to turbulence, which would remain constant, but is a reflection of greater residential, commercial and industrial activity during the week which produces both heat and condensation nuclei in the atmosphere.

Acid rain

Acid rain is a phenomenon which has received enormous media attention in the last twenty years. It is acidic precipitation caused particularly by sulphur emissions from thermal power stations and from the burning of fossil fuels, but also by vehicle and industrial emissions of nitrogen oxides, carbon monoxide and hydrocarbons. The damaging effects on freshwater supplies, vegetation and buildings were recognized in Scandinavia in the 1950s.

Estimates of emissions, depositions and effects are as yet relatively unreliable because of their temporal variability. However, it appears that emissions of SO_2 have more than quadrupled in Europe since 1900, and approximately doubled between 1950 and 1970; they have since levelled off (Dovland, 1987). France was one of the worst offenders for all types of noxious emissions in 1978, ranking fifth in Europe in the production of SO_2, third in the production of NO_x and second in the production of non-methane hydrocarbons (Smith, 1984). However, France has also been one of the most successful European countries in reducing emissions of SO_2 (Table 1.1) (Dovland, 1987: 13). This reduction is mainly attributable to the development of nuclear power in place of thermal sources of energy (see pages 130–3). The emission of NO_x, particularly caused by combustion of fossil fuels and by vehicle fumes, has also declined but less markedly (Table 1.1). Reductions in these emissions have been brought about by industrial closures, by more stringent pollution controls and by the more widespread sale and use of unleaded petrol.

Table 1.1 Noxious emissions in selected countries of western Europe, 1980 and 1984/5

	1980	**1984/5**
Sulphur dioxide emissions in million tonnes per year		
France	3358	1845
West Germany	3200	2400
United Kingdom	4670	3540
Nitrogen oxide emissions in million tonnes per year		
France	1867	1693
West Germany	3100	2900
United Kingdom	1916	1690

Source: Dovland, 1987: 13.

Modelling of European emission and deposition indicates that France contributes significantly to the acid rain that falls over Belgium, the Netherlands, Germany and Czechoslovakia, whereas most of the acid rain falling in France is home-produced (Smith, 1984: 10). Deposition of chemical airborne pollutants is lower in France than in the English Midlands, Germany and central Europe. France receives relatively little airborne pollution from elsewhere, because the prevailing westerly airstreams bring relatively pure air from the Atlantic. As a result, only 32 per cent of sulphur deposition over France originated abroad in 1980, compared with 10 per cent in the United Kingdom and 46 per cent in West Germany (Dovland, 1987: 15). There are also marked regional variations within France in the deposition of sulphur from precipitation, decreasing from northeast France southwards, with an outlier of higher levels at the mouth of the Rhône. This pattern reflects proximity to pollution sources from northern Europe and from within France.

The impact of acid rain was first noticed on forests in northwest Europe. Recently, some of the nation's cultural treasures have been damaged by accelerated weathering. The most notable example is Chartres Cathedral, where stained glass, some of which dates back more than a thousand years, is deemed to be vulnerable (Park, 1987). The cost of air pollution damage to buildings and materials was estimated at FF 600 million per year as early as the 1960s. Although France has suffered less from the effects of acid rain than other countries of northwest Europe, significant evidence of increasing rainfall acidity has accumulated during the 1980s (Massabuau *et al.*, 1987). The impact of increasingly acid rain has been demonstrated by the acidification of freshwater, the decline in fish stocks in some rivers of the Vosges and in damage to French forests and vegetation (see pages 25–6).

Forest damage was estimated to have occurred to more than 10 per cent of the total growing coniferous forest stock of 605 million m^3 in 1986; 74 million m^3 were estimated to be moderately damaged (defined as >20 per cent defoliated) (Nilsson and Duinker, 1987). Comparative data for deciduous forests were 945 million m^3, of which 182 million m^3 were damaged (including slight damage, defined as >10 per cent defoliated). France has proportionally less damage in relation to growing stock and in relation to annual fellings than the United Kingdom, Germany and much of central Europe and Scandinavia, but has a relatively high proportion of conifers at present showing only slight damage (a further 153 million m^3) which may represent areas of further risk (Nilsson and Duinker, 1987). Forest damage may be caused by factors other than acid rain. Prinz (1987) pointed to the importance of climatic stresses such as the 1976 drought as a triggering factor, as well as to the increase in atmospheric ozone and the loss of soil nutrients as a result of acid deposition. Spatial variations in forest damage may be attributed to differences in soil nutrient supply, and to altitudinal variations in ozone concentrations (Prinz, 1987: 35).

The impact of acid rain and air pollution, and the political tensions between the pollution exporting and importing nations, have brought about increased international co-operation to control the problem. The emission, transport and deposition of air pollutants has been monitored since 1977 by a European and North American Co-operative Program for the Monitoring and Evaluation of the Long-Range Transmission of Air Pollutants in Europe (EMEP), under the auspices of the United Nations Economic Committee for Europe (ECE) and the World Meteorological Organisation (WMO). In 1984, France announced its intention of reducing emissions to 50 per cent of its 1980 level by 1990; this goal was achieved before the set date.

Vegetation

The natural vegetation of France is primarily a function of climate and soils, although it is arguable that very little natural vegetation remains. The temperate climate of most of France would have produced a climax forest vegetation over almost the whole country except where topography and soils prevented this. The natural forest taxa vary according to climatic conditions, ranging from beech, birch and oak in the northwest, to chestnut, maritime pine and Mediterranean oak in the warm temperate conditions of the Mediterranean.

However, as with soils, one of the most significant factors affecting the nature of the present-day vegetation has been human activity. Clearance of forest has been evident in the Mediterranean region for at least 2000 years, but most palynological and sedimentological evidence points to a dramatic increase in clearance from about AD 1200 (see, for example, Billard *et al.*, 1985, and Desbordes and Valadas, 1979). Sediments of Petit Lac d'Annecy in the Alps show dramatic changes in chemistry, palynology and magnetic mineralogy, indicating clearance of vegetation and

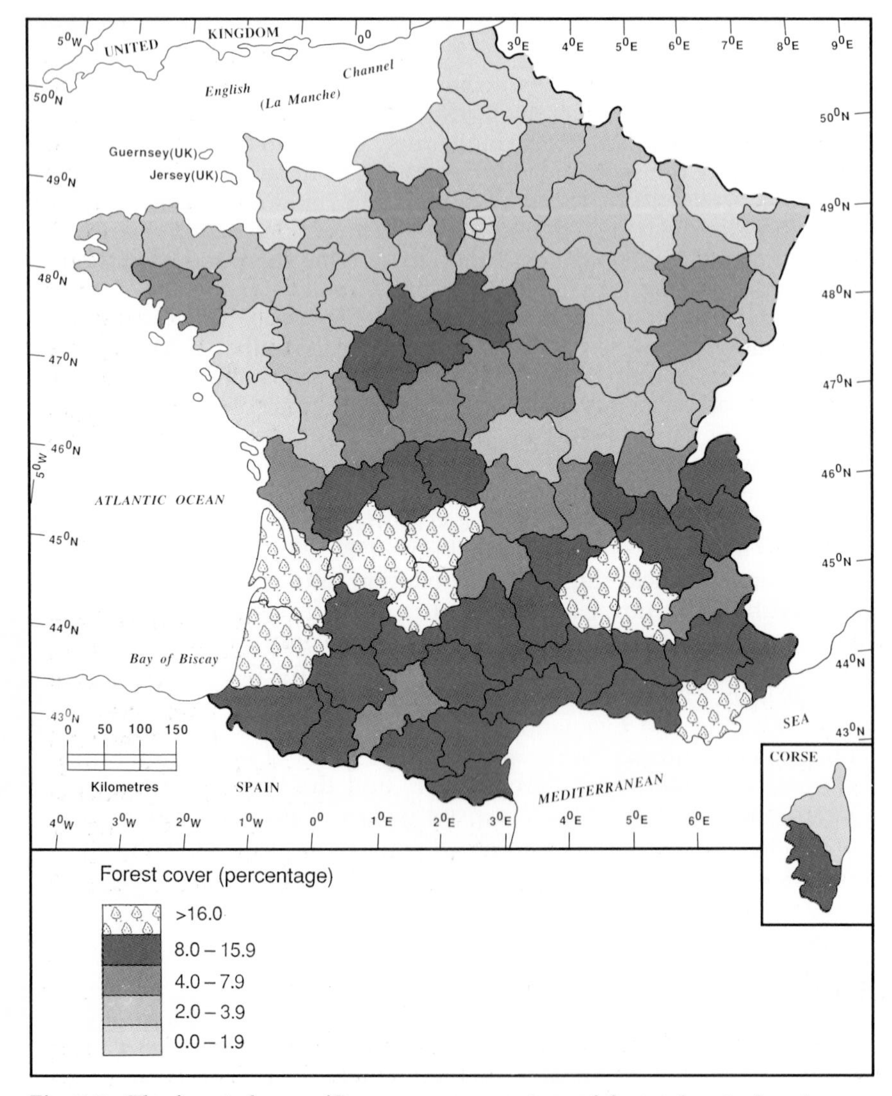

Fig. 1.5 The forested area of France, as a percentage of the total agricultural area. (Data from Ministère de l'Agriculture, 1984)

subsequent erosion of topsoil. Clearance and erosion date from the Middle Ages, with a further increase in intensity of human activity at or around AD 1800 (Higgitt *et al.*, 1991). The most complete record of vegetation change, from cores in the Alsatian peat bog of La Grande Pile, dates back 140 000 years and demonstrates the impact of both climatic change and human activity. The sediments contain evidence not only for past glacials and interglacials, but also of historical changes in vegetation and cultivation (Woillard, 1978).

Clearance and reduction of the forest cover reached its greatest extent around 1920, as a result of agricultural extensification and an increased demand for timber during the First World War. Subsequently, however, rural depopulation and the retreat of agriculture from marginal areas has allowed some vegetational recolonization of uncultivated land. Only about 6.5 per cent of the land is deemed to be uncultivated, and in these areas the vegetation consists of heathland in cool temperate climates, and *maquis* or *garrigue* in the Mediterranean. In parts of the Mediterranean, where abandoned terraces with their thicker soils have provided an excellent medium for renewed growth, recolonization has been very effective (Wright and Wanstall, 1977). Forest cover has increased since the 1920s as a result of planned reafforestation, a policy implemented to combat soil erosion in the Alps and the Landes since the late nineteenth century (see pages 8–9; 91). Reafforestation has more recently enabled commercial exploitation of poor grade land to occur in the Landes, Sologne, Vosges and parts of upland France; approximately 25 per cent of the national land area is now forested (Fig. 1.5).

Natural and induced hazards

Natural hazards, such as floods, avalanches and forest fires have been greatly intensified by human activity in the last two hundred years. In global terms, France suffers less from environmental hazards than most other countries, reflecting high levels of income and technology, and low levels of population density (see Thompson, 1982, for an international comparison). Between 1947 and 1981, France suffered 13 major environmental disasters with the loss of 1048 lives, significantly fewer than the United Kingdom (21 disasters, 4962 deaths) but proportionally higher than West Germany (13 disasters, 592 deaths) (Thompson, 1982: 11). As in most other countries, floods still represent the major hazard to life and property (Chardon and Castiglioni, 1984).

Floods

Floods are a major hazard to life and property, partly because floodplains are desirable places for farming and settlement. The densely occupied valley of the Seine includes Paris, which was flooded in January 1910 after two weeks of extreme rainfall, resulting in 30 000 people being made homeless. In France, susceptibility to flooding is greatest in the southeast where the effects of relief and climate operate in tandem. For example, flash floods occurred in the city of Nîmes in October 1988, after 228 mm of rain fell in heavy storms in only six hours; eight people were killed. Flooding in Alpine rivers is caused predominantly by extremes of rainfall and by spring snow-melt; flooding of the Rhône damaged Lyon several times in the nineteenth century, for example, in 1840 and 1856, while the unpredictability of flow prevented settlement of the upper valley of the Rhône at that time. The series of dams constructed on the upper Rhône since the 1950s, at Génissiat, Seyssel and Miribel St Clair, has provided

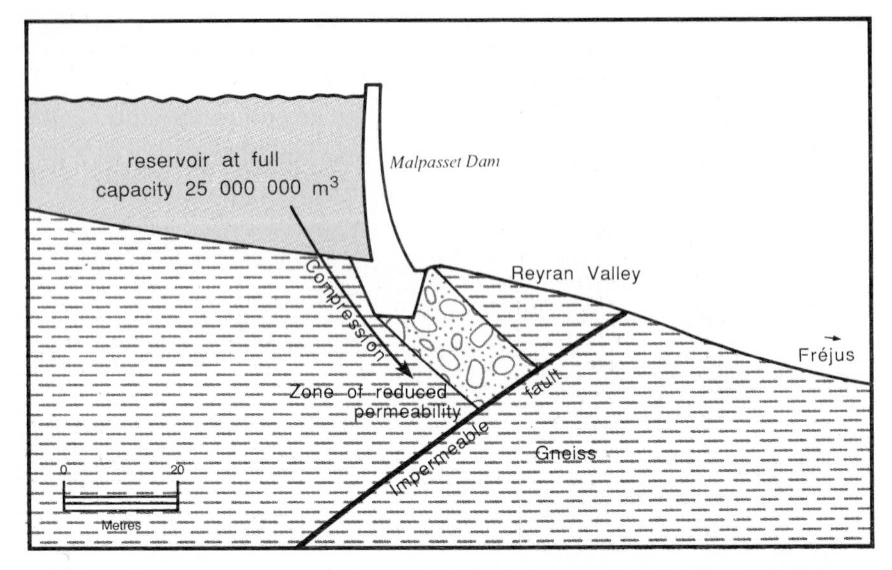

Fig. 1.6 Geological structure beneath the Malpasset Dam on the River Reyran; conditions prevailing just before the disastrous flood of Fréjus 1959. (Source: Waltham, 1978: 101)

flood control and has facilitated land reclamation and settlement on the valley floor (Bravard, 1987).

The most disastrous flood of recent times occurred in Fréjus on the French Riviera, on 3 December 1959. This was caused by the collapse of the Malpasset Dam on the River Reyran. This flood caused 421 deaths and was a disaster significant on a national and even a global scale (Beyer, 1974). The collapse of the dam followed a period of very heavy rain when the reservoir was at full capacity of 25 million m^3. The reason for the dam failure did not become clear until several years after the event, but appeared to be due to geological factors which were at that time unknown (Ward, 1978: 54). Subsequent hydrogeological research established that the gneiss upon which the dam was constructed became significantly less permeable under compression, but more permeable under tension. The key to the failure was the location of a fault containing an impermeable layer of crushed and crystallized rock immediately downstream of the dam (Fig. 1.6). The filling of the reservoir, for the first time since its construction in 1953, provided sufficient pressure to reduce the permeability of the gneiss between the dam and the fault, so that the water was almost completely contained by impermeable material. Initial release of pressure was noted in the form of seepage from the west bank of the dam about three weeks before the disaster. The final increase in pressure, caused by five days of prolonged rainfall, resulted in the instantaneous shattering of the dam. The disaster devastated Fréjus, demolished a naval air base, cut the motorway (N7) and main railway, and left 2524 families homeless. Damage to property, including public utilities and transport routes, was estimated at FF 11 million; rehabilitation and reconstruction cost twice that amount and donations towards the work came from the United Kingdom and other countries. The dam failure could not have been predicted at the time, but subsequent construction of similar dams has included drains beneath the dams to relieve water pressures (Waltham, 1978: 100–2).

More recent floods in southeast France in 1972, 1973 and 1976 have also caused loss of life and property, although on a less serious scale. In each case the flooding occurred as a result of heavy and prolonged rainfall, but the extent of damage has also been attributed to the indirect effects of urbanization (Gabert and Nicod, 1982). In the 1973 floods, 164–81 mm of rain, a significant proportion of the mean annual rainfall of 623 mm, fell in 48 hours in the upper basin of the River Arc. Despite the amount of rainfall, the floods were relatively unexpected, as they occurred at the end of the dry season in October. The extension of the built-up area, which had tripled in extent since 1960, was considered to be a possible contributing factor. However, Gabert and Nicod (1982) estimated that the impermeable area made up no more than three per cent of the total valley catchment and only one per cent in the upper valley. They attributed the flooding to the indirect consequences of urbanization rather than the direct sealing of permeable surfaces. In particular, their study suggested that the river channel had become increasingly inefficient because of the lack of upkeep of the Arc and its banks, the illegal dumping of abandoned vehicles and other objects in the river bed, and uncontrolled land sales, canalizations and water abstraction along the river banks and flood plain.

Disastrous floods, in which 34 people died and a further eight were reported missing, occurred on 22 September 1992 in the town of Vaison-la-Romaine in the *département* of the Vaucluse. The flooding followed torrential rain and electrical storms, with over 265 mm of precipitation reported in 24 hours from one weather station. A massive surge in the Ouvèze River was caused by a build-up of water, mud and debris behind the town's Roman bridge. The debris included numerous cars and caravans, some containing whole families, which were swept from a riverbank campsite upstream of the town. When the Ouvèze finally overtopped the bridge at a height 17 m above normal, the flood wave swept through the narrow streets of the old town and demolished many new houses on the flood-plain below. The peak velocity of the river was estimated at six metres per second. The bridge itself remained intact, although massive damage was caused to other property. As a result of the disaster, enquiries were set in motion to determine why the weather predictions were ignored by the rescue services and other authorities. Serious allegations were made of corruption by town officials who allowed building on flood-prone land in contravention of established planning procedures, despite the fact that these areas had been flooded at least four times this century (Hurel, 1992). The normal requirements for houses built in flood-prone areas had not been enforced; these include deep foundations rather than concrete bases, and ground floors devoted to non-residential uses such as garages, games rooms and laundries. The planning of land uses by local authorities is designed to prevent such disasters (see pages 160–1), but in this case human activity clearly exacerbated the impact of natural hazard.

Floods have been associated with change in rural as well as urban land use. For example, severe flooding in a number of Breton towns in the 1980s may be attributed to the removal of the traditional *bocage* landscape of hedges, banks and ditches. The change to larger, more open fields brought about a series of biometeorological changes; a reduction in bioturbation and evapotranspiration, a greater susceptibility of the topsoil to wind erosion, and a propensity for the formation of hard pans below the soil surface. These changes combined to reduce the permeability of the ground surface and to increase the volume of overland flow, resulting in a very short lag time between precipitation and peak discharge, and contributing significantly to the scale of flooding. However, in a study of floods in the western Pyrenees, Viers (1984) dismissed the role of human-induced land use changes, instead attributing most floods in that region to natural conditions of extreme rainfall.

Flooding due to entirely natural causes has occurred as a result of the sudden release of ponded glacial meltwater. The Tête Rousse glacier in the French Alps was the source of a catastrophic flood in 1892, caused by meltwater accumulating behind a rock sill in the long profile of the valley (Tufnell, 1984: 47). The flood is estimated to have brought down the phenomenal load of 80 000 m^3 of sediment in 200 000 m^3 of water, and raised the water level in the Bonnant gorge almost to the height of the Pont du Diable, 62 m above average water height (Vivian, 1974, quoted in Tufnell, 1984: 48). This violent flood, which caused 177 deaths, appears to have resulted from a special combination of physical circumstances, although regular floods of glacial origin are common in other parts of the Alps.

Mass movements

Mass movements, including falls, slumps, and slides of rock, mud and debris have been of increasing concern to the French authorities since the 1960s. The risk of mass movement is often greatest in areas of high relief and extreme climate; these are areas which are also becoming increasingly attractive both for tourism and permanent residence. Moderate-scale debris flows occur widely within the French Alps, with a recurrence interval of 10–40 years (Van Steijn *et al.*, 1988). Roads are particularly vulnerable to damage and closure caused by mass movements. For example, the Corniche linking Cannes and Nice, in the *département* of Alpes-Maritimes, is a vital road link which winds above the coast perched on a huge calcareous cliff; the road gives stunning views but is also the site of numerous slides. A major slide at Eze in 1977 destroyed both the road, and the nearby railway (Martin, 1984). The physical conditions of relief and climate present a semi-permanent risk of rock falls in many of the gorges of the south and on innumerable mountain roads. The risk of slippage and falls is compounded in areas of thinly-bedded incoherent rocks (Martin, 1984). Chardon (1987) described an example of rockfall in the Romanche valley in the Alps where physical conditions along the glacial trough pose

Plate 1.1 Snow avalanche in springtime in the Alps.

a major hazard. The valley sides, composed of fractured schistose rock, rise over 400 m at angles of 60–80° in an area where the annual rainfall exceeds 1000 mm. Nineteenth-century iron-mining in the area may also have contributed to the potential hazards. A series of rock falls and movements has provoked the imposition of protective measures against a possible calamity. The protection measures have included the re-routing of the N91 trunk road (Chardon, 1987: 108).

The risk of mass movement is not confined to the Alpine regions. Marre (1987) reported a rotational slip and mudflow in the sands and clays of Rilly La Montagne (Marne) in August 1986, following drought and subsequent heavy rain, which caused some damage to property. Similarly, a sequence of slides has been recorded on the coast of Normandie on either side of the town of Villerville as a result of rotational slips in Kimmeridgian marls (Flageollet and Helluin, 1985). The impact of events such as these is very localized. In general, however, avalanches of ice and rock are one type of natural hazard which is having an increasing human impact, killing on average several dozen people each year (Chardon and Castiglioni, 1984). Most avalanches undoubtedly used to occur in uninhabited areas, but now these same areas are increasingly used for skiing, winter recreation and semi-permanent occupation, thus intensifying the hazard potential (Smith, 1988) (Plate 1.1).

A variety of glacier-related hazards has been chronicled in the Chamonix valley in the Alps, including glacier advances over buildings and crops during the Little Ice Age of the seventeenth century. The valley has also been the scene of floods, falls of ice and debris, and an avalanche from the Glacier du Tour in 1949 which killed six people (Tufnell, 1984: 20–2, 34–6). Glacier advance caused the abandonment of villages such as Les Tines in the Chamonix valley during the Little Ice Age; ice movement still poses a hazard to settlement, hydro-electric power stations and communications. It is estimated that an increase in ice thickness of 50–95 m would return the glacier snout near to its maximum Little Ice Age extent near Les Tines (Mahaney, 1987).

Forest fires

Fire is a serious hazard in Mediterranean and southeast France, especially in late summer. Forest fires are most prevalent in conditions of heat and drought; the flames move further and faster in autumn than in spring (Malanson and Trabaud, 1988). They spread rapidly in the resinous secondary forest and scrubby ground cover of the *maquis*. The severity of fire damage is also related to wind direction and intensity; a strong consistent *Mistral* blowing from the north fans flames and makes fire-fighting more difficult (Wrathall, 1985a). Paradoxically, the impact of fires has been accentuated both by rural depopulation and by urban development. Rural depopulation has resulted in less clearance of flammable litter and understorey vegetation; furthermore, fires in remote areas may cause considerable damage before they are observed. The spread of urban and tourist development has increased the number of fires started by carelessness, vandalism or malice.

Forest fires have affected over 25 000 hectares per year in Provence-Alpes-Côte d'Azur, Languedoc-Roussillon and Corse, with damage to over 50 000 hectares in 1979 (Wrathall, 1985b). This represents a loss of 1.25 per cent of forest in those regions each year. The area most vulnerable to fire lies inland from the coast between Cannes and Toulon, where forest loss reaches four per cent per year. This appears to be related to the concentration of holidaymakers in the area, and to the presence of camp-sites and other recreational facilities. The island of Corse is also severely affected; in the period 1973–83 it suffered from 8800 fires affecting 106 000 hectares. In 1980, the cost of these fires was estimated at FF 58.5 million for Corse alone. It seems highly likely that at least some of these fires have been started deliberately by farmers to clear *maquis* in order to improve grazing conditions.

Most fires for which the causes are known are caused by human activity: by dropped cigarette ends and by children playing with fire (Nicod, 1964, quoted in Wright and Wanstall, 1977). More recent data on causes of fires for which the origins are known (30 per cent of all fires) indicated that leisure-related causes are less important than occupational causes associated with agriculture and forestry (nine per cent of all fires) and accidental causes relating to defective power lines and the burning of rubbish at rural tips (eight per cent of all fires) (Gouiran, 1974, quoted in Wrathall, 1985b). Recently numerous fires have been deliberately started as a means of clearing land, prior to seeking development and planning permission. However, this underhand practice has been curtailed by recent legislation which maintains fire-affected land as forest or *maquis* (Wrathall, 1985b).

Forest fires have had a profound and lasting impact on the degraded nature of the Mediterranean forest and are thought to have been an important factor in the production of *maquis* and *garrigue* from Mediterranean forest. The ecological impact appears to be dependent on the heat attained and the intervals between burning (Malanson and Trabaud, 1988). Fires have also caused considerable damage to property and some loss of life, as in the deaths in 1983 of the crew of a Canadair aeroplane which crashed while being used to drop water in the path of a severe fire (Wrathall, 1985b). Exceptionally damaging forest fires occurred in the Gironde and 12 other *départements* of southeast and southwest France in August 1949 following a prolonged drought. The fires in the Gironde started 24 km south of Bordeaux and raged for six days, fanned by strong southwesterly winds, until they threatened the suburbs of the city. In the fires, 134 500 hectares of pine trees, which at that time were important in the domestic softwood industry for the production of turpentine and pitprops, were destroyed; the economic loss to the timber industry was estimated at FF 5000 million. The fires also destroyed 269 houses, killed 84 people and injured a further 100.

Concern over the scale of losses by fire and the lack of a co-ordinated statistical base on fire outbreaks and damage led in 1972 to the organization of *Opération Prométhée*. This is a co-operative scheme operated by 14 *départements* which not only compiles statistics on fires, but which has also established fire prevention and control measures. The prevention measures include firebreaks, emergency water tanks, small hillside reservoirs and helicopter landing pads in dispersed locations. Fire rehabilitation measures include the introduction of fire-resistant species, reafforestation plans and research and development in experimental areas (Wrathall, 1985b). In Corse, a regional nature park has been established, one of the main aims of which is to prevent and control fires, and thereby offer some protection to the natural environment. In the nature park, a *maquis* management system has been designed to control vegetation growth by means of 'mechanical mule' rather than by fire, with the aim of minimizing fires and soil loss.

Resources of the physical environment

Water resources

The water resources of France are intimately related to other physical characteristics and, inevitably, are unevenly distributed. Furthermore, the demand for water does not always correspond to its availability. France has substantial water resources; the former Yugoslavia is the only other country of Europe better endowed in quantity, but France ranks only sixth in Europe for the availability of water in relation to surface area or population (Margat, 1982). Despite the relative abundance of water and low level of usage, there still exist chronic problems of distribution, and increasingly of water quality. Major problems of distribution occur in the Mediterranean and in northeast France. In the Mediterranean, summer shortage of water may occur in areas which have a rapidly-growing summer population, where the effective rainfall is only 50–100 mm per year (Margat, 1982). This problem has been addressed by the development of comprehensive water management schemes and by the use of karstic aquifers for large-scale water storage (Nicod, 1980). Northeast France suffers from a similar but year-round shortage of water; this region, which is the most urbanized, industrialized and densely settled in France, is also the area of lowest rainfall and possesses the most polluted rivers.

The integrated management of river basins helps ensure the regularity of water supplies which is vital for population and industry. This regularization of supply also provides better control over drought and floods (for a discussion of flood hazard see pages 17–20). Further benefits of integrated management include more regular supplies of irrigation water, cooling water for nuclear power stations, hydro-electric power and improved navigability along rivers. The reservoirs created by management schemes are a major resource for leisure and recreation. In the case of the Rhône, channelling of the river in its former braided sections has stabilized the floodplain and has encouraged land reclamation and settlement (Bravard, 1987). The major effect of the management scheme has been a dramatic regularization of daily and monthly discharge regimes in the Rhône as in other Alpine rivers (Edouard and Vivian, 1984). The control of the Loire, formerly noted for its irregular flow and braided course, was achieved in 1982 by the construction of the Villerest Dam near Roanne, and by the building of a series of further dams. These dams have made obsolescent numerous hump-backed bridges along the river's course and have removed the need for semi-continuous dredging of its channel (Scarth, 1983). A more regular flow will be maintained at >50 m^3 per second (compared with a former low of >10–20 m^3 per second) by controlled release of water from the reservoirs. This regular flow is an important asset in the provision of cooling water for the nuclear power stations constructed along the Loire. The discharge from the reservoirs will also generate 167 million kW hours of electricity, meeting 60 per cent of demand in the Roanne area.

The use of water resources for power generation has been important in France since the late nineteenth century. The first use of water to generate electric power occurred in 1885 at Vizille near Grenoble, and since then the water resources, particularly of the Alpine areas, have been developed almost to their full potential. The exploitation of almost all the suitable sites for the production of hydro-electric power occurred after the Second World War, associated with the integrated management of river basins. At this time, the demand for electricity was increasing from growing population and industry, and the control of supply was facilitated by the nationalization of the electricity industry. By the late 1960s, hydro-electric power provided 50 per cent of France's requirements for electricity, a proportion which dropped dramatically in subsequent years (see pages 130–3 and Fig. 4.4). The diminishing relative importance of hydro-electricity in France stems from the fact that all economically-viable sites have already been developed (Electricité de France, n.d. (b)).

Hydro-electric power is still particularly important in supplying electricity at times of peak demand, because of the impossibility of storing electricity at an industrial scale, and the speed at which hydro-electric power stations may be brought into production. This capability is in marked contrast to the base-load nature of production from nuclear power stations, and emphasizes the complementary nature of the two methods of power production. The flexibility of the hydro-electricity schemes has been further refined by the development of pumped storage systems. These schemes, such as the 720 000 kW installation at Revin in the Ardennes, consist of an upper and a lower reservoir. Water released from the upper reservoir produces power at peak times; off-peak electricity is used to pump the water back to the upper reservoir for re-use. A famous example of the use of renewable water power for electricity generation occurs at the tidal barrage of the Rance estuary in Bretagne, a scheme unique in France. Much of the contemporary development of water power resources consists of technical and environmental improvements, automation and increases in capacity (Electricité de France, 1983; 1984).

The quality of water resources became an increasingly prominent political issue in France in the 1980s, when environmental damage from acid rain (see pages 13–15) and pollution became measurable. Until that time there had been little evidence of acid rain damage to water or forests in France. However, by the mid-1980s, pH levels in the rivers of the Vosges had reached minimum values of 4.1, with the worst effects evident in low flow conditions in summer (Massabuau *et al.*, 1987). The acidity has severely affected wildlife, and in the region of Cornimont, Vosges, trout have completely disappeared from streams.

The sources of water pollution include not only atmospheric pollutants but also sewage, industrial waste and artificial agricultural fertilizers (see page 88). The nitrate content of underground waters has been significantly increased in many areas, especially where cropping and stock breeding has been recently intensified (Delavalle, 1983). The high concentration of nitrates in the water supplies arises from a doubling of fertilizer applications in the last 20 years; the level of fertilizer usage would have been even higher but for its rising price during the oil crises of the 1970s. In the early 1980s, concentrations of nitrate in solution reached approximately 16 mg per litre at both Montjean and Thouaré on the River Loire (Manickam *et al.*, 1985: 734–5). These levels, which occurred at various times in the winter and spring, may be compared with the safety standard of 50 mg per litre established by the EC. Apart from making drinking water noxious and being particularly dangerous to infants, high concentrations of nitrates can also be detrimental to wildlife as a result of eutrophication of water resources.

Pollution from phosphates tends to be more constant in intensity than nitrogenous pollution, as the phosphates become bonded to particulate matter which is then eroded to form suspended sediment in rivers. Measurements of phosphorus concentrations in the Girou River by Probst (1985) over a ten-year period demonstrated that 71 per cent was carried in suspension, with peak phosphate concentrations in excess of three mg per litre. The majority of the phosphorus was derived from human activities rather than from natural environmental processes. Thus, only seven per cent of the phosphorus was released by natural biogeochemical processes, mainly weathering, while 32 per cent could be attributed to non-point pollution and 61 per cent to point pollution sources. Most phosphates are derived from point concentrations of domestic sewage, while the non-point sources are mainly agricultural fertilizers.

A comparison of the amount of phosphate and nitrate transported by French rivers showed the River Seine to be the most polluted of the major rivers, with median concentrations of phosphate of 1.6 mg per litre and nitrate of 17.5 mg per litre during 1976–9 (Meybeck, 1982: 444). The River Rhône carried more phosphate than either the Loire or Garonne (the median concentration during 1971–4 was 0.4 mg per litre, compared with 0.1 and 0.3 mg per litre respectively for the Loire and Garonne during the same period). However, the Rhône was less polluted by nitrates (3.9 mg per litre during 1971–4 compared with 5.4 for the Loire and 4.9 mg per litre for the Garonne). These variations reflect significant differences in the nature and impact of settlement, industry and agriculture in each of the catchments. These values may be compared with those measured in small streams in the remoter areas of the pre-Alps. The lack of pollution here means that phosphate concentrations of 0.01 mg per litre and nitrate concentrations of 0.8 mg per litre are typical (Meybeck, 1982: 443).

Industrial pollutants include polychlorinated biphenyls (PCBs) which are by-products of the manufacture of goods such as plastifiers, paints and hydraulic fluids. Levels of such industrial pollution are highest in the Rhône, Seine and the rivers of Lorraine (LeFrou and Bremond, 1975). Substantial PCB concentrations have been found in the discharge of the Rhône at levels from 45 to 121 ng per litre (Raybaud, 1972, quoted in Bernhard, 1981); these concentrations also extend to considerable distances off the coast (3–5 km) and to significant depths in off-shore sediments. Even higher concentrations of PCBs have been recorded in the Seine at and below Paris (Chevreuil *et al.*, 1987); PCB concentrations found in mussels taken from the Seine estuary are five times higher than the French average. As the production of PCBs has been restricted since 1975, their continued high level in the Seine at particular points such as at a car factory at Poissy (Fig. 1.7) suggests that serious discharges of pollutants into the river system are being made in violation of regulations (Chevreuil *et al.*, 1987). The concentration of PCBs at 50–150 ng per litre in high water and 500 ng per litre in low water ranks the Seine in the class of 'moderately to highly polluted' rivers according to the World Health Organization.

The waters of the Seine also receive pollution from industrial by-products and from sewage, even when treated. The drinking water of the Seine at the main extraction point of Choisy-le-Roi has mean levels of cadmium,

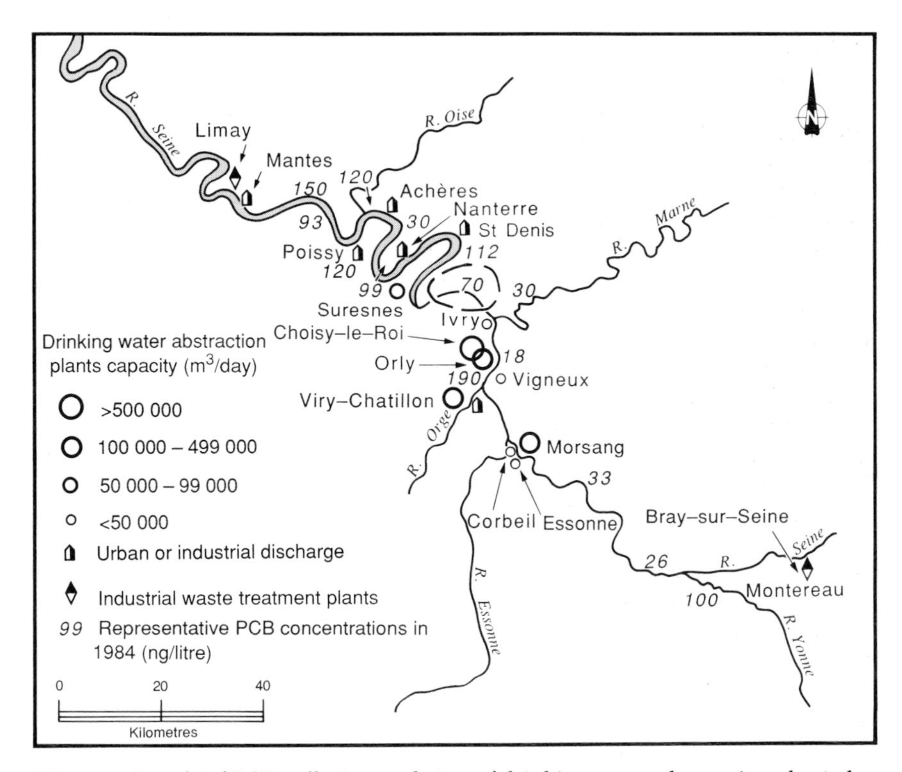

Fig. 1.7 Levels of PCB pollution and sites of drinking water abstraction along the River Seine. (Sources: Fiessinger and Mallevialle, 1978: 246; Dézert, 1981: 270; Chevreuil *et al.*, 1987: 428)

mercury and total coliforms which exceed EC minimum guidelines for drinking water quality. The guide levels for phenols, ammonia and faecal coliforms were all exceeded at some point during a year of study (Fiessinger and Mallevialle, 1978). The pollution is significant because drinking water is abstracted from the river particularly at periods of low water. The presence of water-borne viruses and heavy metals may also pose a danger for swimmers and for consumers of fish (Soudan, 1975).

Partial responses to the problem of polluted water supplies have been the development of monitoring schemes and artificial recharge of aquifers. Aquifers in urban areas have been over-exploited so that water yields are not only declining but are increasingly liable to contamination (Vigouroux *et al.*, 1983). The quality of water from the Rhine aquifer, deemed to be over-used and vulnerable to pollution, is now monitored by a data bank established for management purposes (Risler *et al.*, 1986). Both the problems of yield and pollution can be overcome by the artificial recharge of aquifers in winter with surface water. This has been undertaken since 1959 at Croissy-sur-Seine, near Paris, through the use of lagoons. A substantial recharge rate of 20 million m^3 per year was possible by the late 1970s and serves both to augment water supplies and to improve their quality (Edworthy and Downing, 1979).

The problem of water pollution affects not only rivers but also coastal waters. The most common type of pollution is by toxic heavy metals which are released not only by natural processes such as weathering, but also by industrial processes. Levels of mercury, cadmium, zinc and copper around the coast are generally comparable to those found in the open oceans, except for localized concentrations near the mouths of the Seine and the Rhône. In these waters, potentially hazardous concentrations of heavy metals of approximately 14 μg per litre have been recorded. Heavy metals such as mercury accumulate and do not move far from the point of discharge; hence accumulations are very likely in stationary fauna such as molluscs. In some cases the levels of mercury in fish have been found to be above EC recommended levels, but this is not thought likely to pose a general health hazard (Bernhard, 1981).

Mineral resources

France is not well endowed with mineral resources; processing is more important than extraction. The primary energy resources are notably lacking, especially in the case of oil, the known reserves of which are nationally insignificant. Natural gas was discovered in the Lacq region of southwest France in 1951, and exploited from 1957 (Gaz de France, 1985). Other deposits of natural gas have subsequently been exploited in the same region, while there are two smaller commercial deposits, one near Strasbourg, the other in Marne. The deposits at Lacq occur at great depth (3300 m), and are composed mainly of methane with a significant quantity of sulphur which is extracted as a by-product. Estimates of the size of the deposits have been continually re-evaluated upwards, but France still has insufficient natural gas for national requirements. These are met by imports from Algeria, the North Sea fields and the countries of the former

USSR. There is significant variation in the seasonal demand for natural gas; Paris uses five times as much gas in December as in August (Gaz de France, 1985). Storage of gas is therefore essential, and major developments have occurred in underground storage either in natural aquifers or in cavities dissolved out of salt deposits by the injection of warm water, for example, at Tersanne in southeast France, where compressed gas is stored 1550 m below the ground surface (Gaz de France, 1987).

For many years the major primary energy source was coal. France formerly possessed three major regions of coal production: the Nord–Pas-de-Calais, a field now essentially worked out; the fields of Lorraine, still producing two-thirds of the nation's coal, but a shadow of their former selves; and the scattered fields of the centre, south and Alps, deemed to be of strategic importance in wartime and still operating on a small scale, some using highly mechanized open-cast technology. The production of coal has declined substantially and significantly from a peak of 60 million tonnes in 1958 to 33 million tonnes in 1971 and only 15 million tonnes in 1987 (see pages 132–4 and Fig. 4.6). There are major structural problems which inhibit the economic extraction of coal, principally the great depth of the deposits, and the alignment of the seams, which are thin, faulted and steeply dipping. The decline in the two major northern coalfields results from the removal of the most accessible coal; the remaining coal is difficult and expensive to work. The recent decline of coal production also reflects the late restructuring of the heavy industries of France (see pages 112–16). New developments are evident in the open-cast mines of the central-south region, for example, at Decazeville in the Aveyron. Open-cast mining is cheaper, and has a quicker lead time than the deep mining of the north, but still accounts for only 10 per cent of national production. The French coal industry appears to have little future and, unlike the British industry, was unable to react positively to the opportunities of the oil crises of the 1970s. The policy of the French coal industry is for continued rationalization of production, involving closures, automation, a switch to open-cast mining and a continued dependence on imported supplies. The reduction in the national supply of coal has been matched by a reduction in demand for coal for electricity production, for the iron and steel industry, and for domestic heating (Charbonnages de France, 1988).

France has a limited supply of non-conventional power sources. France was originally estimated to possess about five per cent of the world's uranium supply; this prospect of self-sufficiency and a reduction in dependence on imported oil was a powerful incentive to develop one of the world's largest nuclear power industries. Uranium is mined in central France and is processed and recycled at a number of sites, especially at Pierrelatte on the Rhône. Geothermal power has become a minor but significant resource, mainly in the Paris Basin. In the Melun region, geothermal energy extracted from >2000 m depth has been used since 1971 to heat apartment blocks, at a cost saving of about 13 per cent on conventional oil-fired heating (Crabbe and McBride, 1978).

Other minerals which are produced at a level which is significant on a world scale include arsenic (19–20 per cent of world supplies), diatomite (14–15 per cent), gypsum (8–9 per cent), potash (6–7 per cent), iron ore and bauxite (2–3 per cent each). France is also generally self-sufficient in basic mineral resources such as salt, limestone and sand and gravel. In recent years, the extraction of all major minerals has been in decline. The trends are exemplified by the production of both iron ore and bauxite. France was the world's leading producer of bauxite before the Second World War; bauxite was originally discovered (and named) at Les Baux in Provence in 1821. The aluminium industry using the bauxite grew up in association with the development of the hydro-electric capacity of the Alpine rivers. Peak production was reached in the late 1960s and early 1970s when over three million tonnes of bauxite were produced annually. In the 1980s, closure of aluminium plants and world oversupply combined to reduce production to less than two million tonnes per year. Similar trends are evident in the production of iron ore, although France was never a world producer. The main source of supply is Lorraine, where low-grade iron ore reserves were exploited from the late nineteenth century when techniques became available to process the impure phosphoric ore. Peak production occurred after the Second World War, with 54 million tonnes of iron ore produced in 1974. Decline since then has been very dramatic but fluctuating, occurring hand-in-hand with the demise of the heavy industry complex of steel and textiles based on the coalfields.

Environmental protection

The adoption of anti-pollution legislation

Concern over the damaging effects of acid rain (see pages 13–15) has stimulated national and international legislation to control pollution. In 1979, an international Convention on Long Range Transboundary Air Pollution held in Geneva declared its intention to reduce noxious emissions. Initially the declaration was seen as a statement of intent rather than an executive prescription (Sand, 1987), but in 1983 the Geneva Convention on Transboundary Air Pollution was ratified by 31 countries. However, it was not until 1987 that a Protocol on the Reduction of Sulphur Emissions by At Least 30 Per Cent came into force. By this time, a number of countries had already made unilateral decisions to reduce their sulphur emissions. In 1984, following a lead set by West Germany, France announced its intention to halve its 1980 levels of SO_2 emissions by 1990. Other countries such as Denmark and Belgium gave a commitment to a 30 per cent reduction. The 30 per cent target figure was arbitrary and a number of countries achieved this target as early as 1986.

The EC also responded to the environmental debate with a directive issued in 1984 to reduce emissions from large combustion plants. Large combustion plants include power stations, chemical works and metal manufacturing and processing plants. All these establishments were required by law to reduce their emissions of SO_2 by 60 per cent, of NO_x by 40 per cent, and of dust by 40 per cent. These measures, applicable from January 1985, are to be achieved by 1995. The same 1984 directive also sought to regulate emissions from vehicle exhausts. This was to be achieved by making unleaded petrol widely available from 1989, and by requiring all cars to have clean motor exhausts by 1995 or earlier. These measures were to apply initially to large cars and subsequently to medium and small cars, and to all new cars by 1993. Many conservationists argued that these measures were too lenient and too late. The EC directives also specify close monitoring of forests, and increased research on acid rain, while 92 measuring stations for the monitoring of air pollution were established by 1987 under the Geneva Convention. In 1988, attention was turned to the control of emissions of nitrogen oxides, and a protocol is being drawn up similar to that in place for sulphur.

France is party to a similar international agreement on the control of water pollution in the Mediterranean which was drawn up with other Mediterranean countries in 1976. This agreement attempts to control noxious discharges into the Mediterranean. Several substances are not permitted under any circumstances; the blacklist includes PCBs, DDT, mercury and radioactive waste. The dumping of other substances requires special permission; these substances include zinc, copper, lead, cyanides and fluorides. In 1981, lists of endangered species were compiled and France is an active contributor to the programme of monitoring.

France itself drew up anti-pollution legislation in 1975. The main thrust of this legislation was to cope with the disposal of industrial waste. The law established collective treatment centres for industrial waste, which were set up near existing industrial centres. Industrialists were compelled to use these for waste disposal, but at subsidized cost. Alternatively, plants could establish their own treatment centres with 50 per cent grants and 20 per cent loans from the national government. Despite the subsidized cost, small firms in particular were bitterly opposed to the legislation on the grounds of increased costs at a time of severe competition and restructuring (Dézert, 1981). Opposition was also felt from industries with older factories which produced considerable levels of pollution, such as the electro-chemical works at Val de Livet in the Alps. Larger firms were in general better able to cope with the new demands, and new factories have taken on board the requirements despite an increase in costs. The extra costs were estimated at seven per cent of total cost at the iron and steel plant at Fos on the Mediterranean, and at 14 per cent of total cost for a new chlorine factory in the same region. The legislation is seen by Dézert (1981) as a factor in the changing industrial geography of France, providing a stimulus pushing noxious industries to the periphery.

Protection of environmentally-sensitive areas

France has a comprehensive land-use planning system, which is considered in more detail on pages 160–1. Special protection is accorded to areas of environmental sensitivity. Areas designated for special protection include National Parks, regional nature parks and the coastline. These areas of mountain and coast are coming under increasing pressure both for permanent residence and for seasonal development for recreation and tourism.

Seven National Parks have been designated since 1963, much later than in many other countries. These zones are extensive areas of mountainous terrain of particular scientific interest, usually of a biological or a geological nature. They are almost totally unpopulated and the central park zones are deliberately difficult of access to maintain protection of the natural environment. There is total control over building in the central park zones, and many prohibitions on use including overnight camping. Tourist access is restricted to well-defined 'honeypots' in an established peripheral zone, such as the small town of Gavarnie at the edge of the Pyrenees National Park. The National Parks are established, financed and maintained by the state.

The National Parks are complemented by a greater number of regional nature parks; the 21 which have been designated since 1967 cover 2.8 million hectares, that is, 5.2 per cent of the national area (Fig. 1.8). They consist mainly of large areas of mountainous, coastal or rural land, usually of exceptional beauty or landscape value, and containing a large proportion of the nation's forest. They are less stringently protected from development than the National Parks, as they are designed not only to be a means of protecting nature but also of preserving rural space from urban encroachment. Since 1975, they have become less and less controlled by the state but are increasingly run as joint ventures between the local *communes* and the *département* and regional governments. The regional nature park in Corse is one such joint venture which contains a large nature reserve, but also caters for tourists with a long-distance footpath and ski resorts, and plays a supplementary role in rural renovation and education (Richez, 1983). In recent years, several new regional nature parks have been designated in the more industrialized lowland areas of France (Fig. 1.8). Wherever they are located, they are increasingly playing an active role in rural regional development, particularly in the exploitation of historic and recreation sites for tourism; examples of this type of development include the proliferation of regional museums, *écomusées*, craft centres, *zones artisanales*, and tourist routes.

Since the early 1970s, special protective measures have also been accorded to the coastline, because of the increasing pressure on the coast from tourist resorts, counterurbanization and large-scale industrial developments such as power stations and iron and steel works. A recognizable coastal policy was first formulated in 1972 (French Ministry of the Environment, 1982). The policy is designed to protect and enhance coastal areas by extending existing land-use planning. Coastal *départements* have been encouraged to designate 'sensitive perimeters', and 22 of 25 *départements* have done so. Some model areas have been designated for integrated coastal management to combine the needs of leisure and the environment; these include Combrit in Finistère, Ecault in Pas-de-Calais, Sallenelles in Calvados and the East of Dunkerque in Nord.

A second arm of coastal policy has been the acquisition of coastal land. This has been achieved through a body established in 1975, the *Conservatoire du Littoral*, which acts in much the same way as the British National Trust. This body can acquire coastal land, by pre-emption or expropriation; the land is then usually managed by other bodies. Once acquired, the land is protected, for example by the provision of access paths across sensitive areas or by the stabilization of dunes. By 1979, the *Conservatoire* had acquired 10 000 hectares, approximately 120 km of coast, but only a small proportion of the land under threat.

In 1979, the French Ministry of the Environment issued a coastal directive to codify practices to maintain the delicate balance between competing land uses. Coastal policy aims to ensure public access to all beaches, and to maintain the agricultural and aquacultural uses of the coast. The policy also aims to control linear urban development and other potentially harmful developments such as marshland drainage. These measures,

Fig. 1.8 National parks and regional nature parks, France. (Data from Laborie *et al.*, 1985: 74)

designed to protect and enhance coastal areas, are made operational through the land-use planning and designation process outlined above (French Ministry of the Environment, 1982).

An example of land-use conflicts on the coast occurred on the sand dunes at Le Touquet, in Nord–Pas-de-Calais (Ernecq, 1988). Conservationists were anxious to preserve and stabilize the dune system, which had been used for many years as the site for a major annual moto-cross competition. This usage also resulted in year-round damage from practice, but the competition was an important event for organizers, competitors and local retailers. The preferred solution to the problem was the relocation of the moto-cross to a nearby area which was less environmentally sensitive.

Limiting and repairing damage

The reclamation of derelict land is a large-scale problem in old industrial areas of the country. The region of Nord–Pas-de-Calais alone has about 10 000 hectares of derelict industrial land, about one per cent of its total area (Ernecq, 1988). Derelict land results from industrial decline and restructuring and from widespread exploitation of land resources; many of these areas exhibit related problems of soil contamination. The decision to reclaim land was made on economic rather than on environmental grounds, and was seen as a measure to absorb some of the unemployment caused by industrial restructuring, rather than as a means of maintaining the ecological balance. The reclamation programme in Nord–Pas-de-Calais has taken an annual investment of FF 73 million, and created 340 jobs. The responsibility for reclamation lies with *commune* councils, but many of these local bodies lack both expertise and conviction, and increasingly expertise is being pooled between the regional government, the *communes* and consultancy firms. Other attempts at damage limitation include the recycling industries which are still very much in their early stages. French industry is considered to be relatively resistant to the introduction of new, clean technologies. Industrialists are wary of the extra costs involved, and they are reluctant to implement new technology unless it has demonstrated technical and economic advantages as well as environmental benefits. The adoption of environmentally-sound technologies may be stimulated by the decision of the EC in 1989 to become involved in policies of waste disposal and elimination, recycling and the management of polluted land and water.

In order to minimize damage it is necessary to have an accurate picture of the areas susceptible to hazard. From 1982, an inventory of areas subject to floods, avalanches and mass movements has been prepared resulting in a series of maps showing hazard-prone areas (Besson, 1985; Humbert, 1987). The maps have been incorporated into the land-use planning system, and zones established where construction is prohibited or controlled. In mountain zones, a series of measures has been taken to minimize hazards. These have been classified as measures of active defence, passive defence and temporary defence. The strategy of active defence involves tackling the physical phenomenon itself, for example, revegetating slopes to prevent landslides or gullying. Passive defence involves protection of the buildings or object under threat, but does not tackle the cause of the problem. Hence, it may be possible to stop, slow or divert falls of rock or ice; this approach is often relatively cheap, but it is recognized that damage may be caused elsewhere. Temporary defensive measures generally include surveillance and evacuation.

Green politics

One of the most dramatic changes in public perception since the late 1970s has been the growth of popular ecological awareness, and of political support for environmental policies. The stimulus to ecological activism came primarily from reaction to a number of environmental disasters, and from opposition to the nuclear power programme (see pages 130–3). Also the success of the Green Party in West Germany was a powerful positive stimulus to like-minded groups in other European countries.

In March 1978, the oil tanker Amoco Cadiz was shipwrecked on the coast of Bretagne, resulting in the worst oil pollution disaster ever recorded at that time. The Amoco Cadiz shed 230 000 tonnes of crude oil, creating a huge slick along 160 km of beaches from Brest to the north coast of Bretagne. The oil caused a massive loss of marine life and brought the fishing, oyster and lobster industries to a halt; the profits from the tourist season were irreparably damaged. The clearing up involved 3000 troops and many volunteers. The costs were estimated at FF 370 million, not including the losses sustained by the tourist industry. This disaster focused public attention on the vulnerability of the natural environment as a result of the normal operations of the oil industry. An interesting footnote to the story is that in 1988, ten years after the event, Standard Oil of Indiana was ordered to pay US $ 85 million in damages and accrued interest to France and Bretagne; this order immediately became the subject of an appeal.

During the late 1970s, popular opposition to the nuclear power programme reached crisis point in West Germany and, to a lesser extent, in France. In 1977, clashes between demonstrators and police at the Super-Phénix site resulted in 100 people being injured, one death, and several demonstrators receiving prison sentences. Further opposition at the Breton site of Plogoff erupted in 1980. These demonstrations followed international concern over French nuclear bomb testing in the Pacific. Since 1966, the French government has exploded 105 bombs surrounded by a veil of secrecy; their high-handed dealings with opposition from groups such as Greenpeace lent credence to scares of nuclear spillage and of radioactive fallout. Their answers to queries over the health and environmental implications for Polynésie Française have been described as 'half-truths, glaring omissions and downright lies' (Danielsson, 1984).

More recently, the impact of potential disasters has come uncomfortably close. The Seveso chemical disaster in Italy, when an enormous cloud of toxic dioxins escaped after an explosion, had repercussions in France. A French firm of subcontractors to Hoffman-La Roche agreed to remove dioxin-contaminated waste from the Seveso site; subsequently 41 'missing' canisters of the waste were found dumped in a disused abattoir near St Quentin. This discovery was closely followed by the arrest of the firm's director. Possibly the most significant environmental disaster was the explosion of the nuclear reactor at Chernobyl in 1986, when radioactive fallout affected much of western Europe. Arguably the repercussions of Chernobyl have been the single most important influence in rallying support to the Green Party.

Concern for the environment was a major campaign issue in the 1989 European elections (Coles, 1989). The Green Party, *Les Verts*, gained nine seats and over 10 per cent of the vote, compared with only three per cent in the 1984 elections. In 1989, the Green Party emerged as an alternative centre force; as a result, both the socialists and the right wing scrambled to show their concern over environmental issues. Local elections in 1989 also resulted in the Greens obtaining seats in large towns such as Rennes (four seats), Nancy and Rouen (three seats each) and Tours and La Rochelle (two seats each). In 18 towns of over 20 000 population, the Greens gained increased support although they engaged in a variety of local political alliances. The French press reported that the elections were notably affected by local issues; for example, the support for the Greens in Bretagne was largely attributed to concern over water pollution from nitrate fertilizers and from the animal wastes produced by battery breeding establishments.

It is not yet clear whether the Greens will continue as a major independent political force, or whether environmental issues will become subsumed in the mainstream economic concerns of politics. Ecological concerns may fade from the foreground under conditions of deep economic recession. The Greens walk a political tightrope between popular ecological consciousness on the one hand and continued consumer demand on the other, often from the same people. Sallnow and Arlett (1989: 14) considered that 'for the Greens to be successful they need to show how environment-oriented policies can produce a cleaner, greener, form of growth'.

CHAPTER 2

A distinctive demography

'And my fridge is here!' Paulette exclaimed, tapping her belly in front of the other housewives in the co-op.
For us to have a fridge we'd need at least triplets. Maman glared at her rival, who was five weeks more advanced than she was.
'And I'm going for a washing machine!' . . .
A young mother of only three, who wasn't expecting her fourth until spring, looked at her elders with admiration.
'Don't worry, Madame Bon', said the grocer 'they'll come one by one without you even noticing'.

Rochefort (1982: 84–5)

Contemporary population patterns

Contemporary patterns of population distribution in France reflect a distinctive demographic evolution. Many countries of Europe experienced rapid population growth from the nineteenth century onwards. This was caused by demographic transition, that is a sequence of changes in fertility and mortality rates from a pre-industrial phase of high fertility and high mortality to a post-industrial phase of low fertility and low mortality. The intervening years, when fertility rates greatly exceeded mortality rates, were a period of sustained population growth and industrialization. While countries such as Britain and Germany underwent this demographic transition, levels of fertility in France fell in parallel with mortality levels. The vast, thinly peopled areas of rural France result primarily from a period of prolonged population stagnation followed by a rapid rural exodus which accompanied economic and urban development after the Second World War. The distinctive features of French population change are its slow growth during the nineteenth and early twentieth centuries and its rapid expansion in the post-war baby boom.

From the 1960s, French demographic trends became more similar to those of the rest of Europe, with a resurgence in natural growth of unprecedented proportions and immigration from abroad fuelling the post-war recovery. Internal patterns of migration before the Second World War showed a huge rural exodus, with the destinations dominated by Paris. After the Second World War, the pace of urban growth quickened, resulting in the suburban expansion of most major cities.

New population patterns have begun to emerge since the 1970s. Natural growth has slowed from its post-war peak and internal migration rates have declined. Migration from abroad has also been reduced in scale; restrictions on immigration have been caused not only by the economic recession of the 1970s but also by a growing national xenophobia (White, 1986). Internal migration since the 1970s, although slower than in former years, has begun to show evidence of both counterurbanization (see pages 53–5) and gentrification (see pages 162; 211). In contrast to earlier patterns of rural-urban migration, these movements are dominated by people of higher socio-economic status. These migrations are clearly related to rapidly changing social structures and to patterns of housing demand and supply. Despite these recent trends the French population still carries the legacy of its historical evolution (Ogden and Winchester, 1986). The effects of the past are particularly evident in the population age-sex structure, in national attitudes to population growth, in the ethnic and national composition of the population of France and in the pattern of regional population distribution.

Table 2.1 Population change in selected French *départements*,1846–1911

	1846	**1911**	**% change 1846-1911**
Seine	1 364 000	4 154 000	+205
Nord	1 132 900	1 962 000	+73
Pas-de-Calais	695 700	1 068 000	+54
Basses-Alpes	156 000	107 000	−31
Hautes-Alpes	133 100	105 000	−21
Lozère	143 000	123 000	−14

Source: Huber, 1931.

The distribution of the population in France has changed in response to the development of the country's resource base and to the dramatic restructuring of the national economy; this development has been spatially uneven. Figure 2.1 shows the distribution of the population in 1982 and almost one hundred years previously. The maps reveal significant polarization of population in the 1980s by comparison with the distribution a century earlier. Out-migration from the rural areas began as early as the 1830s, but throughout much of the nineteenth century this loss was camouflaged by relatively high rates of natural growth. At the end of the nineteenth century, the rural areas in general supported a substantial population maintained by a high birth rate; natural environmental conditions allowed only limited improvement and extensification of agriculture to provide a living for this growing rural population. Areas such as the

Massif Central and Bretagne formed rural reservoirs of population which have subsequently been drained to supply the growing urban and industrial agglomerations. Even by the early years of this century, a massive redistribution of population was under way, as people left rural areas to seek wider opportunities in the cities. This redistribution is shown by the radical alteration in the population of the three most populated and the three least populated *départements* during the second half of the nineteenth century (Table 2.1).

The areas of net population loss at the end of the nineteenth century were the remote, mountainous, rural areas of the Alps, the Massif Central, the Pyrenees, the Jura and the Morvan Plateau, where a progressive reduction in population was caused by a combination of selective out-migration, demographic ageing and a reduction in birth rate. Meanwhile, population growth was occurring in Paris, which by 1911 already contained almost 3 million people, at very high densities in the central city (Fig. 2.1). The industrial coalfield-based agglomerations of Nord and Pas-de-Calais also showed rapid growth, as migrants were attracted to areas of newly-developing industry.

The contemporary pattern of population distribution is strongly influenced by this historical pattern of rural to urban movement. In the 1980s, the areas of lowest population density were the upland rural areas of the southern Alps, the Massif Central and the Pyrenean border, all of which are peripheral to and remote from the major industrial complexes of the north and northeast. These urban-industrial areas, together with the cities of Lyon and Marseille, constitute the major concentrations of population and employment outside Paris (Fig. 2.1). Areas of moderate population density occur in the hinterlands of the major cities, the Paris Basin, the Rhône Valley and the Mediterranean littoral. In all cases, these areas benefit not only from proximity to urban areas, but also from the advantages of coastal, riverine or coalfield locations. At a regional level, there is an imbalance in French population distribution; if a dividing line is drawn between the mouths of the Seine and the Rhône, the areas to the north and east contain 77 per cent of the population on just under 50 per cent of the land area, creating problems of relative underdevelopment in the south and west, and congestion and resource exhaustion in the north and east.

Natural population change

Pre-war demographic stagnation

The major distinguishing characteristic of nineteenth and twentieth century French demographic trends has been that of stagnation, particularly when compared with the significant growth taking place in other European populations. The demographic transition from high fertility and high mortality to conditions of low fertility and low mortality took an early and distinctive form in France (Wrigley, 1985). The early decline in levels of fertility was unprecedented in the nineteenth century (Fig. 2.2). By 1881, the crude birth rate in France was only 25 per 1000, compared with 35 per 1000 in the United Kingdom. France therefore never underwent the period of demographic expansion which characterized her neighbours. During the nineteenth century, the French population grew from 28 million to 40 million, an increase of some 43 per cent, trifling in comparison with the flourishing growth of the German population from 22 to 63 million, an increase of 186 per cent; or the British population which rose from 16 to 40 million, an increase of 150 per cent over the same period.

The causes of the early decline in fertility are far from clear, but have been linked by Bourgeois-Pichat (1965) with social and economic upheaval in the post-revolutionary era and particularly with the desire to hand on land and property intact under conditions of partible inheritance. White's (1982) study of Normandie demonstrated the importance of property as an influence on fertility; the land-owning classes reduced fertility significantly more than the landless. The restriction of fertility was facilitated by a late age of marriage for women, by high levels of celibacy and by societal mores inhibiting extra-marital sexual relations. The low number of children per woman also indicated the effective practice of contraception within marriage, particularly by *coitus interruptus* and by the abortion of unwanted foetuses. Although this pattern of late marriage and low fertility was characteristic of most Protestant proto-industrial Europe, it reached its most extreme form in France, which was largely a Catholic country.

The population stagnation of the period was also brought about by high levels of mortality, which were generally higher than in the United Kingdom. Infant mortality rates were particularly high because of poor nutrition, sanitation and health care; these conditions were exacerbated in the war years of the 1870s, 1910s and 1940s. The First World War of 1914–18 claimed 1.3 million lives as a direct result of military action, and the total population deficit from those years is estimated at about 3 million, including children never born and deaths of poorly-nourished people in the influenza epidemic of 1919. This deficit had a profound effect on socio-demographic structures as the loss of hundreds of thousands of young men had an immediate impact on the domestic labour force and on the marriage market. As a consequence of this loss, women became increasingly important as wage-earners and farmers; many of these women never married. The repercussions of the First World War may still be seen in the national age-sex pyramid (Fig. 2.3).

The Second World War, because of the rapid capitulation of the French army, was less devastating in its death toll, but this nevertheless approached one million, including 250 000 servicemen and 350 000 civilians killed in air raids and during forced labour abroad. There was also an above-average mortality during the wartime period, estimated at about 297 000 (Dyer, 1978). Apart from these major losses, general levels of mortality fell steadily throughout the century. Life expectancies in 1900 were 48 years for men and 52 years for women; by 1946, life expectancies had risen to 62 years for men and 67 years for women (Van de Walle, 1979). Mortality levels were generally worse than in the United Kingdom, particularly deaths from tuberculosis and typhoid. This was the result of poor housing, welfare and medical conditions especially in the urban and industrial areas of the north. In the early part of the century, the practice of

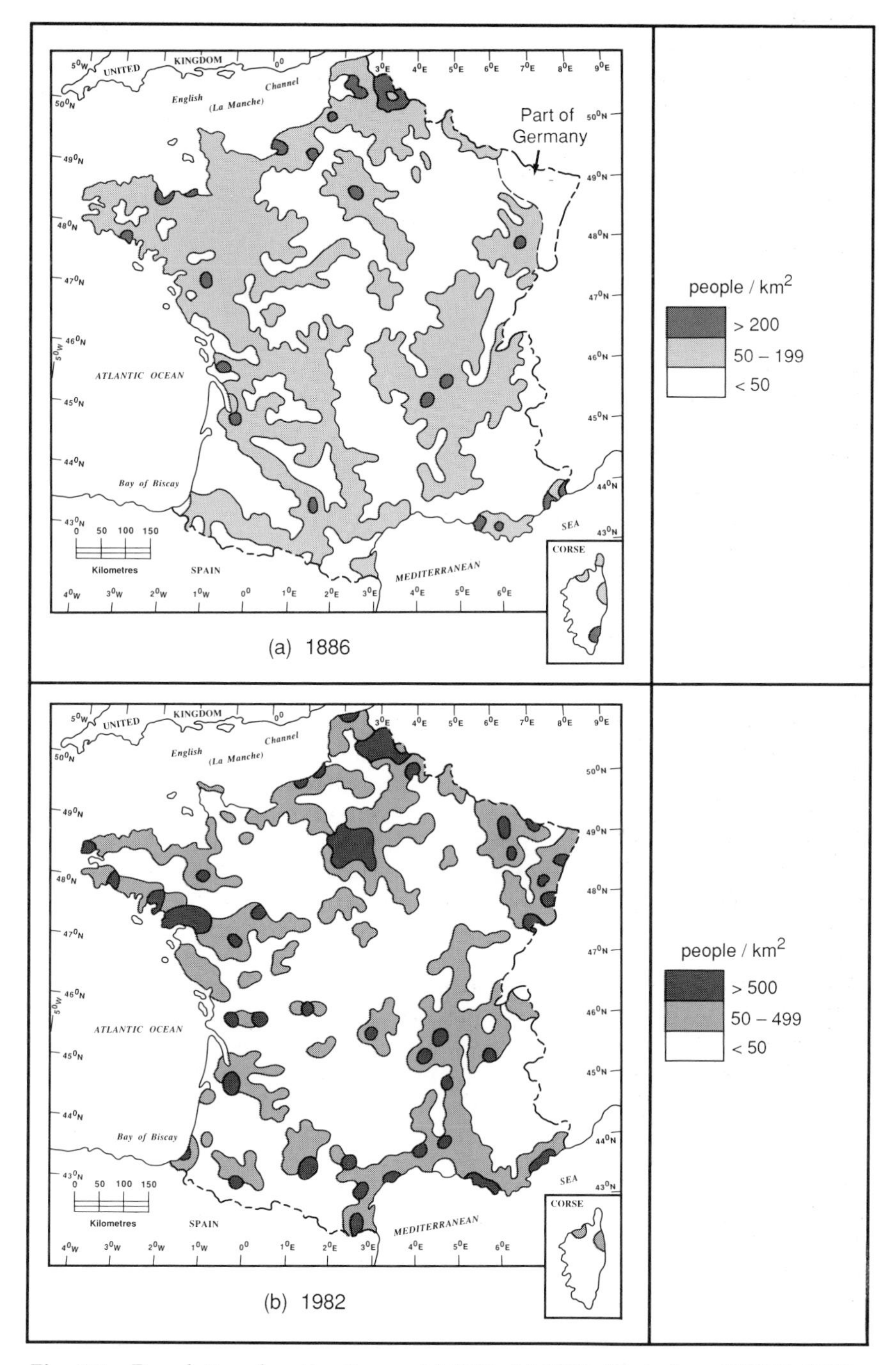

Fig. 2.1 Population density, France (a) 1886; (b) 1992. (Data from INSEE, 1982)

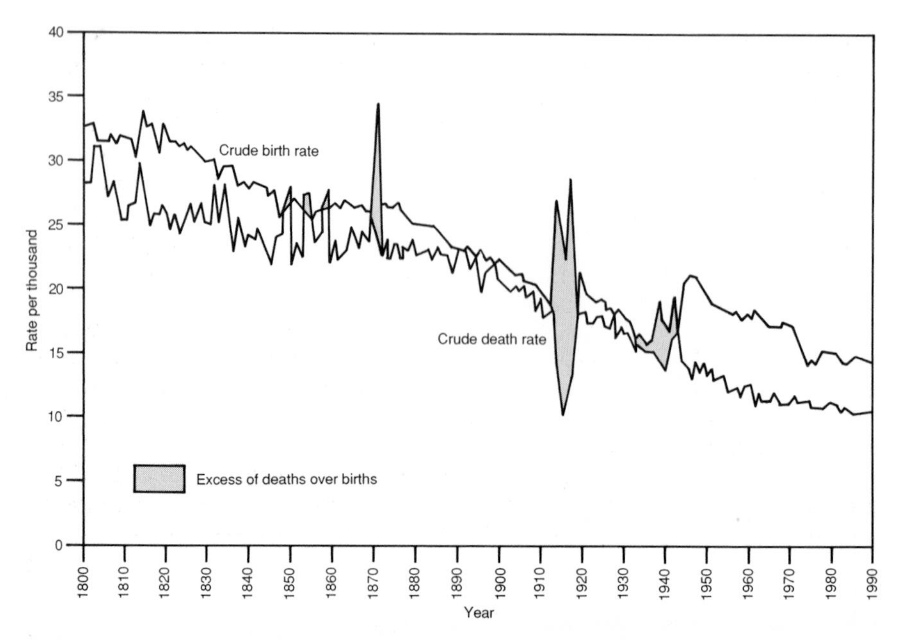

Fig. 2.2 The demographic transition, France, 1800–1990: trends in fertility and mortality. Note the sharp increase in fertility in the 'baby boom' of the 1940s. (Data from INSEE, 1982; OECD, 1991)

wet-nursing contributed to high levels of infant mortality. Van de Walle (1979) estimated that more than a quarter of all infants were sent to wet-nurses in the 1880s; the level of wet-nursing was significantly positively correlated with infant mortality. More than one child in 10 died in the first year of life during the 1920s, a rate which did not improve significantly until after the Second World War (Nadot, 1970).

Although the elevated death rate contributed to demographic stagnation, the most significant factor in slow population growth was undoubtedly the unusually rapid reduction in births. In countries of the Third World, high levels of infant mortality are considered to be a major factor contributing to high fertility, but in the early years of this century that relationship did not exist in France. Fertility in France underwent a slow decline from 1831 to 1931. In 1929, France had a deficit of births, and between 1934 and 1939 there were fewer births than deaths in every year. There was a minor baby boom after the First World War when exceptional numbers of male children and twin births were recorded for a short period, but generally the era was one of slow growth which left an indelible mark on the population, both in terms of attitudes and in terms of demographic structure. The effects of population decline were cumulative, with fewer young people reaching marriageable age. This became particularly obvious in the late 1930s when the unborn babies of the First World War should have been making for the altar and the marriage bed; in fact, only half the numbers of marriages were being celebrated as had been the case a decade previously (Dyer, 1978). In 1920, there had been more than 600 000 marriages, but by 1938 fewer than 300 000, and as over 90 per cent of children were born within marriage, this decline in marital fertility brought about a significant cumulative effect on population decline. Furthermore, after the First World War, marriages were increasingly ending in divorce; in 1913, 13 500 divorces were declared, but 26 400 by 1938. The ageing of the population also reinforced conservative attitudes and children were seen as disruptive influences, causing financial hardship, housing and inheritance problems as well as noise and clutter. The prevailing attitudes of the time, and the examples of parents and peer groups, were not conducive to demographic revitalization.

At a national level, the demographic stagnation and the economic depression of the time were giving cause for concern. In the 1920s, measures were adopted to prohibit abortion and the sale of contraceptives. After the natural decline of the 1930s, the family code, *Code de la Famille*, which provided financial incentives for those couples having children, was introduced in 1939 (see also pages 46–9). Concern was prompted by unfavourable demographic comparisons with France's European

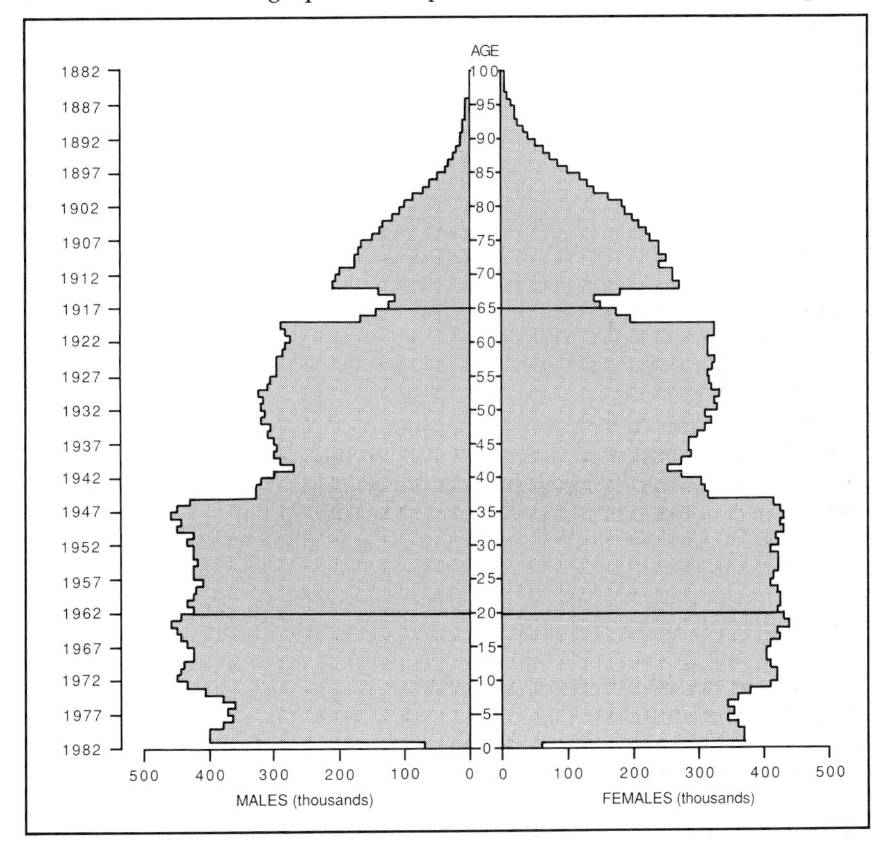

Fig. 2.3 Age-sex pyramid, France, 1982. Note the 'bites' in the pyramid caused by low birth rates during the First and Second World Wars. (Source: INSEE, 1982)

neighbours, particularly Germany: these fears of relative population decline were felt to be demonstrably justified when Germany occupied France early in the Second World War. Militaristic concerns and the more obvious extremes of national chauvinism are now much diluted, but pronatalist policies are an enduring and distinctive feature of French national thinking, despite the demographic miracle of the post-war years.

Post-war population growth

In fewer than forty years after the Second World War, the French population has grown by some 14 million, compared with only 12 million in the preceding century and a half. The major cause of this phenomenal rise in population was a 'baby boom', a period of unusually high fertility which lasted until 1965 (Fig. 2.2). The relative significance and size of the post-war baby boom in France was greater than in any other country of western Europe. This prolonged period of raised fertility was preceded by an increase in the number of marriages. This began in 1942, the very middle of the war for Europe, but after the occupation of France by German forces. By 1943, births exceeded replacement levels and carried on rising into the late 1940s (Dyer, 1978: 128).

For the country as a whole, the increase in births may be seen as a nationalistic response to the German occupation and as an attempt to remedy the deficiencies of the relatively childless years of the Depression. Individually, procreation could be seen as a sensible social and economic reaction to the provisions of the *Code de la Famille*: the material benefits conferred by children included individual ration cards for each child, concessions for large families and a reduced likelihood of the father being called up for military service. Financial advantages could therefore accrue from the presence of children, and in the post-war years further benefits in terms of housing facilities and medical care became available. Attitudes to child-bearing became much more positive and the French people undertook a remarkable demographic expansion.

Although the most significant factor in the post-war demographic expansion was the increase in fertility, a small element of growth is attributable to a reduction in mortality rates. After the elevated death rates of the war years, mortality stabilized at relatively low levels, about 13 per 1000, comparable with the rest of Europe. By 1962, these rates had fallen to about 11 per 1000, aided by a marked reduction in the rate of infant mortality from 56 per 1000 in 1946 to 26 per 1000 in 1962. The continuing improvement in mortality rates has increased life expectancy from 64 years for men and 69 for women in 1950 to 72.5 years for men and 80.7 years for women in 1989. Similarly, infant mortality rates continued to improve, from 10.9 per 1000 live births in 1976–80 to 7.4 per 1000 live births in 1989. The low rate of infant mortality must be attributed, at least in part, to the extra payments made to women who attended ante-natal clinics during the first sixteen weeks of pregnancy, the financial incentive producing an improvement in ante-natal care. The general mortality rate appeared to approach a stable level in the mid-1970s at around 10 per 1000 (Hall and Ogden, 1983). Subsequently, it has improved marginally to

9.4 per 1000 in 1988 and 1989 (INSEE, 1990).

There are, however, still significant transitions occurring in the major causes of death. The infectious and contagious diseases are of generally diminishing importance as preventive medicine has improved. There are a few notable exceptions to this general trend. The incidence of malaria and some other tropical diseases has actually risen, stimulated by increased foreign travel. Deaths from AIDS, although small in number, have received enormous media attention. The major causes of death are the degenerative diseases; circulatory diseases account for 37 per cent of all deaths, while cancers cause a further 24 per cent. Significant variations exist in the causes of death between regions, social classes and the sexes. One of the most persistent regional differences is the elevated levels of mortality from all causes in the north of France from Bretagne to Alsace (Noin and Chauviré, 1987); this includes circulatory disease, cancers and alcohol-related deaths from disease and accidents. The death rate for men from alcohol-related deaths and from violence is significantly higher than for women, reflecting differences in the social environment and way of life for women and men.

The period from the mid-1940s to the mid-1960s was therefore one of unprecedented population growth as a result of elevated fertility and reduced mortality. The demographic miracle of 1946–62 resulted in a net gain of 6.6 million people, from 14.0 million births and 9.0 million deaths, the balance being made up by 1.6 million immigrants. Since the mid-1960s, however, the situation has altered, with a fall in the crude birth rate, which is also evident throughout Europe. Generally, the fall in the birth rate may be associated with the changing role of women in the paid labour force and with an increase in the perceived costs of child-rearing; this fertility reduction has been facilitated by the greater availability of oral contraceptives and abortion. In France, the legalization of contraception occurred with the *Loi Neuwirth* of 1967 and the legalization of abortion with the *Loi Veil* of 1975. The increase in the perceived costs of families may be related in part to a decline in the buying power of family allowances relative to salaries by 1967. By the 1970s, however, economic depression and a prolonged period of uncertainty and rising unemployment were also affecting attitudes to child-rearing. In the late 1970s, fertility rates fell temporarily below replacement levels, but by the early 1980s they had recovered and were higher than any other country of northwest Europe, except for Eire, and were only marginally lower than some countries of Mediterranean Europe (Hall, 1986). The French fertility rate was still 13.6 per 1000 in 1989, higher than in most other European nations (INSEE, 1990).

Short-term fluctuations in birth rates are a characteristic of post-industrial societies and may possibly be related to economic cycles of prosperity and depression. Furthermore, just as the effects of population decline are cumulative, so periods of demographic expansion show an 'echo effect' as the products of each baby boom reach the age of family formation. Recent upswings in the birth rate may be attributed in part to this echo effect of large cohorts and to the appearance of infants postponed from the 1970s when potential mothers joined the workforce outside the home; this latter factor is demonstrated by an increase in births to women aged over thirty. The relatively high level of natural growth has been related to the continuing importance of the family, supported by high family allowances, a relatively low divorce rate and the low level of adoption of contraceptive practices (Sullerot, 1978). Nevertheless, overall, the intercensal period 1975–82 revealed a reduction in the level of natural growth and total population growth when compared with the preceding intercensal periods (Table 2.2). This gave rise to renewed concern over population levels during the 1980s. Preliminary results of the 1990 census showed an increase in the total population of France from 54 300 000 in 1982 to 56 556 000 in 1990, an increase of four per cent since 1982, and an annual rate of increase of 0.5 per cent, marginally higher than that experienced in 1975–82 (Table 2.2).

Table 2.2 French population change, 1954–82

	1954–62 (% p.a.)	**1962–8 (% p.a.)**	**1968–75 (% p.a.)**	**1975–82 (% p.a.)**
Total population change	1.09	1.14	0.81	0.46
Natural growth	0.69	0.67	0.58	0.40
Net migration	0.40	0.47	0.23	0.06

Source: INSEE, 1982.

Regional variations in demographic characteristics may be related to country-wide variations in demographic trends. In particular, the intercensal period 1975–82 showed a continued regional imbalance in natural growth between north and south. The north was demographically buoyant, with a young and expanding population (Fig. 2.4b). The Paris region, which has attracted young migrants for over a century, accounted for a third of national natural increase from less than a fifth of the population total (Boudoul and Faur, 1982). By contrast, five planning regions of the south and the Massif Central exhibited an excess of deaths over births as a result of a prolonged history of depopulation and an ageing population structure (see also pages 49–51).

Malthusianism and pronatalism

The early fall in fertility, so distinctive in French demographic evolution, has encouraged a pronatalist tradition based on fears of cultural, economic and military decline (Huss, 1980). Concern about the decline of the French population became evident in popular and academic journals in the late nineteenth century. Early pronatalist policies were restrictive: abortion was outlawed and the sale of contraceptives was banned. Nevertheless, it was estimated that the abortionists, *les faiseuses d'anges* (angel-makers), produced hundreds of thousands of angels in each of the inter-war years (Dyer, 1978). Measures aimed at encouraging births were introduced in the early part of this century, mainly by private organizations formed by concerned individuals. State encouragement for higher fertility began in

1932, when family allowances were introduced; family support schemes were codified in 1939 as the *Code de la Famille*. The Vichy regime, which established a Ministry of Population in 1940, acquired a hard-line reputation, for example in the application of the death penalty for angel-makers (Ogden and Huss, 1982), while continuing family allowances. By 1949, allowances, tax incentives and family incentives took an astonishing 22 per cent of Gross Domestic Product, showing the importance attached to population in the reconstruction of the war-torn economy. The post-war governments adopted a softer style of pronatalism and the urgency of their task was lessened by the upsurge of fertility between 1946 and 1965, such that by the mid-1960s the value of family allowances had been reduced in real terms.

The fall in fertility since 1965 continues to engage public opinion in France. The spectrum of opinion ranges from those seeking to increase fertility, the *natalistes*, to those in favour of birth limitation, the *malthusiens*. Huss (1980) has argued that the middle-point of French opinion is not one of indifference, but is positively pronatalist (Plate 2.1). This provides a stark comparison with other west European countries; the tag of *malthusien* is applied to almost any one believing in free choice (see page 128). This widespread interest in demographic issues has been shown by a series of opinion polls, which have recorded a notable increase since 1975 in the perceived ideal family size. While the majority of families actually have two children (see Table 2.8), over 50 per cent of those interviewed saw three children as the ideal. Nationally, as fertility fell below replacement levels in the late 1970s, 41 per cent of the population were in favour

Plate 2.1 Pronatalist poster. The baby is saying 'it appears that I am a socio-cultural phenomenon' (rather than a fact). The message on the right is 'France needs children'.

of population increase, compared with only 23 per cent four years previously (INED, 1975; 1979). This public concern over population issues is reflected in frequent media coverage, particularly at times when the birth rate drops.

Present government policies continue to be markedly pronatalist, while being concerned to shake off the militaristic overtones of earlier policy and to demonstrate the country's evolution into a liberal advanced society, particularly by the liberalization and legalization of contraception and abortion. The introduction of these 'malthusian' measures in 1967 and 1975 has been counterbalanced by a series of related family support measures introduced in the mid-1970s. These include, first, medical improvements aimed particularly at reducing infant mortality, which have been remarkably successful, leaving little scope for further improvement. Secondly, the package of family allowances has been made exceptionally large, the combination of inducements and tax relief often forming between one-third and one-half of the take-home pay of the average worker. In the late 1950s, when family allowances were extremely high, Thody (1982) quoted an example of a family with 11 children, the father of whom held an unskilled job in a mustard factory earning 33 690 old francs a month, but who received an additional total of 81 240 old francs a month in family allowances. This was made up of 9000 old francs for the mother staying at home and child allowances at varying rates for the 11 children; it would have been further supplemented by a housing allowance, equivalent to 61 per cent of the rent and public transport subsidies. These benefits were not subject to income tax. Levels of family allowances in the 1990s are relatively lower, but with a large jump in allowances for the third and subsequent children providing incentives for the family to indulge in fertility beyond replacement levels. However, couples still seem loath to have families larger than two, so the policies themselves may be less influential than the prevailing climate of social opinion.

As a third measure to encourage reproduction, there is some official recognition of the job of mothers at home with children, known as the *statut social de la mère de famille*, which gives mothers increased pension rights. The introduction of a maternal salary has been floated, but has not yet been implemented, mainly because of the expense. This type of mother's salary is also seen as contrary to the creation of equal opportunities, as it has the effect of keeping women tied to domesticity and out of the paid labour force in times of unemployment. Similarly, the introduction of measures which nominally allow the combination of work and family life, may in fact reduce the employment opportunities available to women. Finally, the commitment to a family dimension in other government policies is most evident in housing allocation policy, recent immigration policy and, less obviously, in the traditional support offered to the family farm. Huss (1980: 64) has seen French family policies as a 'qualified success', but has pointed to the inevitable time-lag between changes in fertility rates and the intensity with which policies are applied.

French pronatalism and the French perception of population levels as a national issue, can best be understood by an appreciation of the country's distinctive demographic history. The French emphasis on raised fertility levels is distinctive among the countries of western Europe, most of which have lower fertility levels but a more laissez-faire attitude to growth. There are, however, some interesting comparisons to be made with early pronatalism in Sweden and with attitudes to fertility prevalent in eastern Europe in the 1970s and 1980s (Jones, 1981). French attitudes do bear some resemblance to the traditional pronatalism of the eastern bloc, although without its coercive and militaristic overtones. Despite the pronatalist attitude, France registers a large number of abortions; in 1986 there were 213 000, estimated at 22 per cent of conceptions. This number is comparable with abortions registered in England and Wales and in West Germany (Blayo, 1989), but is much lower than in countries of eastern Europe. The sensitivity of the French to issues of domestic population growth is also shown by their concern over the national dependence on foreign immigrants both as members of the labour force and as contributors to natural population increase: in the 1930s immigrant groups accounted for over 70 per cent of natural growth, from 1966 to 1973 for 30 per cent, and in the early 1980s for 13 per cent.

Internal migration

Internal population movement within France is intimately connected with deep-seated economic change, which has transformed the rural peasant society of the nineteenth century into the advanced urban-industrial capitalist economy of the late twentieth. This economic transformation occurred later in France than in Britain or Germany and necessitated huge transfers of people and property between regions and between sectors of the economy. Until the 1960s, migratory movements were predominantly from rural to urban areas and from agriculture to industry and to service occupations. The growth of urban and industrial functions required the input of labour, initially from the provinces and later, as this source became exhausted, from abroad. More recently, migratory movements have been reversed and there has been a substantial rural revival and a decline in population concentration in the large cities.

The rural exodus

Undoubtedly the most important feature of twentieth century population movement has been the rural exodus, *l'exode rural*, which has had a profound impact on areas of both population origin and destination (Merlin, 1971). Paris and its region formed the major focus for internal migrants, drawing on a hinterland which has extended over virtually the whole of France. In the period up to the Second World War, the rate of out-migration was positively associated with both population density and with the rate of natural growth, suggesting that population pressure on rural resources may have been a significant factor in internal migration. Many rural areas established patterns of seasonal migration as a mechanism for stretching limited resources (Van de Walle, 1979). Early this century, pressure on resources may have been especially significant in stimulating rural out-migration from areas of France less favoured for agricultural production (Winchester, 1986). Rates of natural growth were at their highest in Bretagne and the Massif Central, but these were areas of polyculture in which agriculture could not easily be intensified, although some extensions to the cultivated area were possible. As a consequence of the limited opportunities available in these areas, rural depopulation was particularly marked; the migrants were positively attracted to the towns, where alternative employment could be found. Migrants from Bretagne and the Massif Central were principally attracted to Paris, as there were few towns of sufficient size in the home regions to act as loci of intervening opportunity. Patterns of permanent movement replaced the seasonal migration streams which had been established when rural workers left their home regions to earn money in the winter season, for example as peddlers or labourers. Within Paris, these migrants often formed distinctive occupational and spatial groups. Thus, young women from Bretagne were stereotypically in domestic service in the expensive districts of western Paris, while young men from the Massif Central formed groups of masons and café proprietors in areas close to the railway station from which they were disgorged (Ogden and Winchester, 1975).

Rural migrants from the south and east of the country arrived in those cities which were large enough and sufficiently distant from Paris to form distinct regional foci. These included Marseille, Lyon, Bordeaux and Toulouse; they were often the final destination of migrants who had moved in a step-wise fashion up the urban hierarchy from the countryside (Winchester, 1977). Similarly, the growing industries of Alsace–Lorraine and Nord–Pas-de-Calais absorbed local population growth and drew in migrants from their surrounding areas and from abroad.

This pronounced rural-urban movement was stimulated by many interrelated developments. The construction of the railway network from the early nineteenth century onwards allowed movement and aided the integration of the economy and the breakdown of rural isolation. Other factors, such as the widening net of primary education in a common language and the new horizons opened up by military service also facilitated mobility (Weber, 1977). However, the primary factor was the push from the poverty and the overpopulation of the rural areas, and Zeldin (1973) has argued that the pre-war waves of migration were essentially caused by the stagnation of agriculture; they have therefore been considered to be 'migrations of poverty'. The poverty caused by low economic returns from agriculture was compounded by the grinding drudgery of farm work and by the lack of services and alternative employment opportunities, particularly as employment in local crafts and industries declined rapidly from the 1920s (Pinchemel, 1980). By contrast, jobs in factories, shops and offices and even in domestic service were seen to be less arduous and better paid than farm work and had the added benefit of an urban location with enhanced public facilities, better services and more enticing marriage prospects. Gradual changes in economic organization and the social order resulted in continued urban growth even in a period of slow overall population increase.

The impact of rural-urban migration was cumulative, leaving an increasing proportion of the elderly and less mobile in the rural areas, while the flow of migrants to the cities was reinforced by a supportive network of friends and relations in the destination areas. The rural migrants, as with their foreign counterparts in later years, usually moved in at low socio-economic levels to provide manual labour as construction workers or domestic servants, or as unskilled non-manual labour. Provincial migrants formed the labour force of the growing industries of Paris and the north, the resort areas of the Mediterranean and the industries and services of the larger cities. Until the Second World War, the new urban poor endured conditions worse in many respects than those experienced in their rural origins, especially in terms of housing price and quality and in many aspects of health and welfare, indicated by such measures as rates of infant mortality. At a regional level, the dichotomy between the industrialized north and east and the less developed south and west became even more pronounced. The regional industrial development of core economic areas relied on the immigration of an army of reserve labour drawn from agricultural areas (White and Woods, 1983).

Suburbanization and the rural revival

The years of post-war reconstruction brought about complex and dynamic patterns of movement in response to the quickening pace of industrial and economic growth. Urban development was not only fuelled by the movement of labour from the agricultural sector, but was maintained by an increasing proportion of inter-urban and intra-urban moves. Urban growth necessitated expansion of the urban area, and the housing needs of the post-war period resulted in the construction of large housing estates on the edges of existing cities. By the late 1960s, new patterns of internal migration were evident, particularly the beginnings of out-migration from declining industrial areas and of suburbanization from major urban centres. The centre of Paris had been losing population since 1921 (see pages 170–2), but other major cities did not feel the impact of deconcentration until the late 1960s. The growth areas of the late 1960s were the *communes* adjacent to the large cities where new suburban districts were expanding. These areas were often legally defined as rural districts or small towns, but they rapidly became part of the functional regions centred on the urban agglomerations. In this way, suburbanization stimulated population growth in many rural areas and small towns and villages. This trend towards the revival of rural areas became even more pronounced after 1975 (see pages 53–5). Internal migration in the 1960s therefore became increasingly complex, with continued urbanization, but also with concurrent patterns of suburbanization and some out-migration from regions which were beginning to show industrial decline (see pages 55–60).

The individual motives for mobility are inevitably even more complex than the aggregate patterns of migration. Important studies of the motivations of French migrants, especially migrants to Paris, were carried out by Pourcher (1964; 1970). His evidence indicated that moves were primarily economically motivated, particularly to find employment or to improve the type of occupation and the financial rewards accruing from it. This was a stronger motive for men than for women (Table 2.3). However, economic, family and other motives, such as housing, are often difficult to disentangle, especially when a family unit moves together. Furthermore, the individual decision to migrate is highly constrained by the structure of the economic opportunities available.

Table 2.3 Motives for migration of French migrants in the 1960s

	Migrants from: Provincial France		Paris
	Men %	**Women %**	**Total %**
Motive for migration:			
Work	59	41	50
Family	21	42	32
(of which, marriage)	(14)	(30)	(22)
Retirement	8	4	6
Other	24	23	23
TOTAL	112*	110*	111*

Source: Pourcher, 1964; 1970.

*Totals exceed 100 because of multiple replies

Of the people who had moved to Paris in the 1960s, 70 per cent felt that they had been able to fit into their new lifestyle. This adaptation was often helped by the presence of friends or relations in Paris or by frequent contact with their former home communities in the provinces, which alleviated isolation. Most people enjoyed the relative liberty, independence and activity which the city provided, while a minority felt disorientated by the sheer size of the city, by the urban way of life and by their alienation from a familiar home community.

In a study of migrants to the provinces, it was found that the motives differed from those of Parisian migrants while their problems of assimilation were less marked. Migration to the Côte d'Azur was particularly influenced by the employment available in manufacturing and services in the Mediterranean towns. It was also attractive for its climate (see page 11) and for the facilities available for migrants of retirement age, especially those from the financially more favoured classes of society. The Côte d'Azur has long been renowned for its warm winter season, seaside amenities and sophisticated pleasures which have proved attractive to the French as well as to foreign visitors. The impact of retirement migration is seen in the age-structure of the region and particularly in the imbalanced age-structure of certain resort towns, where the proportion of elderly residents rises to over 20 per cent, compared with a national average of 14 per cent. During the 1970s and 1980s, retirement migration also became increasingly important to the less industrial regions of the south, particularly Aquitaine, Languedoc and even the Auvergne, while other predominantly rural regions such as Bretagne have also regained population, in some cases as a result of former out-migrants returning on retirement.

This movement back to the rural regions has in part taken the form of second-home ownership. In 1990, it was estimated that there were 2.8 million second homes in France, mainly in rural and 'retirement' regions. Such retirement or pre-retirement moves help maintain links with rural origins, especially in those areas with particular climatic and scenic advantages which are easily accessible from major urban centres. Ownership of a second home may, however, be purely for vacation purposes and does not necessarily imply a desire to return to rural living at any stage. The location of the majority of second homes in coastal and mountain regions, and the loosening of financial restrictions on their purchase since 1986, indicates that for many families they are a popular form of holiday investment, rather than part of a genuine rural revival.

Counterurbanization

The most recent change which has occurred in patterns of internal migration in France has been the trend to counterurbanization, *rurbanisation*. This first became evident in most countries of the developed world in the late 1960s, and was clearly apparent in France by the time of the 1975 census. Counterurbanization is best defined as a growth of population in small settlements, and is a net movement or deconcentration of population down the urban hierarchy, rather than merely a spatial expansion of urban areas (Champion, 1989). Several theoretical explanations for counterurbanization have been put forward by Fielding (1982). The traditional explanation for counterurbanization, and the one favoured in most of the French literature, is that families and individuals who have experienced at first hand the disadvantages of living in large urban areas have chosen to move to the country to improve their quality of life (Bauer and Roux, 1976). It is argued that more and more people are choosing to escape from the stress of urban living and are rediscovering some of the benefits of living in the country or in small towns, such as simplicity, spaciousness and closer links with the environment. This explanation stresses that the individual decision to relocate is the motive force for the rural revival. However, it implies a complete freedom of choice which is not available to most people, who are constrained by the operation of the job and housing markets. It may well apply, however, to the elderly who are no longer employed and who own their own homes, and could also apply to people in footloose well-paid professional occupations. This explanation for counterurbanization is probably most usefully applied to choices of suburban locations, and to retirement migrations.

Fielding's other explanations for counterurbanization all redirect attention from the decisions of the individual to the operation of the job market. First, neo-classical economic models indicate that job opportunities in a given region will result in migration to balance employment and wage levels. This explanation is generally more useful in the interpretation of urbanization rather than counterurbanization processes (Winchester, 1989). Secondly, the state intervention model emphasizes government incentives and decentralization policies which result in the relocation of firms to the periphery. These models of the counterurbanization process are useful in that they focus attention on the role of employment in migration, and on the decision-making of individual firms and the state. The movement of jobs away from Paris, and particularly to areas within a 200 km range of the capital, has provided job opportunities which have attracted new population to those accessible areas.

Fielding (1982) has integrated these considerations of industrial relocation into an argument which relates counterurbanization to the new spatial division of labour. The movement of population to rural areas is viewed as a response to the restructuring of employment opportunities as monopoly capital is replaced by global capital. He considered that manufacturing industry has become progressively released from its traditional locational constraints, such as raw materials and proximity to the market. Industry, increasingly owned by multinational firms rather than individual entrepreneurs, has chosen to relocate production plants in peripheral regions where the labour force is plentiful, cheap and non-unionized. The managers and professionals running the plants have to be drafted in from urban areas, a move which they generally welcome because of the perceived benefits of rural life. Counterurbanization is therefore viewed as a movement predominantly of middle-class people following industrial trends. Logically, however, at least part of the labour required in these factories would also make the move to areas of employment, cheaper housing and pleasant lifestyle.

Most studies of counterurbanization in France have focused on the decisions of individuals rather than on the impact of economic restructuring (Bauer and Roux, 1976). It may be argued that in France individual decisions regarding the desirability of the rural lifestyle may be more significant than in other developed countries. The recency of urban development means that many people who are first- or second-generation urban dwellers have retained close family and property links with rural areas. A further factor which differentiates French counterurbanization from that of other countries in Europe is the extreme severity of the post-war housing shortage (see pages 150–3). As a consequence, the search for good quality housing in a time of increasing affluence led people to build detached houses, *pavillons*, outside the existing city boundaries. In the period 1975 to 1982, over three million new dwelling units were built in France, the majority of them constructed by the private sector on the fringes of existing built-up areas. A major factor in French counterurbanization, therefore, has been both the demand for and the supply of good quality housing. Fruit (1985), in a study of counterurbanization in the Pays de Caux (Normandie), found that the most significant motives for migration were housing and land, while reasons of amenity and the quality of life were cited by only 20 per cent of respondents. Similarly in the Paris region, Belliard and Boyer (1983) found that 88 per cent of recent in-migrants lived in individual houses, and 40 per cent in houses built since 1967. From these data the authors inferred that housing requirements were an essential ingredient of the decision to move. Similarly, 1982 census data for Isère showed a preponderance of recent in-migrants in new owner-occupied *pavillons* (Winchester, 1989).

However, the significance of the employment motive cannot be overlooked in areas which are close enough to Paris to have benefited from industrial decentralization. Ganiage (1980), in a study of the Beauvaisis region of the Ile de France, considered population growth since the 1950s to be a direct result of industrial decentralization which brought major international firms such as Nestlé and Lockheed into the area, providing thousands of new jobs. In some cases, the actions of entrepreneurs and local government officials helped secure jobs for particular localities such as the small Breton town of Loudéac (Limouzin, 1980).

It therefore appears that no one explanation of counterurbanization can be applied to the whole of France (Winchester and Ogden, 1989). State decentralization policies have resulted in the relocation of industry from the centre of Paris mainly to the wider Paris Basin (see also pages 123–5), stimulating population movement into that area. However, there is little evidence of the wholesale relocation of production-line processes to the periphery by the forces of global capital. Indeed, many French firms are looking beyond their own national borders for peripheral locations with the benefits of cheap factors of production (see page 121). In parts of the French rural periphery, counterurbanization takes more the form of diffuse urbanization fuelled by housing need (David *et al.*, 1986; Noin, 1987), while in the very remote areas depopulation is still occurring. Table 2.4 shows that most *communes* in truly rural areas are still losing population, while the rural *communes* at the edges of the urban-industrial agglomerations are gaining most rapidly. It is clear that, in France, counterurbanization trends are spatially differentiated between the core, the periphery, and the remote regions of the country.

Table 2.4 French rural population change by type of *commune*, 1975–82

	% Increasing		% Decreasing	
	Total	by >15%	Total	by >15%
All rural *communes*	55.5	23.4	44.5	9.0
Communes within ZPIU*	74.1	37.8	25.9	3.1
Communes not within ZPIU:				
(a) near to ZPIU⁺	55.1	22.9	44.9	8.8
(b) entirely outside ZPIU	39.1	11.0	60.9	14.7

Source: Boudoul and Faur, 1982.

*ZPIU: *Zone de peuplement industriel ou urbain*, an urban or industrial agglomeration
⁺ where at least one *commune* in a *canton* belongs to a ZPIU

The regional impact of recent population changes

In the 1950s and 1960s, the dominant regional pattern in the demography of France was a fundamental division between urban growth, overwhelm-

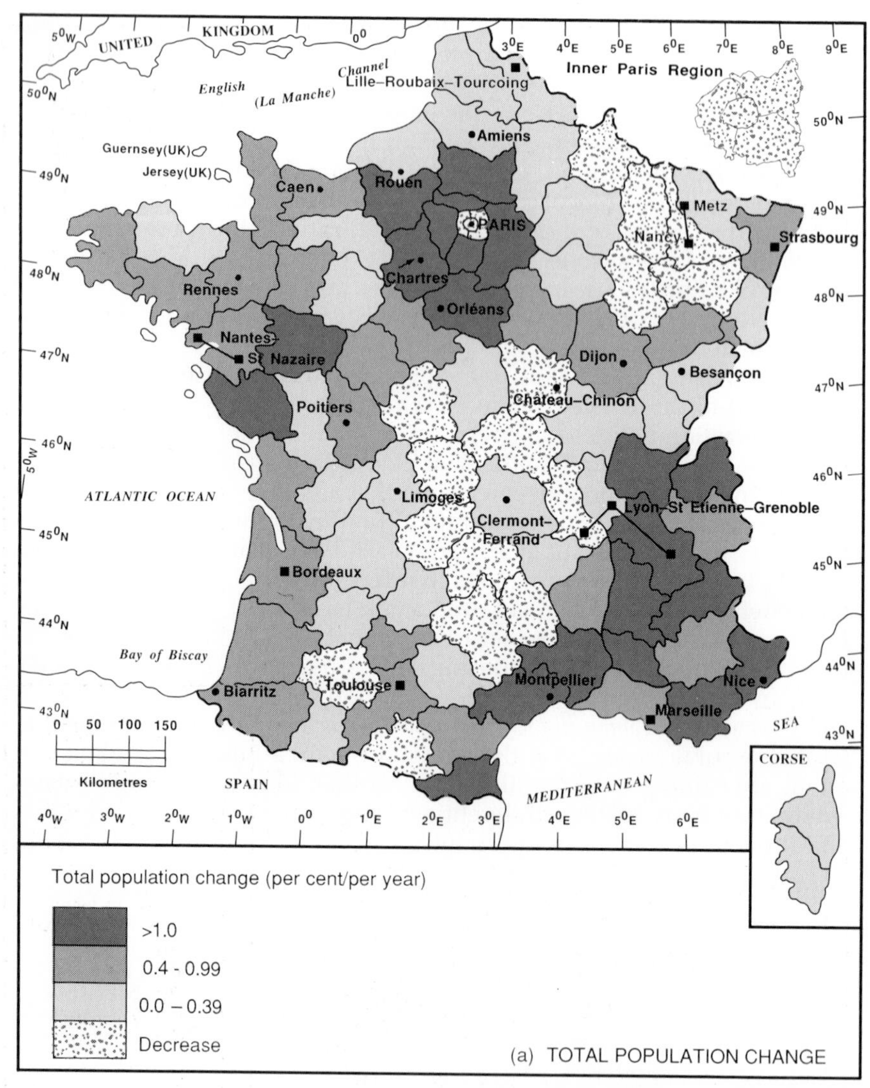

Fig. 2.4 Population change, France, 1975–82 (a) total population change; (b) natural population change; (c) net migration. (Source: INSEE, 1982)

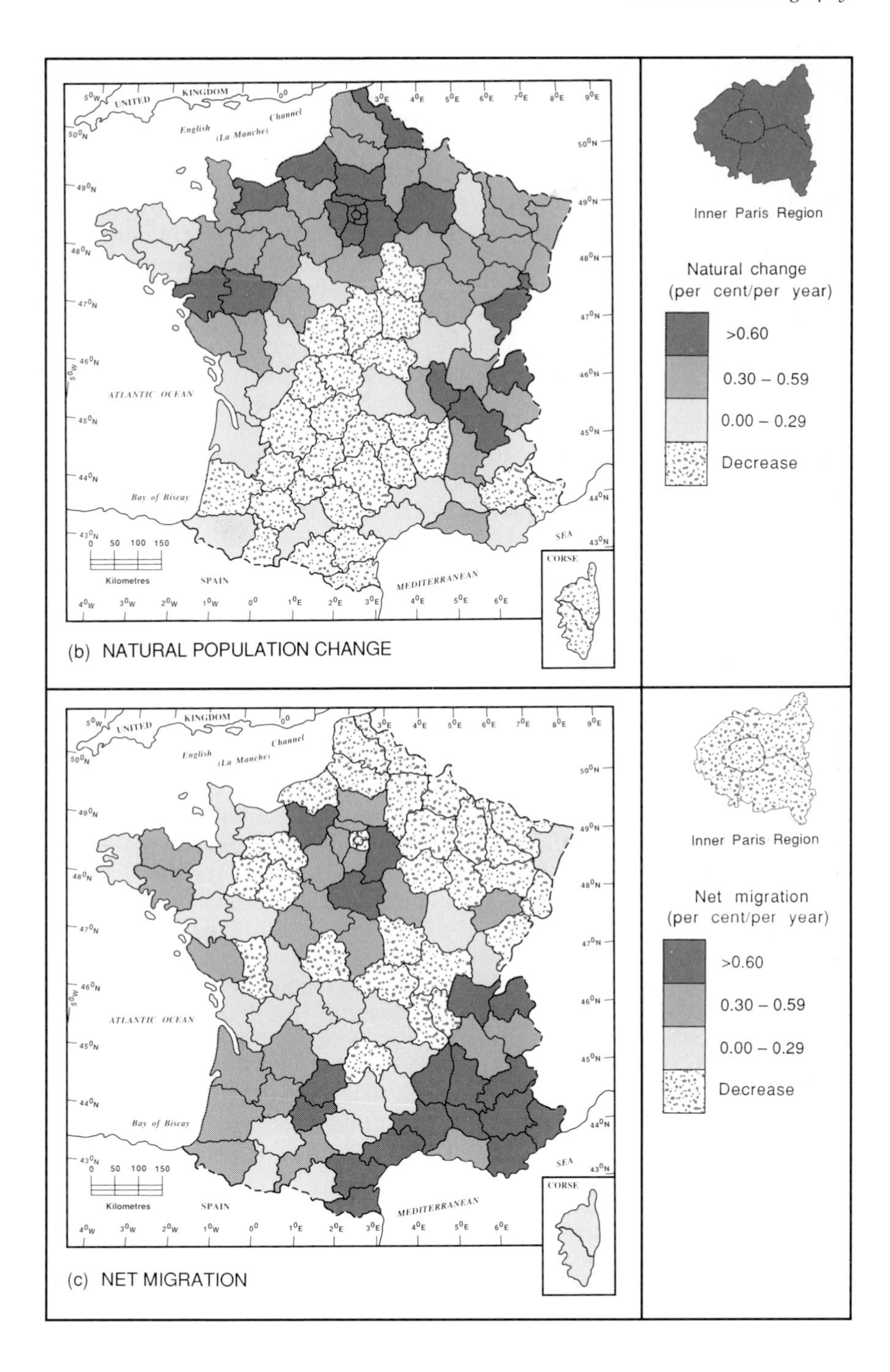

(b) NATURAL POPULATION CHANGE

(c) NET MIGRATION

ingly concentrated in Paris, and a declining countryside. This pattern has recently been greatly disrupted by the phenomenon of counterurbanization, which has brought growth to rural *départements* which had been declining for a century or more. Overall rates of population growth nationwide have slowed substantially since the late 1960s. Since 1975, the highest rates of overall growth have been found in a number of *départements* in the southeast and in the Paris Basin, although not in the City of Paris (Fig. 2.4a). Population growth, at a rather less spectacular level, has also been a feature of most of the west of France, including Bretagne and Normandie. The City of Paris and the three inner *départements* of the Parisian suburbs lost population between 1975 and 1982 (see also pages 172-4). Other areas which suffered population loss are concentrated in the regions of Lorraine and the Massif Central, as well as two *départements* in the Pyrenees.

Figures 2.4b and 2.4c show the relative contribution of natural population change, brought about by fertility and mortality, and of population change brought about by net migration, to total population change in the period 1975–82. The pattern of natural change (Fig. 2.4b) shows a marked regional variation, with a demographically buoyant north, where rates of annual natural increase exceeded 0.3 per cent, and a declining south. Figure 2.4b also shows 28 *départements* where deaths exceeded births, heavily concentrated in the centre and the Pyrenees. The large number of areas experiencing natural decline is significantly greater than in the 1960s and early 1970s, and reflects national trends in fertility reduction. However, the location of these areas is indicative of the cumulative effects of long-term population decline, especially substantial out-migration, the ageing of the residual population, and its low fertility.

The migration pattern (Fig. 2.4c) is almost the inverse of the pattern of natural growth and decline. Many of the southern areas, which have a long history of population decline, have experienced a reversal of migration flows, and in 1975–82 gained migrants. This is even the case in some of the *départements* of the Auvergne which exhibited a tiny margin of net in-migration over out-migration. In-migration was particularly marked in the *départements* of the southeast and was substantial in much of the southwest. Traditional areas of migration attraction which maintained their net inflow were the outer *départements* of the Paris Basin and the lower Seine, while Bretagne continued the demographic revival begun in 1968–75 (Dean, 1987). The regional pattern shows clearly the influence of the national slow-down in natural growth. It also shows the spatial impact of urban decentralization, with growth in outer suburban areas, and of counterurbanization, with migration affecting the areas of traditional population decline.

The pattern of population change in the period 1975–82 shows some significant alterations from the regional trends of earlier intercensal periods. Before 1968, the western regions exhibited migration loss, but since that date the migration balance has become and has remained positive. The Mediterranean regions are now the most attractive for migrants, with a gain at the regional level of over 1.0 per cent annually. On the other hand, the regions of the north and east have seen their situation deteriorate, and growth has slowed in the Ile de France. Recent migration change

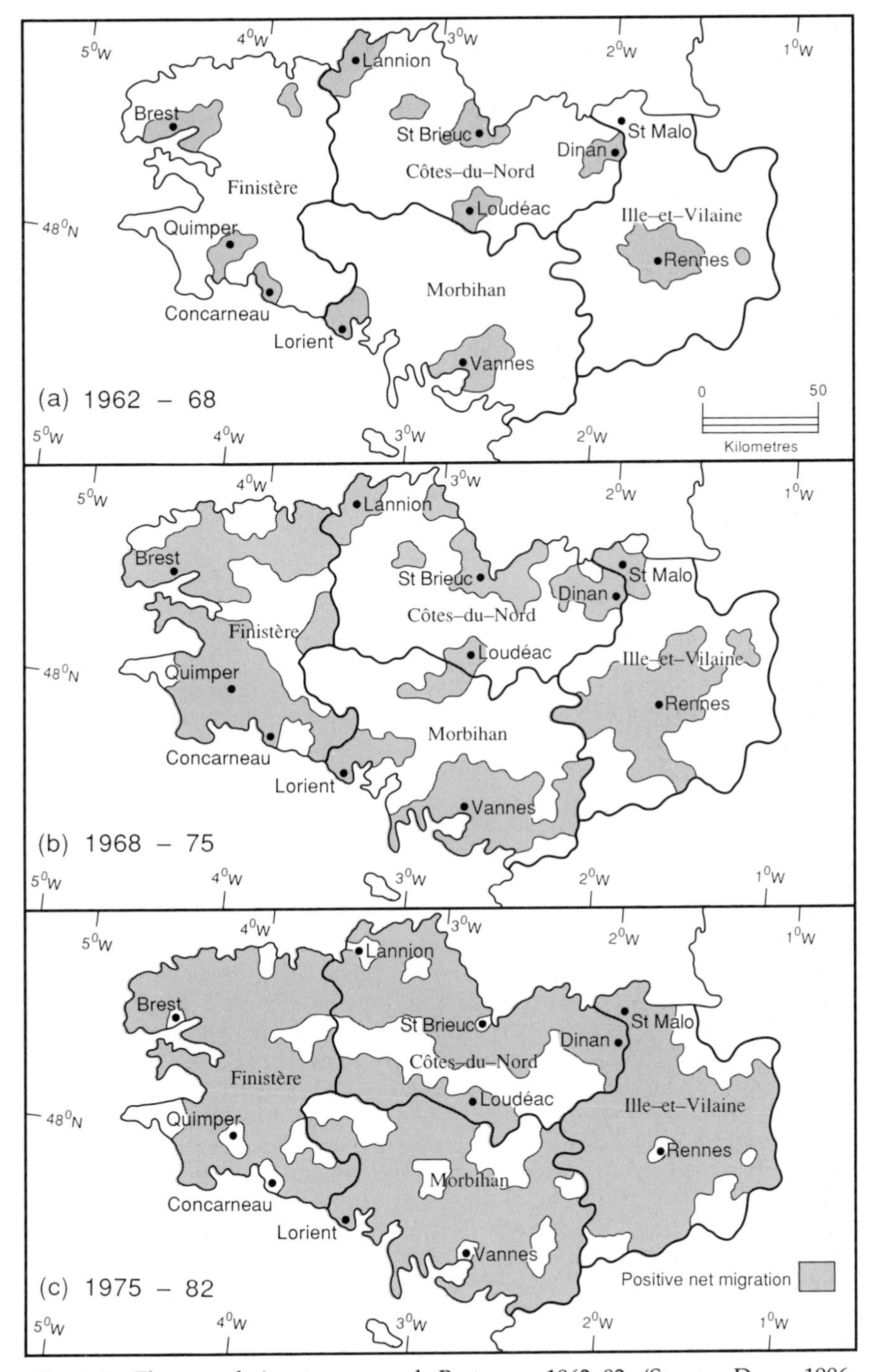

Fig. 2.5 The population turnaround, Bretagne, 1962–82. (Source: Dean, 1986: 152)

is complex, and the patterns are much more clearly revealed at the scale of the *canton* or *commune*, as in Fig. 2.5, which shows the unmistakable turnaround of migration between 1962 and 1982 in Bretagne. In the 1960s, in-migration took place to only a handful of major urban areas; by 1975–82, the pattern had been reversed, with the larger urban areas losing population and the rural areas gaining (Dean, 1986; 1987). This population turnaround has resulted from the processes of suburbanization, retirement migration, second home ownership and counterurbanization outlined above.

In summary, overall patterns of population change and the relationships between natural change and migration in the 1975–82 intercensal period are complex. The basic regional pattern may be succinctly summarized as natural growth in the north and migration growth in the south. Paris provides an added complication (see pages 170–2), stagnant in the centre and growing principally in the outer suburbs, by both migration and natural increase.

Foreign immigration

Pre-war foreign immigration

The long period of French demographic stagnation in the early twentieth century brought about a labour shortage which was supplied first from the provinces and subsequently from abroad. France early became a net recipient of foreign immigrants, as it was one of the few large European countries which did not make a significant contribution to the trans-Atlantic migrations. Various estimates have indicated that between 1881 and 1910 about 40 000 people left France each year, but that the level of emigration was far exceeded by the wave of foreign immigrants which moved into France as replacement labour from the end of the last century. By 1911, there were over one million foreigners in France, almost three per cent of the population total. There were also a large number of French colonial citizens, predominantly Algerians with some Tunisians and Moroccans, in the French army in France at this time (Bennoune, 1975). Most of the foreign nationals in the country came from France's immediate neighbours in western Europe, with Italians and Belgians comprising over half the total. Once having entered the country, most foreigners settled either near their homeland or in the capital city. The region of Nord–Pas-de-Calais contained over two-thirds of the Belgian immigrants in France, while the Italians dominated the Mediterranean *départements*, and the Spaniards the southwest from Bordeaux to Montpellier. *Départements* distant from the frontier, such as those of Bretagne, were home to relatively few foreigners, as were the economically unattractive areas of the Massif Central. Moreover, around the turn of the century, both these areas experienced considerable population growth and pressure on resources, and would have been unlikely to have provided major opportunities for migrants seeking work.

Table 2.5 Changing birthplace composition of the foreign-born population of France, 1936–82

	1936	1962	1982
Total population	41 907 000	46 459 000	54 273 200
Total foreign-born	1 823 800	2 169 665	3 680 100
Country of birth:			
Italy	720 900	629 000	333 740
Poland	422 700	177 200	64 820
Spain	253 600	441 700	321 440
Belgium	195 400	79 100	50 200
Portugal	28 000	50 010	764 860
Algeria	n.a. } 87 000	50 500	795 920
Morocco	n.a. } 87 000	33 320	431 120
Tunisia	n.a. } 87 000	26 570	189 400
Turkey	n.a.	n.a.	123 540

Source: Dyer, 1978; INSEE, 1982.
n.a. not available as an individual country

The First World War created a vacuum of labour in France, and in the inter-war years many thousands of foreign migrants provided a valuable contribution to demographic and economic growth. By the 1930s, there were three million foreigners in France, spread over about half the country, particularly in the capital city, the frontier districts and, increasingly, the rural southwest. The Italians maintained their predominance, but the French government also brought in Polish workers on contracts to work in the mining districts of the north, and an increasing number of Spaniards came as agricultural workers to the demographically-declining southwest. Both these groups often endured worse conditions and lower wages than the French. At the same time, the number of Belgians in the population was declining, predominantly by naturalization, as there was relatively little adjustment in language or culture to be made by this group of immigrants. Table 2.5 shows the numbers of foreign-born (excluding naturalized citizens) in France in 1936.

The foreign population, then as now, was demographically and economically distinct from the host population. In particular, the age and sex ratio was imbalanced, with a high proportion of men of working age. The excess of foreign men was in sharp contrast to the depleted numbers of French men after the First World War. Similarly, despite the high proportions of single men, the fertility rate of immigrants far exceeded that of the French population. In economic terms, the immigrants were heavily concentrated in manual work and extractive occupations at the lower end of the occupational spectrum; they were also more frequently in gaol and in hospital than their French counterparts (Dyer, 1978: 75). Nonetheless, assimilation of these predominantly European groups by intermarriage and by naturalization occurred relatively easily, except in the major urban centres where concentrations were high, and where expatriate organizations, such as Polish libraries or Italian holiday camps, could become established.

Post-war foreign immigration

Immigration since the Second World War has maintained its demographic and economic significance, and further major changes have occurred in the origin of foreign groups. The Italians, who had dominated the migration stream to the war-torn towns of northern France in the late 1940s and the 1950s, were exceeded numerically from the early 1960s by new groups of migrants. In particular, there were massive increases in the numbers of north Africans from Algeria, Morocco and Tunisia, and in the numbers of Portuguese (Table 2.5). Between 1962 and 1968, the number of Portuguese increased from 50 000 to almost 300 000; the Moroccan population more than doubled from 33 000 to 80 000; and the numbers of Algerians and Spanish each increased by over 100 000. These numbers taken from the census are likely to be underestimates. Figures from the Ministry of the Interior are almost invariably higher, and there are estimated to be many thousands of illegal immigrants whose presence has been completely unrecorded. It is known that the 1968 census underestimated the national population by over one per cent, and this underestimation is likely to be higher for immigrant groups, especially in urban areas.

The explanation for these changes in the nationality of immigrant groups must be found mainly in the changing conditions of the countries of origin. From the mid-1950s, Italians found wider and better-paid opportunities in the rapidly-industrializing areas of northern Italy, and also as guest workers, *gastarbeiter*, in Germany and Switzerland. Spain, with a surplus rural population and an inadequate number of jobs, positively encouraged emigration to its nearest European neighbour. Although Portugal banned emigration because of the war effort in Africa under Salazar and Caetano, nonetheless many illegal Portuguese emigrants made their way to France, confident that their position would be legitimized on arrival. Hundreds of thousands of emigrants came also from the former French colonies in Africa once the principle of freedom of movement was established by the Evian agreement in 1962 after the end of the Algerian war. By 1982, there were three quarters of a million Portuguese and about 1.5 million north Africans in the country. The movement was essentially a labour movement from the Mediterranean countries with surplus labour to a receiving country eager to fuel economic growth, and unable to do so from its own rural domestic labour force (White and Woods, 1983).

The foreign population of France shows marked geographical concentrations, which are particularly evident among the newest arrivals. Over 90 per cent of the foreign population lives in urban areas (Hémery, 1986). Furthermore, foreigners are heavily concentrated in Paris and in the large cities; 65 per cent of foreigners live in agglomerations of more than 100 000 people, whereas only 43 per cent of French-born live in these cities (Guillon, 1986). In 1982, foreigners made up over 10 per cent of the population in 13 of the largest 36 cities in France, and 17 per cent of the population of Paris (Ogden and Winchester, 1986). Recent migrant groups are particularly concentrated, with 65 per cent of Algerians and 45 per cent of all Portuguese in towns of over 200 000 population. Over half the foreign population of France lives in only three regions, the Ile de France, the Rhône-Alpes, and Provence–Côte d'Azur, while there are very low proportions of foreigners resident in Bretagne, in the west of the country and in the rural centre (Fig. 2.6) (see also page 238).

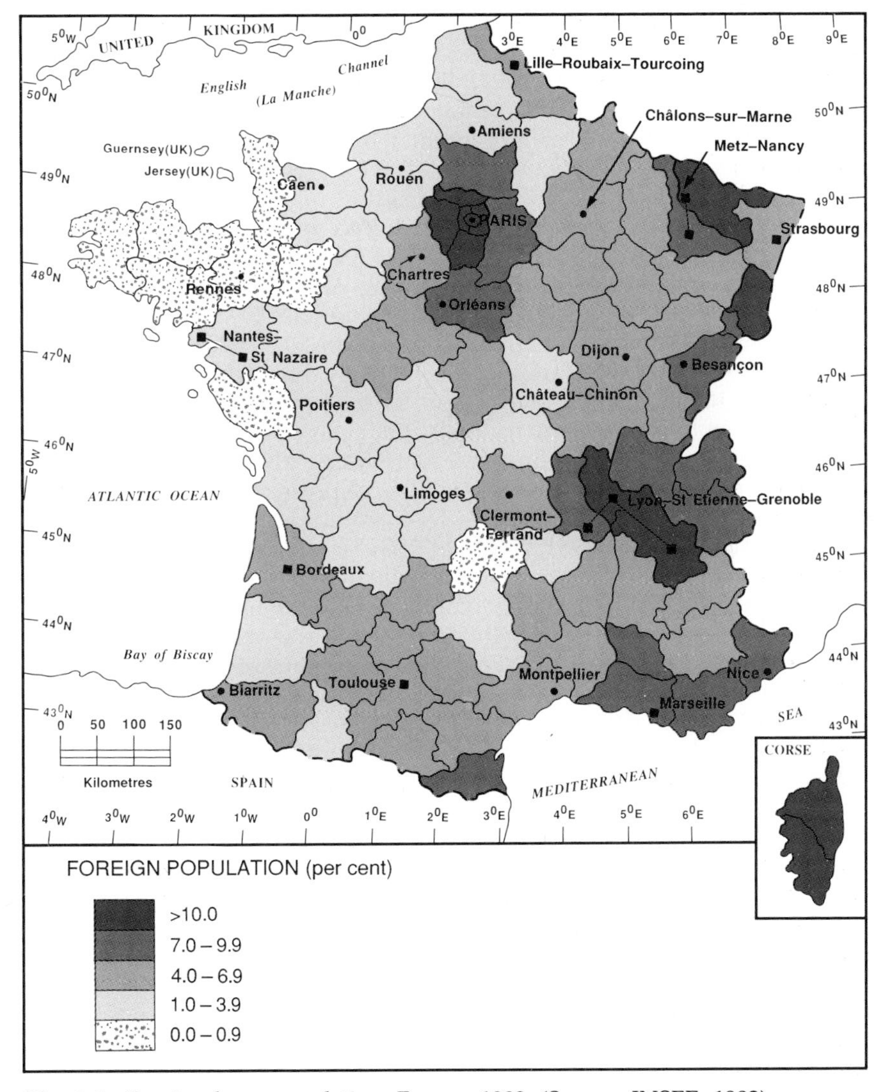

Fig. 2.6 Foreign-born population, France, 1982. (Source: INSEE, 1982)

hange in migration trends occurred in 1974, when the was reversed. The oil crisis, and the economic recession caused most west European countries to reassess their , and to curb the flow of immigrants. White (1986) has nstrated that the change in migration policy may have been due more to rising xenophobia, fanned by active political groups of the far right such as the National Front, than to purely economic considerations (see pages 220–2; Fig. 7.3). A total ban was placed on immigration in 1974, and France has effectively restricted immigration to family reconstitution since that date. Despite the stricter immigration controls which have been in force, France has accepted large numbers of 'boat people' as refugees from southeast Asia. France, because of its history of colonial involvement in the region, gave asylum to about 87 000 refugees in the period 1979–82, compared with about 16 000 taken by Britain (UNHCR, 1983). The refugees were taken initially to reception centres in the provinces, to areas with low concentrations of immigrants, but many of them have subsequently made their way to the capital (see pages 173–4; Plate 6.2).

The significance of the social and political impact of immigration in determining migration policy is demonstrated by the imposition of some hard-line measures introduced in the late 1970s. Voluntary repatriation for migrants was introduced in 1977, and grants of FF 10 000 were offered to each worker, male or female, and a supplementary FF 5000 for working children. From 1977 to 1979, only about 70 000 people, mostly Spanish and Portuguese, took up these grants, and it is likely that the takers were those who would have returned in any case, as the level of grant in itself was insufficient incentive. In January 1980, there was a clamp-down on illegal migrants, and many whose presence had been tacitly accepted in times of economic necessity were threatened with expulsion. Mounting racial tension in the early 1980s resulted in a relaxation of attitudes by the Mitterrand government in 1981. During 1981 and 1982, the process of 'regularization' of the presence of an estimated 10 000 migrants without papers was begun, repatriation payments were stopped (until 1984 when they were resumed on a slightly different basis), and some further concessions were made in allowing in families of existing migrants, while not relaxing the general policy of halting all new labour immigration.

The assimilation of migrant groups

Despite the ban on immigration in 1974, France still has a problem in adapting to the presence of its very large foreign population. The country now has approximately four million foreigners, about seven per cent of the total population, who are distinctive in their origins, spatial distributions and socio-economic status. Many came to France during the boom years of the 1950s and 1960s as temporary contract workers, estranged from their families and from their host community. Their continued presence in times of unemployment and recession has given rise to heightened prejudice and racial tension, on occasions flaring into violence, as at the Talbot car factory at Poissy in the Paris suburbs in 1984, when the threat of job losses particularly affected immigrant workers. Over half the workforce originated from north and west Africa, and after three days of rioting between striking and non-striking workers, many immigrants were demanding repatriation allowances of FF 200 000. Tensions are most evident in areas of immigrant concentration, particularly in the large cities; in these areas, such as Marseille, the political activities of the National Front became prominent in the 1980s (see pages 220–2; Fig. 7.3).

Immigrants are distinctive as a result of their demographic and socio-economic characteristics. Nearly two-thirds of the foreign labour force is employed in the secondary sector of the economy; the foreign population accounts for six per cent of the labour force, but eight per cent of those working in industry, and 17 per cent of those in building and public works. Foreigners are still exceptionally concentrated on the lower rungs of the socio-economic ladder, with high proportions found in labouring jobs, domestic service, mining and routine factory work. The importance of immigrant labour in menial jobs was brought to the fore in 1980 and 1981 when Parisian commuters on the *Métro* endured accumulations of rubbish and dirt which built up during a protracted strike by immigrant cleaners. The type of jobs performed by migrant workers means that they are rarely competing directly with French for jobs, but instead form a 'replacement' labour force, doing work which the indigenous French are no longer prepared to do. In this way, immigrant labour can be seen as a 'fundamental element in the economic structure of European capitalism, and not simply an extra source of labour in conditions of rapid growth' (Castles, 1985: 39), and so may coexist with unemployment of indigenous French.

Plate 2.2 Demonstration by migrant workers in Paris against forcible assimilation of migrant workers in Bulgaria. Note the age and sex of the demonstrators.

The immigrant groups are distinctive too in their demographic characteristics and community structures. The age-sex pyramids of some immigrant groups are heavily distorted from the normal age-sex structure. Men of working age may outnumber women by more than two to one (Plate 2.2). The longer-established migrant groups, such as the Italians, have more balanced age-sex structures, while recent groups of migrants, particularly the Turks and west Africans, exhibit gross distortions from the normal pyramid. The age as well as the sex structure is imbalanced, with a preponderance in the 20–45 age group, and relatively few elderly people. There has been, however, a recent increase in the number of young women as a result of family reconstitution migration, and consequently there are more young children who are forming a widening base to the pyramid. Indeed, the average household size of recent immigrant groups is significantly larger than the national average, and has grown larger since 1975, contrary to national trends (Table 2.6) (Guillon, 1986). The number of children under 16 is also higher than the French average, and there are more single men and fewer one-parent families. However, the household composition of the Spanish and Italians approximates to that of the French as a whole (Table 2.6), and the family character of contemporary immigration is causing a growing alignment of foreign-born families to the national norms (Guillon, 1988).

Table 2.6 Demographic characteristics of the foreign-born population of France, 1982

	Number	% change 1975-82	Mean size of household*	Mean number of children age 0–16+
Total population	54 273 200	+3.2	2.70	0.90
Born in France	49 167 180	+2.9	2.66	0.85
Naturalized French	1 425 920	+2.4	2.77	0.75
Foreign born:				
Total	3 680 100	+6.9	3.34	1.54
Italy	333 740	−27.9	3.06	0.91
Spain	321 440	−35.4	2.99	0.93
Portugal	764 860	+0.8	3.63	1.51
Algeria	795 920	+12.0	3.99	2.46
Morocco	431 120	+65.8	3.91	2.52
Tunisia	189 400	+35.5	3.60	2.03

Source: INSEE, 1982.

* Household refers to 'ordinary households' where the 'reference person' (formerly the head of household) was born in the country in question
\+ dependent children in ordinary households

The most recent migrant groups form spatially and socially segregated communities, while the older groups are tending to become naturalized and assimilated. The second generation north and west Africans are arguably the most difficult to assimilate. Born and raised in France, but differentiated by religion, culture and education, they are caught between two cultures, and faced with unemployment and discrimination in a society which has aroused in them higher expectations than those of their parents. The relatively high birth rate of migrant groups, caused by their distinctive age and class structure, means that the problem of second and third generation migrants may further intensify. However, Faidutti (1986) considered that the Algerians are becoming assimilated in a manner similar to the Italians and Spanish, and that there is substantial evidence of multigenerational integration. Newer immigrant groups are still confined to manual jobs traditionally reserved for immigrants, Moroccans as farm labourers, Portuguese in construction and Turks in manufacturing (Faidutti, 1986). Levels of discrimination and abuse are not, however, solely related to recent migrant status, but must partly be attributed to more deeply entrenched racism. The French Jewish population, a long established and large minority group (about 700 000), suffered persecution during war time and has been subjected to numerous acts of violence and terrorism in recent years, including the bombing of a cinema and restaurant, and the desecration of a Jewish cemetery.

Table 2.7 Selected housing indicators for all French households and for foreign-born households, 1982

	% of all ordinary households	% of foreign-born households
Tenure:		
Owner-occupiers	50.7	20.9
Renting privately	26.0	39.3
Renting public housing	13.5	23.6
Amenity:		
With telephone	74.4	50.3
With inside w.c.	85.0	76.1
Crowding:		
In buildings with >20 dwellings	13.8	22.8
Severely overcrowded	2.5	11.9
Size of dwellings:		
Mean number of persons per room	0.7	1.1
Mean number of rooms per dwelling	3.7	3.1

Source: INSEE, 1982.

The so-called immigrant problem is most obvious in the social and spatial segregation of the immigrants in areas of poor housing (Table 2.7) (Hémery, 1986). In the 1960s, the rapid influx of migrants of low socio-economic status exacerbated the pre-existing housing crisis, and many were forced to live in shanty towns or *bidonvilles*, often located on suburban fringes close to new construction projects where many of the immigrants were employed. In the late 1960s, at least 100 000 foreign migrants were estimated to be living in *bidonvilles* without basic amenities such as heating, electricity, running water, drains or sanitation. The largest *bidonvilles* were in the Parisian suburbs of Nanterre and Champigny, while other major cities such as Marseille, Toulon, Nice, Lille and Rouen also housed substantial immigrant communities in sub-standard accommodation. Most of the shanty dwellings disappeared during redevelopment schemes in the 1970s, but *microbidonvilles* still exist on urban wasteland and in hidden corners. Many older-established and family groups of migrants have been moved to older public housing, predominantly in the relatively undesirable *grands ensembles* (see pages 166–7), while the French have been allocated to the newer housing blocks (Sporton and White, 1989). Single migrant workers often live in dormitory-type accommodation provided by firms, or in special hostels where the living is spartan and regimented and lacking in privacy but has the advantage of cheapness (Jones, 1989). Large communities of immigrants live in marginal housing, and more immigrants than French find accommodation in the private rented sector (Table 2.7), where sub-standard conditions are common. Exploitation at inflated rates for makeshift facilities particularly affects recent migrant groups and illegal immigrants. On any housing indicator (Table 2.7), foreign households are disadvantaged.

Migrant groups are deprived in other ways besides housing. Immigrants are over-represented in inner urban and suburban areas where services are already overburdened. Poor nutrition and limited access to expensive health care services have detrimental effects such as a higher than average incidence of infant mortality and morbidity. Education facilities are similarly limited and hampered by language difficulties, and it has been estimated that Algerian children lag a year behind their French counterparts in terms of educational attainment (Ogden, 1982). In the immigrant district of Marseille known as La Cage, one in six children leaves school illiterate. Since 1975, the number of unemployed immigrants has tripled. Conditions of multiple deprivation, magnified in urban areas, form an unacceptable and increasingly obvious consequence of the mismanagement of foreign immigration in France.

These recent trends in foreign immigration in France are also evident in other industrialized countries of northwest Europe (White and Woods, 1983; Castles, 1985; Salt, 1985). The immigration of cheap foreign labour, often on a short-term contract basis, has filled a labour vacuum in the economic development of western Europe, particularly in manual occupations. The costs and benefits of this transaction are notoriously unequal, and labour immigration of this type has been seen as a form of development aid from poor countries to rich. The benefits of the transaction have accrued primarily to the employers who have received a docile non-unionized labour force; and to the state which has incurred little expense in the provision of services for a relatively healthy, adult and predominantly unmarried labour force. The costs have been borne more on an individual basis, as the migrant workers have had to endure a variety of deprivations, but have rarely acquired marketable skills. Many, however, feel some benefit in obtaining employment which would not be available in their home countries, which has enabled them to send remittances to their families, and eventually to return (Ramos, 1981), or which offered hope of a brighter future for their children.

The use of migrant workers in this way has been likened to the concept of the reserve army of labour, whereby a segment of the labour force is brought in to work when needed, and discarded in time of economic recession (White and Woods, 1983; Castles, 1985). This argument can apply not only to migrant labour, but also to agricultural workers and to women. However, recent changes in immigration patterns have made this comparison less appropriate. The recent emphasis on family reconstitution has helped to settle the immigrant populations permanently rather than temporarily and repatriation has been largely unsuccessful; the immigrants cannot therefore be discarded or sent home because economic conditions have deteriorated. Furthermore, increasing numbers of migrant workers are engaged in occupations which are not vulnerable to recession, particularly manual public services, such as in hospitals or on public transport, and as such occupy a necessary place in the occupation distribution; they are 'here for good' (Castles, 1984).

Household structure

One of the most rapidly changing aspects of population geography in France and throughout Europe is the structure of households and families (Hall, 1986). From the sixteenth to the early twentieth centuries, France conformed to the European marriage pattern of small families, low fertility and high celibacy. Since the Second World War, the size and the structure of the family has changed dramatically, with a marked reduction in household size, and new types of family and household formation.

French historical demography is characterized not only by a distinctively early decline in fertility, but also by a characteristic structure of families and households known as the 'European marriage pattern'. This pattern developed from as early as the sixteenth century (Hajnal, 1965). The household and family structure of the French was greatly affected by demographic stagnation and by the very early decline in fertility. The major features of the European marriage pattern from the sixteenth to the early twentieth centuries were a very small family size (averaging less than five) brought about by a late age of marriage for women, and a deliberate reduction of fertility within marriage. The average family size was also kept small by a very high rate of celibacy, particularly among people who were unable to inherit land or the means to support a family. Extended families, with more than two generations living under one roof, were relatively rare, because of low life expectancy and the late age of marriage; if a woman married at 28 and had her first child at 29, it would be relatively unusual for her parents, probably themselves approaching 60, to be alive for many years longer. However, celibate siblings, lodgers and servants often did make up quite complex households until well into this century.

The most important recent changes in households have been a very marked increase in the number of households, a corresponding decrease in their size, and an increase in the number of non-family households. In 1975, there were 17.7 million households in France, each containing an average of 2.9 people. During the period 1975–82, the number of households increased to 19.6 million with an average size of 2.7 (Table 2.8). This trend to more but smaller households has been confirmed by the preliminary results of the 1990 census, which show a further rise in the number of households to 21.6 million at an average size of 2.6 people (INSEE, 1990). Proportionally, the number of households has increased much more rapidly than the increase in total population; in 1975–82 and 1982–90, households increased by 10 per cent, whereas the total population grew by three per cent (1975–82) and four per cent (1982–90). Only two per cent of the population live in non-household arrangements, that is, in institutions such as prisons, workers' hostels, student hostels or religious communities.

Table 2.8 Household structure, France, 1975 and 1982

Number of persons per household	**1975** %	**1982** %
1	22.2	24.6
2	27.8	28.5
3	19.2	18.8
4	15.4	16.1
5	8.2	7.4
6 or more	7.2	4.6
Mean number of persons per household	2.9	2.7
Number of children aged 0-16 years per family	%	%
0	48.3	50.5
1	23.0	22.7
2	16.7	17.7
3	7.3	6.5
4	2.7	1.7
5 or more	2.0	0.9
Complex households	1.3	0.8

Source: INSEE, 1982.

The change in household size is shown particularly in the increase in one-person households, which in 1982 made up a quarter of all households, over 65 per cent of which are female (Chauviré, 1986). The high proportion of women living alone is a function of their increasing longevity; the proportion of female single households consequently rises with age. In 1982, there were two million single-person households of people aged over 55, of which 77 per cent were female; the proportion rises to over 81 per cent for those over the age of 75. Some of these women may live with their children or siblings (the census definition would still class them as a separate household), but an increasing proportion of the elderly live on their own. In fact, one third of all people aged over 65 years live alone (De Saboulin, 1986). Not all the single-person households are elderly, and men form the majority of them between the ages of 25 and 54. This is related to a decreasing rate of marriage, a later age of marriage and an increase in divorces which usually result in a single man and a single mother (Fouquet and Morin, 1984). The regional distribution of one-person households shows the dominance of the elderly component; there is a marked concentration in the rural areas of the Centre and parts of the Alps and Bretagne, areas which have been greatly affected by rural out-migration, leaving a residual elderly population with high rates of celibacy (Chauviré, 1986).

At the same time as single-person households have increased, there has been a reduction in large households of five people or more (Table 2.8). This is clearly related to the decline in the fertility rate and a consequent decrease in the number of dependent children (Table 2.8). The number of complex households has also diminished; these are households with more than one family, and include a couple with children and grandparents, or families of two siblings living under one roof. This type of household arrangement was never the norm even in agricultural households, and is now extremely unusual. The regional distribution of two-family households is extremely distinctive, being highly concentrated in the southwest (Chauviré, 1986).

A consequence of recent social changes has been a marked upsurge in the number of one-parent families, from 720 000 in 1975 to 850 000 in 1982 (Thumerelle and Momont, 1988). One-parent families are mainly headed by women (85 per cent). The cause of this increase is principally related to a rising divorce rate, which has doubled since 1968, facilitated by changes in the law in 1975 (Boigeol *et al.*, 1984). Only 17 per cent of the women who headed one-parent families were unmarried mothers; although the number of conceptions outside marriage has been increasing since 1965, the number of births outside marriage did not begin to match this until after 1975 (Desplanques and De Saboulin, 1986). It is suggested that the character of non-marital births is changing from one of clandestine unwanted births to young women in low-paid jobs to one of a positive choice for an older and higher-status group of women (Deville and Naulleau, 1982). Support for this view comes from the evidence of Pailhé (1986) who found an increasing number of non-marital births; these constituted only six per cent of all births in 1966 but 16 per cent of all births in 1983. Many of these births were to older women, and the proportion of the children acknowledged by both parents doubled between 1972 and 1983. The increase in lone parenthood and the changing status of non-marital births are paralleled by a similar trend for couples to live together without going through a marriage ceremony, but statistics on this are unavailable, as unmarried couples are not differentiated from married couples in the census family statistics.

The household changes outlined here may be related to recent changes in fertility and to changes in the labour force. Lower fertility levels are producing smaller families and households, and are releasing more women for work in the paid labour force. At the same time, these changes are producing some new marginal groups in the population, particularly one-parent families and the isolated elderly, both of which are predominantly female, and often impoverished and relatively immobile. The massive changes in fertility and labour force participation, and rising trends in celibacy, divorce and illegitimacy are changing traditional household and family structures and challenging the patriarchal norm. Paradoxically, these recent social changes are resulting in family structures very similar to those of the European marriage pattern of centuries ago, with small families, low fertility and high celibacy, but with a diminished emphasis on the institution of marriage itself.

The characteristics of French population are distinctive within Europe. The two most significant underlying causes for these patterns are a very early fertility transition and a very late mobility transition. The early decline in fertility in France was the principal reason for a century and a half of demographic stagnation. This has resulted in a wide awareness of population issues, and has helped bring about a generally pronatalist attitude. Most French people believe that it is good for France to have many children, but their behaviour indicates that they would prefer those children to belong to someone else. Demographic stagnation was also significant in bringing about early and substantial foreign immigration, a flow which rapidly expanded in the 1960s when domestic labour sources could not satisfy industrial demand.

French patterns of internal migration have been profoundly influenced by a very prolonged period of rural population loss. The major internal redistribution of population from rural areas to urban started in the mid-nineteenth century, with local and temporary movements giving way to more permanent long-distance migration over time. The rural to urban flows peaked in the period of rapid urbanization and economic growth from 1945 to 1968. Urbanization was then followed by suburbanization and counterurbanization processes. Although the flows of internal migration have apparently been reversed, there appears to be no clean break in the process, which is one of the continued growth of core areas, but of their simultaneous deconcentration.

There have been other significant demographic changes in the last twenty years, notably continued improvements in life expectancy, a dramatic turnaround in immigration trends; and an atomization of household structure. All these demographic changes have profound implications for social policy in the fields of gerontology, race relations and welfare. Temporal trends and regional patterns in population are profoundly influenced by wider patterns of structural economic change. The following chapters consider the changes in the agricultural and industrial sectors with which population changes are so fundamentally interrelated.

CHAPTER 3

The transformation of rural France

Should not in this best garden of the world,
Our fertile France, put up her lovely visage?

Shakespeare, *Henry V*, V, ii, 36

Patterns of agricultural development

In the post-war period, France has become Europe's major agricultural producer and exporter. During this time, and especially since the 1960s, agricultural change has been profound, resulting in widespread mechanization, the loss of millions of agricultural jobs and the replacement of a peasant system of farming by a highly commercialized and semi-industrial agribusiness. There are significant regional variations in patterns of agricultural development which reflect the range of physical and cultural conditions within France, the types of farming systems which have developed over many centuries and the influence of the core area on all forms of economic development.

French agriculture has experienced a rapid post-war boom after a long period of stagnation (Pautard, 1965). The economic stagnation of the first half of this century was exacerbated by the impact of two periods of war and occupation, which left agriculture in a parlous state at the start of post-war reconstruction. For many years, at least until the 1960s, agriculture constituted a lagging sector in the French economy, providing both human and financial resources for the development of other sectors. The stagnation of agriculture until the 1960s provided a building block for the construction of the French industrial miracle (see pages 107–8); agriculture then drew on the fruits of that industrial success to enable its own 'silent revolution' to take place (Debatisse, 1963). These trends in agriculture reflected similar patterns of demographic change (see pages 39–46).

The relationship between agricultural development and population change is a crucial one which has changed over time. A direct relationship between population pressure and agricultural development can be demonstrated in the early years of this century, when areas of growing population such as Bretagne and the Massif Central first extended their cultivated area and then intensified to the available technological limits (Winchester, 1986). By contrast, in the post-war period, intensification of agricultural productivity occurred at the same time as rural populations experienced dramatic decline. In recent years, rural population decline has been an enabling factor allowing an increase in labour efficiency, an enlargement of farm holdings and an intensification of agricultural productivity. Agricultural development was first stimulated by population growth, but continued progress occurred as rural population declined. This changing relationship reflects advances in agricultural technology, which form part of wider changes in the French economy and its institutional structures.

The population exodus from rural to urban areas has traditionally been described as a push-pull phenomenon (see pages 49–51). The pull was provided by growing industrial and urban areas and the opportunities they offered. The push factors were the low incomes, poor living and working conditions and the drudgery of unmechanized farm work. However, this push-pull model may be too simplistic. The supply of rural labour to the cities was not only a free choice for individuals to better their own conditions, but was a structural imperative for economic change (Cleary, 1989). It is arguable that the low economic returns to agriculture and the movement of rural labour were necessary elements in providing the capital and labour required for the massive growth of manufacturing industry in the French economy.

Regional variation in agricultural development in France corresponds in essence to a core-periphery structure. This structure may be related both to physical factors of relief, altitude and climate, and to human factors (cultural, economic and institutional) which have reinforced the dominance of Paris. The core area consists of the Paris Basin and much of northern France; areas of lowland, good quality soil and moderate climate (see pages 3–6; 8–11). The accessibility of this region stimulated early settlement and the development of the national capital. However, its agricultural development as an area of intensive wheat and arable production was delayed until the mid-nineteenth century when machinery became available to break up the heavy soils. Nonetheless, the concentration of transport routes and markets on Paris, and the industrial growth of Paris and the north provided economic stimuli to agricultural production (Weber, 1977). The leadership of the core area of France was enhanced by processes of cumulative causation, and was further assisted by the post-war economic growth of Europe and by the location of the European political heartland around Brussels.

The more difficult agricultural environments of the mountains and the Mediterranean are on the geographical and the economic periphery of France and Europe. These areas suffer from poorer soils and harsher climates, as well as being of greater altitude and relief than the rest of the country (see pages 4–11). Furthermore, they are far removed from the major national and European markets and from industrial innovation. As a result, small-scale peasant agriculture has persisted longer in the periphery, often accompanied by outdated and inefficient marketing and processing procedures, resulting in wasteful practices such as the ploughing in of over-produced fruit and vegetable crops. Until the expansion of the EC to include more Mediterranean countries in the mid-1980s, it could be argued that the southern peripheral areas gained least from national and European subsidies and policies. The intensification of the agriculture of the core since the 1960s occurred, at least to some extent, at the expense of the periphery and accentuated its backwardness. The diffusion of innovations in machinery and marketing from the core has dramatically changed the face of French farming, but it has not changed its uneven regional structure; rather it has served to reinforce it. In recognition of the continued marked regional discrepancies, both France and the EC have instituted agricultural policies to benefit the lagging regions.

The development of early agriculture

Physical geography and climatic conditions serve to divide France into a large northern temperate lowland zone and a smaller coastal Mediterranean zone, flanked and separated by significant mountain areas which impose their own limitations on agricultural development (see Chapter 1). These environmental differences produced significant regional variations in agriculture many hundreds, even thousands, of years ago. Two fundamental agricultural systems developed. These were the 'Mediterranean complex' to the south and the temperate crop and livestock system to the north; both were types of polyculture, incorporating a variety of crops and animals (Grigg, 1974). The high plateau and mountain areas of France, where agriculture was restricted by climate and topography, were able to sustain only extensive pastoral systems of more limited variety.

Mediterranean agriculture supports a characteristic range of crops and livestock. The typical crops are cereals, particularly wheat, grown on the plains; tree crops, especially olives, vines and figs, grown on the lower slopes; and a variety of fruits and vegetables ranging from onions and garlic to oranges and lemons. The cropping pattern is complemented by the grazing of animals, particularly sheep and goats. Given the availability of sufficient water, sub-tropical crops may be grown in the hot summers, while temperate crops may be grown in the warm winters. In poor areas, Mediterranean agriculture is a basic combination of wheat and sheep giving limited financial returns. It is hampered by long, hot drought-ridden summers, by the sparse pasture of steep difficult terrain, and by soils eroded from years of grazing. This rather impoverished system of farming is enriched by the tree crops which provide saleable produce. It is a system of agriculture which developed in response to environmental constraints; remnants of it are still much in evidence and in many areas agriculture is still poor.

Mediterranean agriculture is greatly influenced by climate and topography. The precipitation regime, with its long summer drought (see Fig. 1.3), limited the types of crops which could be grown and reinforced a dependence on tree crops and on extensive grazing. In post-war times, the effects of drought have been alleviated by irrigation which has allowed the introduction of a greater variety of crops, including rice. The Mediterranean topography, where coastal plains are backed by foothills and then high mountains, has also influenced the development of complementary agricultural systems using the mountains and plains at different times of the year. The system of movement of people and animals up and down the mountains, known as transhumance, developed as an efficient use of the variation in landscape and climate (Cleary, 1986). In summer, the mountains were used for grazing, while the plains were used for cereal growing; in the autumn, the animals were brought down from the mountains to graze the stubble.

Temperate agriculture, by contrast, has a different range of crops and livestock, although it was also a system of mixed farming. The characteristic combination was wheat and cattle. In the higher plateau and mountain areas, extensive pastoralism with subsistence cultivation was the norm. Complementary systems of crops and livestock developed at the scale of the individual farm, with much of the cropland used, directly or indirectly, to feed the animals. Soil fertility was initially maintained by leaving land fallow and later by the introduction of crop rotations (Bloch, 1966). Small peasant farms were widespread throughout much of northwestern Europe by the fifteenth century (Grigg, 1974) and still dominate much of France.

Intensification of traditional agriculture

Over time, regional variations in the basic two-fold division of agriculture have become apparent. Some of these variations were purely local, in response to particular physical or socio-economic conditions. Such local specializations included the use of the chestnut and the cultivation of lavender in the eastern Massif Central, the breeding of geese for *foie gras* in southwest France and the cultivation of silkworms around Lyon. At a national scale, however, the growth of Paris brought about a major spatial differentiation in agriculture. The concentration of people and political influence in the Paris Basin became very marked from the mid-nineteenth century, but had been occurring for over a thousand years before that. Parisian growth stimulated not only an increased demand for agricultural produce, but also fostered the industrial development which would provide new methods of transport and new agricultural technology. The cumulative process of regional growth brought with it an intensification of agriculture around Paris (Pautard, 1965). The new ideas and methods which developed in the Paris Basin gradually moved out from the core to the periphery by a process of diffusion. The diffusion pattern can be traced both in the adoption of machinery and in organizational methods such as the growth of co-operatives (see, for example, Baker, 1980).

The core-periphery structure of the country significantly changed the existing pattern of temperate and Mediterranean farming systems. Particularly during the nineteenth century, the Paris Basin developed a much more intensive, commercial and specialized agriculture than that of the rest of the country. This development was in response to demand, but was only possible after the manufacture of heavy ploughs with which to break up the soil. As a result, within the temperate zone a major distinction arose between, on the one hand, the *grande culture*, or large-scale commercial monoculture of cereals, which dominated the Paris Basin; and on the other hand, the *petite culture* of the rest of the country, where small-scale farmers produced a variety of crops and livestock for their own consumption.

The system of small-scale polyculture forms a long-standing and pervasive image of French farming, a caricature image characterized by unmechanized and inefficient smallholdings, each with a few cows and hens and a patch of vegetables. The term 'peasant', now often used in a derogatory way, is one which should correctly be applied to a social system which has now substantially disappeared (Mendras, 1964). A peasant farming system is not one which would be considered as economically rational in the usual sense of maximizing profits or minimizing labour (Franklin, 1969). A peasant system is rational, but in the sense of maximizing labour input in order to provide work, food and shelter for a whole family unit. Such a farming system would therefore be considered inefficient if it were viewed as a commercial enterprise. However, generally the food produced under peasant farming is for the use of the family unit rather than for sale to the wider community; the peasant is therefore essentially self-sufficient and isolated from the mainstream of trade and also from changing ideas and methods. This type of family-run subsistence polyculture relied heavily on property as the basis of security, land rarely being sold or relinquished except in cases of dire necessity.

Most peasant smallholders in France owned their own land: at the turn of the twentieth century, approximately 60 per cent of French soil was in owner-occupation. However, owning land under the Napoleonic Code was a mixed blessing, as the code required land to be divided between all heirs. After a number of generations, the farms became ever smaller and more fragmented as the land, and often each field, was subdivided. This system decreased the efficiency of the farming enterprise as much time was wasted travelling between field and farm, and the tiny plots were unsuitable for mechanization (Baker, 1973). A more rational allocation of land has been a major concern in the reform of post-war agriculture (see pages 96–8).

Many farmers rented more strips or fields in order to eke out a living from the land. However, tenancy was by no means simple and some forms of tenancy were heavily weighted in favour of the owner. The most inequitable form of tenancy was share-cropping, *métayage*. In this system, there were two partners, one of whom owned the land and the other who provided the labour. Any profits from the produce were divided between the partners. However, the landowner had two major advantages. First, the peasant had no security of tenure and could be evicted at any time. Secondly, although the profits from the produce were divided, any profits from the land itself accrued to the landowner. These profits from land might come from improvements such as draining, or from inflationary increases in land prices. There was therefore little incentive for the peasant to improve the land or even to farm it well. At the turn of the century, approximately 13 per cent of French land was held under *métayage*, and the proportions were much higher in the Mediterranean and the southwest of the country.

Despite the tiny amounts of land held by many farmers, the farming community was very numerous and could exert a significant influence in national politics, which was especially evident in the late 1950s (Boussard, 1990). The French have traditionally adopted an ideological stance of support for the peasantry as a vital unifying element in national social and political stability (Barral, 1968). Franklin (1969) explained that the French have viewed the system of small-scale unproductive polyculture, overburdened with labour, as the backbone of the nation, a view intimately linked with pronatalist attitudes. This attitude to peasant farming helps to explain early French policies towards agriculture, which were supportive and protective. Even today the French farming lobby still has significant political influence and attains international media coverage, for example in the successful activism by Mediterranean farmers over competition from Spanish winegrowers.

Agriculture in the early twentieth century

The early pattern of Mediterranean and temperate polyculture (see pages 75–6), with more intensive cereal production in the Paris Basin, was further modified in the early twentieth century. Increases in population and urbanization, and improvements in transport facilities stimulated further changes. Specialized production of agricultural goods occurred in areas accessible to Paris, such as Normandie where dairy products became a major commercial enterprise, and the Morvan and Nièvre in central France, where beef production became important. The regional pattern established at that time essentially continued until the middle of this century. The specialist livestock areas were either those which had grown up to serve Paris, or were mountain zones where little cultivation was possible.

The early part of this century was a period of agricultural depression, exacerbated by a number of factors external to the farming system (Lowe and Buller, 1990). In the late nineteenth century, the vineyards of the Mediterranean regions suffered from ravaging attacks of *phylloxera*, a fungus which killed vines and put many small farmers out of business (Stevenson, 1980). At about the same time, wheat-producing areas suffered from foreign competition, particularly from cheap wheat produced under mechanized monocultural systems in north America. The French government reacted by imposing tariffs on imports to make foreign goods as expensive as home-grown. This protectionist policy not only shielded the inefficient farmer from foreign competition, but also from the need to modernize. International affairs had a further devastating effect on agriculture during the first half of this century; the First World War (1914–18) ruined land, drastically reduced the labour force and decimated livestock (Demangeon, 1920). In the inter-war period, the Western world suffered from a period of acute economic depression, which affected agriculture as well as other sectors of the economy. At the same time, France experienced a demographic depression and there were more deaths than births. The labour force tended to be rather elderly and conservative, and farming modernized little. Again, in the 1940s, in the period of German occupation, agricultural productivity was severely impaired.

Amid all these tales of gloom and stagnation there were some glimmerings of progress, even before the Second World War. The use of artificial fertilizers, of agricultural machines and the development of scientific breeding methods for livestock were all gradually becoming common, providing pointers to further progress and a basis on which post-war intensive commercial agriculture could be built. Similarly, the growth of scientific agricultural knowledge and the establishment of agricultural training schools and of credit schemes for farmers improved the organizational basis of French agriculture.

Post-war change: the 'second agricultural revolution'

In the immediate post-war years, the agricultural sector was ripe for change. In 1946, farms were small and fragmented, inefficient, under-resourced, poorly served by marketing organizations and run down by prolonged German occupation. The limited improvements and innovations which had filtered through to the provinces were much less widespread than in the United Kingdom. The rural exodus was accelerated by social dislocation and industrial change after the war. The level of food production had fallen to only half the pre-war figure and did not again achieve pre-war levels until 1950. Agriculture was in need of reconstruction and radical reform.

In the First National Plan, 1947–53, the French government established its priorities for reconstruction. The national priorities selected were the reconstruction of industry and the mechanization of agriculture. Mechanization alone, however, could not bring about the long-term change that would transform agriculture into a leading export sector. Far-reaching change did not really occur until the 1960s when the state intervened in the restructuring of agriculture, using direct sectoral measures such as pension schemes for elderly farmers, as well as wide-ranging powers of land appropriation. Meanwhile, during the 1950s, the agricultural sector struggled both with its historical legacy and with the effects of the nation's industrial transformation. One significant adverse impact on the agricultural sector was that of price inflation. Ardagh (1970) has described the disparity between agricultural and industrial prices, which by 1959 had widened to a three to one ratio. Farming families, while receiving low returns from the sale of their agricultural products, were required to pay high prices for their increasing agricultural expenses, such as fertilizer and machinery, and also for their domestic requirements and household goods. As industrial sectors of the economy modernized, so disparities in levels of living between rural and urban areas widened. This difference was reinforced by a second impact of industrial transformation: the exodus of rural workers for better paid jobs and improved living conditions in the cities. Meanwhile, discontent simmered in the agricultural community, which saw itself threatened by, yet isolated from, the external changes occurring in the rest of France.

The 1950s were years of serious agricultural unrest and rural riots. In Bretagne and Paris, farmers blocked the roads with tractors and heaped vegetables or truckloads of manure into the street to draw public and political attention to low agricultural prices. Confrontations with the CRS (riot police) in Bretagne in 1961 resulted in one death and thirty wounded (Gervais *et al.*, 1976). In response to the unrest, government measures were introduced to provide guaranteed prices for some agricultural products. In the late 1950s, rural dissatisfaction found new expression through the leadership of the Christian Agricultural Youth movement or *Jeunesse Agricole Chrétienne* (JAC). This organization co-ordinated demands for financial investment in agriculture and demanded structural reform rather than price support and protectionism from the government. The government reacted to this pressure by introducing legal measures to modernize farm structures in 1960 and 1962.

Major elements of post-war agricultural change

Several major elements of agricultural change in the post-war period can be distinguished. First, there have been substantial reductions in the agricultural labour force and changes in its composition. Secondly, there have been alterations in farm structure, producing fewer but larger farms, together with changes in land tenure. This restructuring was paralleled by changes in the organization of industrial firms (see pages 117–19). The results of these structural changes are visible in the landscape: large, open, prairie-like fields have replaced small fields bounded by ditches and hedges. Thirdly, there have been significant changes in farming practice, for example the use of tractors and artificial fertilizers has become widespread. There have also been significant increases in farm output (OECD, 1991).

The impact of these changes has varied regionally; a major contrast still exists between large-scale commercial agriculture and small-scale peasant farming. New practices have been most rapidly adopted in areas such as the Paris Basin, where large rented farms are entirely commercial in outlook. They have been least rapidly adopted in remoter areas, such as the Massif Central, where elderly and part-time farmers, who have little incentive to innovate, are in the majority. Overall, the combined impact of the changes has been to make France's rural landscape more exclusively agricultural and empty of people, services and artisanal activity (Buller and Lowe, 1990), although these changes have been counteracted in more accessible regions by the impact of counterurbanization (see pages 53–5; 59–60).

Changes in the agricultural labour force

Since the Second World War, the agricultural labour force has been substantially reduced and has undergone major changes in its age, sex and social structure. From a pre-war peak of six million working in agriculture, the total has declined sharply to less than two million in the 1980s (Table 3.1). The loss of agricultural labour since 1955 has been continuous and prolonged, reaching three per cent *per annum* in the crisis years of the late

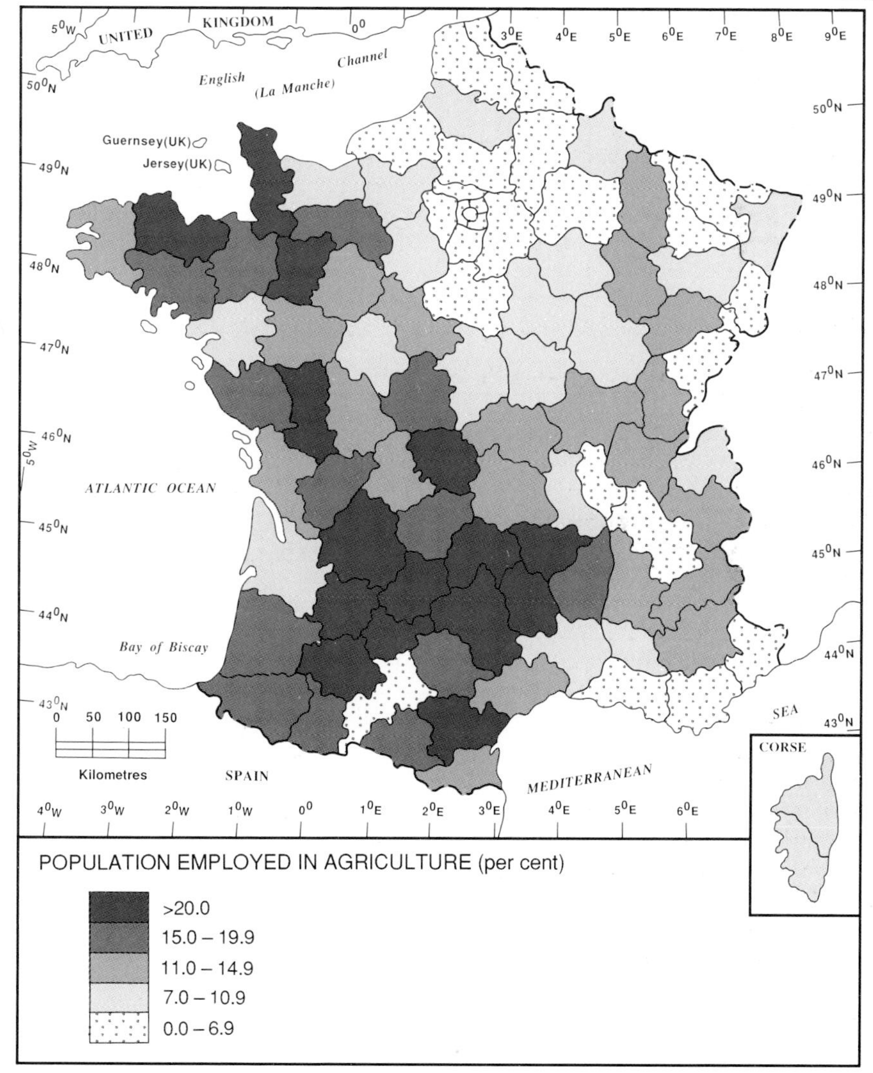

Fig. 3.1 The population employed in agriculture as a percentage of the total labour force, France, 1982. (Source: INSEE, 1982)

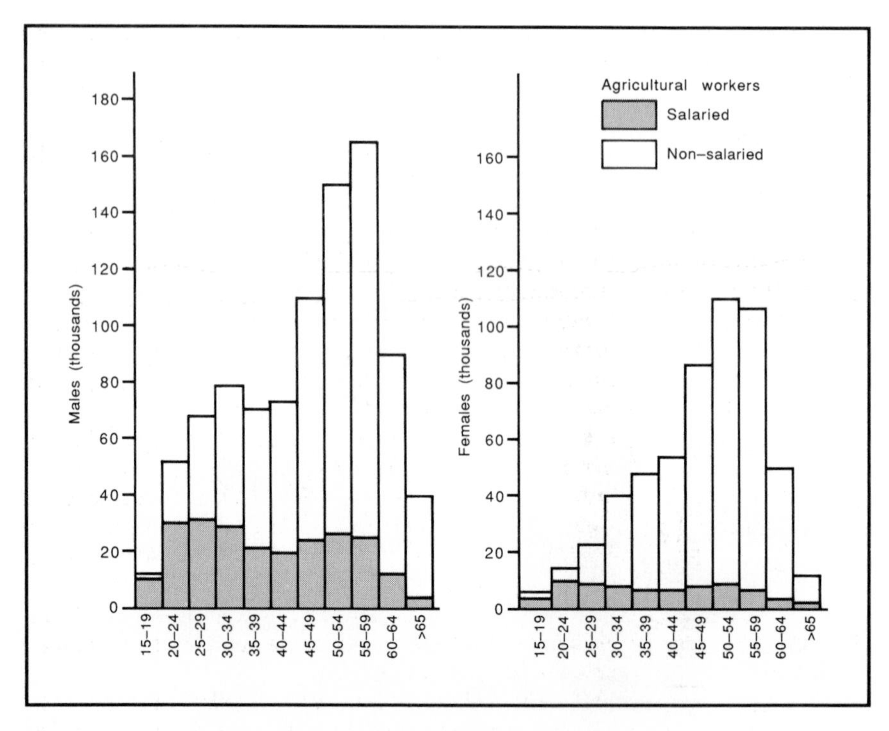

Fig. 3.2 The structure of the agricultural labour force, by age, sex and status, France, 1982. (Source: INSEE, 1982)

1950s (about 160 000 people per year). This exodus has occurred in all agricultural regions, but its impact has been uneven. However, France still has three times as many people working in agriculture as in the United Kingdom, a country with a very similar total population.

Table 3.1 Agricultural labour force, France, 1929–89

	Total agricultural labour force	% of total labour force in agriculture
1929	6 098 500	35.7
1955	5 135 400	27.6
1970	3 460 700	14.9
1982	1 769 900	7.5
1989	1 556 100	6.4

Source: Ministère de l'Agriculture, 1984; OECD, 1991.

Regional variations in the agricultural labour force are still marked (Fig. 3.1). In the Paris Basin, less than 10 per cent of labour is employed in agriculture, whereas areas of the west and centre have over 20 per cent of their labour force in farming. In particular, the southern Massif Central has a concentration of agricultural workers, a legacy of the small-scale peasant agriculture of early years and a result of the lack of alternative employment opportunities.

The agricultural workforce is very varied and is rather different in age and sex characteristics from the labour force as a whole (Fig. 3.2); there are far more men and a higher proportion of elderly people than in other sectors of employment. There are also major differences in the employment status of agricultural workers compared with the rest of the labour force; notably, there are relatively few salaried or waged employees, who are greatly outnumbered by unsalaried family workers and owner-managers. Significant proportions of farm owner-managers work only part-time. These differences in labour force status emphasize the continuing tradition of the small family farm with few paid employees.

The regional distribution of the agricultural labour force aged 65 and over shows a significant concentration in the southeast of France, Aquitaine, the Pyrenees and parts of the Alps and Alsace. As the workforce in this age group retires or dies out, so the total population in agriculture will continue to decrease as many elderly farmers will sell out and farm labourers will not be replaced. By the year 2000, this natural wastage is likely to remove at least another half a million people from the agricultural labour force, bringing it into line with the more agriculturally advanced countries of the EC such as the United Kingdom.

The agricultural labour force also shows a major imbalance between women and men, with women recorded in the 1982 census as comprising only one-third of the agricultural labour force (compared with over 40 per cent in the labour force as a whole). Furthermore, over 80 per cent of unsalaried family workers are women, but less than 20 per cent of owner-managers and salaried employees. There is, therefore, a highly significant imbalance in the roles attributed to women and men. The definition of jobs by gender almost certainly under-records and under-values the contributions made to family farms by women, who have traditionally borne the bulk of agricultural labour in peasant farming systems. The proportion of women in agriculture has fallen since the pre-war period when they formed over 40 per cent of the labour force. This change reflects the key role played by women in maintaining farms in the early years of this century when male labour was decimated by war service; it also indicates a subsequent movement of women away from farming, particularly to jobs in the tertiary sector. Increasingly, traditional family structures and family farms are on the wane and women are working in paid jobs away from the home and farm, rather than working on the farm without pay or status.

The growth of part-time farming has been an important recent phenomenon in western Europe; Frank (1983) estimated that a quarter (over 1.3 million) of farmers in the EC in 1980 had a gainful activity outside farming. In the United Kingdom and Germany many part-time farmers are 'five o'clock' farmers who have another business such as a guest house or a rural craft, or else are hobby farmers who have well-paid urban jobs, but engage in low-intensity farming as a form of recreation or tax-deductible activity. France has not been exempt from these influences, but very many part-time farmers are worker-peasants or are elderly and semi-retired. In France in 1979–80, about one-third of all farm owner-managers did not work full-time in agriculture (about 300 000 people). The regional distribution of part-time farming is very similar to the pattern of elderly farmers, being concentrated in the southeast, Aquitaine and Alsace. Over 85 per cent of the farms run by the retired and by farmers with another occupation are small and unproductive, producing less than the equivalent of five hectares of wheat per year. Although some of the elderly and the worker-peasants may represent the last vestiges of peasant farming which may disappear within the next generation, this is not likely to be the case for hobby farmers, and the development of part-time farming is considered likely to be a permanent feature of the European rural landscape, especially in areas accessible to major metropolitan centres (Ilbery, 1985).

The employment status of workers in agriculture is markedly skewed towards owner-managers (about one million) with few waged employees (about 300 000), themselves out-numbered by unpaid family workers (480 000). Many small farms are run by one person or by one family; one-third of all paid workers were the only employee on the farm. At the other end of the spectrum, 7000 of the largest farms in France employed 80 000 agricultural workers. These were heavily concentrated in the large rented farms of the commercial cereal-growing areas of the Paris Basin. The low number of agricultural workers has resulted from the massive post-war rural exodus of labour to the cities and their replacement by agricultural machines.

The reorganization of farm holdings

The decline in the agricultural labour force has been accompanied by a major change in farm structure and organization. Since 1955, the total amount of agricultural land has remained constant at about 32 million hectares, although some land has gone out of production in areas of suburban expansion and in most economically marginal areas; at the same time other agricultural land has been gained by farm consolidation. The agricultural land has, however, been redistributed to form fewer but larger farms. Table 3.2 shows the reduction in the total number of farms, the increase in average farm size and the declining proportion of workers per unit area. The total number of farms has decreased dramatically since the turn of the century, and since 1955 over a million farm holdings have disappeared. Since 1955, the average farm size has more than doubled from 14 to 32 hectares, while the number of workers per 100 hectares has diminished from 14 to five.

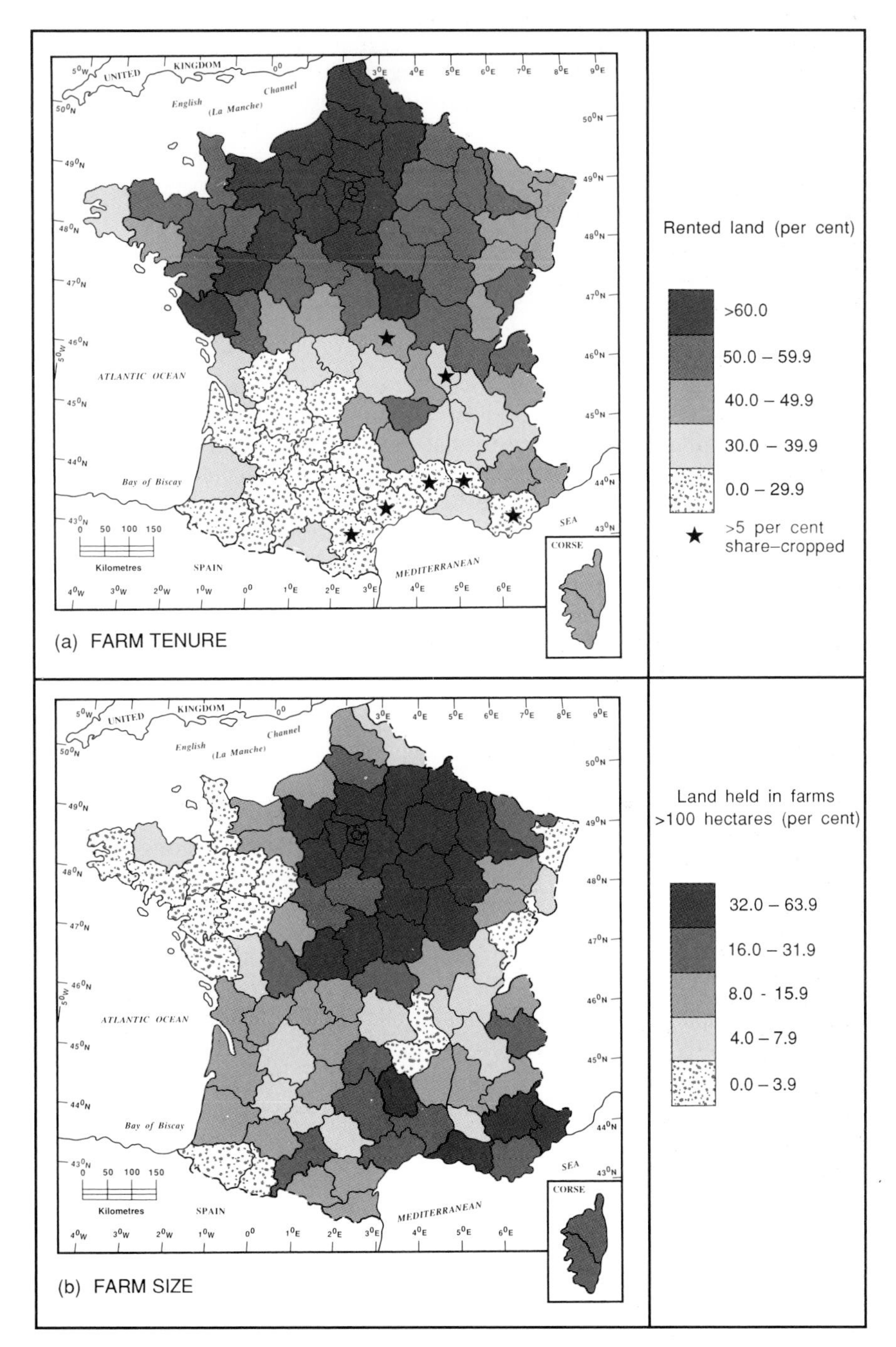

Fig. 3.3 Farm structure, France, 1982 (a) farm tenure; (b) farm size. (Source: INSEE, 1982)

Table 3.2 Farm numbers and size, France, 1929–87

	Total number of farms	Average farm size (hectares)	Workers per 100 hectares
1929	3 966 400	11.6	17.9
1955	2 285 700	14.1	14.0
1970	1 587 500	20.8	9.4
1980	1 262 700	26.7	5.2
1987	981 700	32.0	4.7

Source: Ministère de l'Agriculture, 1984; INSEE, 1989.

French farms are still small and over 60 per cent of all farms are less than 20 hectares, but there has been a notable general increase in farm size. Most of the farms which no longer exist as separate entities were relatively small, between one and 20 hectares. Changes in farm size since 1955 show that the proportion of large holdings (over 50 hectares) has increased dramatically. There has therefore been a redistribution of land holdings caused by the amalgamation of small holdings into units of more economic size. One anomaly in this trend to larger farms is the relative stability of the number of farms of less than one hectare; this is mainly due to problems of enumeration and definitional changes between the agricultural censuses of 1955 and 1980.

Despite the general increase in farm size, marked regional variations in the size of holdings still existed in the 1980s (Fig. 3.3b). The largest holdings, averaging more than 50 hectares, were to be found in the Paris Basin. The *département* of Lozère in the southern Massif Central also contained very large farms averaging over 60 hectares, but in this case it was the result of extensive pastoral agriculture and prolonged rural depopulation. At the other extreme, farms of below average size were to be found in the regions of Bretagne, Alsace, Aquitaine and parts of the southeast.

Changes in land tenure have also occurred since 1955; in particular, the proportion of rented land has increased, while the proportion of land owner-occupied or held under share-cropping arrangements has decreased. Many farmers have some land which they own and other land which is rented. In general, smaller farms tend to be owner-occupied, a legacy of the peasant farming tradition based on property (see page 77), while larger farms are more likely to be rented in whole or in part. The regional pattern of land holding (Fig. 3.3a) shows that in the 1980s, over 60 per cent of agricultural land was rented in the industrialized areas of the Nord and the Paris Basin. Land ownership in these regions has also become increasingly dominated by companies, banks and trust funds rather than by individuals. Conversely, very low levels of rented land and high proportions of individual owner-occupation were typical of the small-farm areas of the south, Aquitaine and the Massif Central. Even in the 1980s, share-cropping was still locally important in parts of the Massif Central and Languedoc–Provence–Côte d'Azur (Fig. 3.3a). The few farms still held under share-cropping arrangements tended to be small and unimproved. Share-cropping is now controlled by government legislation and the worst excesses of exploitation of share-croppers are now consigned to history. Share-cropping is likely to die out almost completely by early next century as the existing tenants die, retire or move away.

Changes in farming practice

Major regional variations remain in farming specialization. Many of these originated hundreds of years ago (see pages 75–6). Figure 3.4 shows the distribution of cereals, concentrated in northern France and, to a lesser

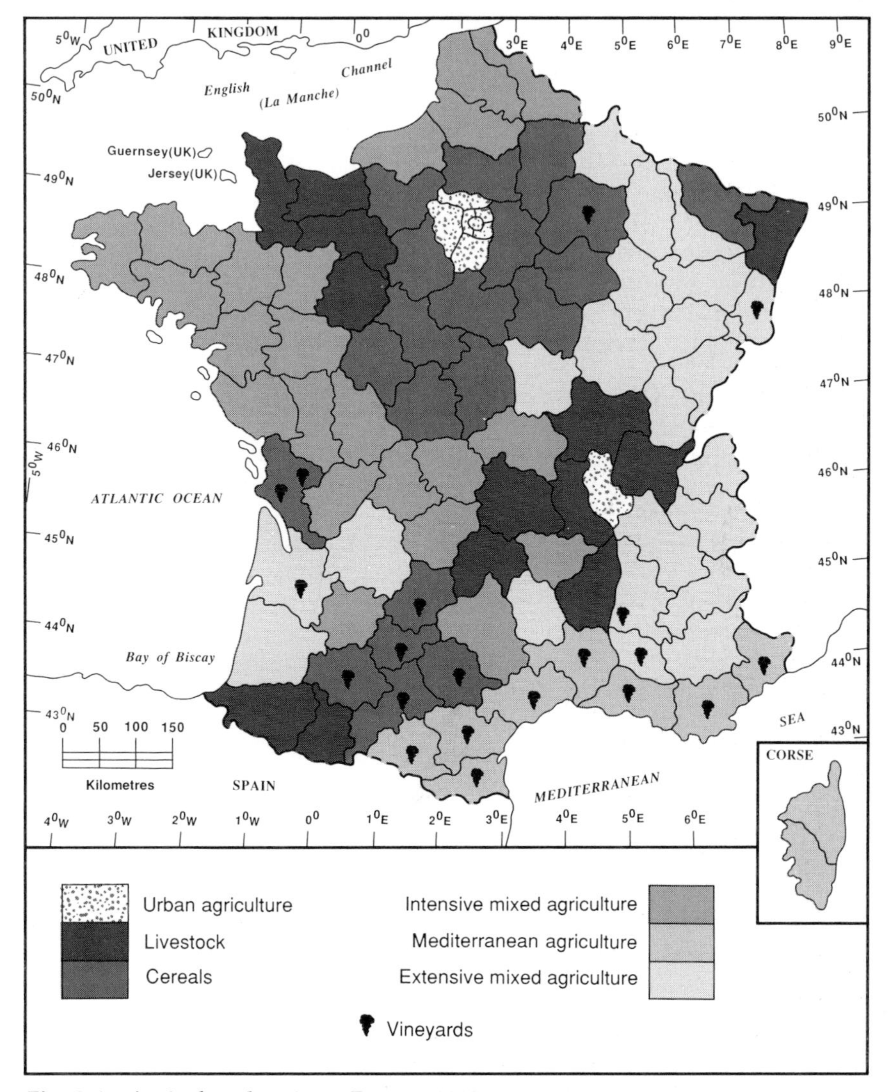

Fig. 3.4 Agricultural regions, France, 1982

extent, in the Garonne Valley. The inverse pattern is shown by the distribution of dairy cattle, which are predominantly raised in Bretagne, Normandie, parts of eastern France and the Massif Central. Specialized crops, such as vines, tree fruits, soft fruits and horticultural products are concentrated in traditional areas of viticulture and horticulture, such as the Mediterranean coast, Aquitaine, the Rhône and Loire Valleys and the farming areas adjacent to the industrial and urban concentrations of the Nord and Paris.

The modernization of agriculture has involved land improvements, mechanization and an increased use of chemical fertilizers and pesticides. Together these have resulted in spectacular increases in agricultural productivity. Areas of specialized viticulture and horticulture have experienced the greatest amount of land improvement, especially in irrigation and land drainage. These improvements have been particularly concentrated in the market gardening zones around Paris and the Nord, where demand has brought previously unimproved areas into intensive use. Land improvements have also influenced the development of new crop specializations, such as rice in irrigated areas of the Languedoc.

Mechanization of farming is now widespread and over 70 per cent of farms possess a tractor; more specialized machinery is, of course, distributed according to the specialization and level of development of the area concerned. The proportion of farms with tractors, the most basic item of all-purpose farm equipment, nonetheless varies widely throughout the country. Only one-third of the intensive small farms close to Paris have tractors and only 15 per cent of the small peasant farms of southern Corse.

The mechanization of agriculture has been accompanied by many other technical improvements, the most notable of which has been dubbed the 'chemical revolution'. The use of chemical fertilizers doubled between 1967 and 1987, a substantially greater increase than in Germany, Belgium or the United Kingdom. Their increased use has been a contributing factor in the major increases in farm productivity which have been evident since 1960. The chemical revolution has played its part in the transformation of French agriculture from a system of low inputs and outputs to one of high inputs and outputs. However, some of the adverse side-effects of excessive uses of fertilizers include pollution of groundwater by nitrates (see pages 25–8) and the eutrophication of water bodies.

The volume of agricultural production increased by more than 60 per cent between 1960 and 1980. This enormous increase in output is unparalleled in the history of French agriculture. Total agricultural production rose from a base level of 100 in 1973 to 129 in 1987; comparable data for the United Kingdom are 100 to 118; and for Germany 100 to 115. Greater increases in output were recorded by the Netherlands, Denmark and Ireland over the same period. For certain products, the increase in production has been remarkable, with increases of over 100 per cent in the 20 year period, notably for cereals, sugarbeet and vegetable oils (although the latter started from a very low base). In the livestock sector, dairy produce and poultry have shown above average production increases. The greatest increase in agricultural production has occurred in crops such as cereals, sugar and fruit, which can be processed and used in the production of other goods such as pie fillings and breakfast foods; they have therefore contributed to the development of downstream processing of agricultural produce (Fel, 1984) (Plate 3.1).

The production increases have been relatively less for goods which are used relatively infrequently in processing or for which the demand is otherwise virtually static, such as lamb or vegetables. Other products have experienced particular difficulties in production, as a result of competition with other EC producers, or because of a mismatch with market demand. For example, wine production has suffered from foreign competition, and milk production, having increased very rapidly, became subject to EC quotas in the mid-1980s.

Modern commercial farming requires high inputs and high outputs, whereas peasant farming was characterized by low inputs and low outputs. The results of investment in technology may clearly be seen in the intensive raising of pigs, poultry and veal calves and the development of factory farming. Factory farming has aroused some opposition, both on ethical grounds, because of the deprivation of animals of light, space and stimulation, and on economic grounds, because intensive methods have contributed to national and European overproduction. Increased productivity, especially when it leads to overproduction while using scarce resources, is ecologically unsound and has caused some backlash from 'green' groups. It is, however, undoubtedly true that the technical improvements which have occurred in French agriculture in the post-war period have turned the nation into a major exporter of agricultural products as well as a major producer.

Government agricultural policy

The major and widespread agricultural changes which have occurred in France in recent years have not happened in isolation. They have taken place in response to market forces such as changes in demand and disparities in wage levels between the agricultural and industrial sectors. More generally, they have formed part of the post-war transformation of the capitalist system which is considered more fully on pages 112–14. Furthermore, the agricultural system has been increasingly channelled and regulated by government. Intervention by governments is occurring not only in France, but also in other countries of Europe and throughout the developed world; in France, however, the state has assumed a particularly important role in economic regulation.

The policies which influence French agriculture operate at two main levels, the supranational (especially the EC) and the national. In theory, these policies should work together to fulfil the aims of the EC's Common Agricultural Policy (CAP); indeed, many of the grants given by national governments are subsidized by the EC. However, in practice, the policies of national governments may conflict with those of the EC, especially where national interests are felt to be at stake. At other times, the policies and interests of individual member states may conflict with each other; for example, competition between French and Spanish wine growers, and

Plate 3.1 Downstream processing: a flour mill in Picardie, an area of large commercial cereal farms. (Photo: S J Gale)

French and German livestock producers, reached the point of violence and sabotage on a number of occasions during the 1980s.

The national policies adopted by the French government have been altered and extended over time. Early policy measures tended to be conservative and aimed to improve farm conditions without altering the structure of farms or of the agricultural labour force. Later policies were much more radical and far-reaching and had a significant structural impact. Modern French agricultural policy really began in the early 1960s when a wide range of measures was introduced affecting the development of co-operative systems, retraining and reorganization of the work force and the management of land. In the late 1970s, these sectoral policies were supplemented by specific regional measures and by a range of amendments and improvements to existing laws, so that there now exists a fairly comprehensive package of socio-structural policies. Agriculture in France is thus affected by government policies at all points in the production process from the purchase of land to the processing of food.

French national policies before 1960

The early policies adopted by the French government were piecemeal in nature, were essentially protectionist and were intended to support the peasant subsistence farmer. The most important government influence on agriculture was the long-standing protectionist policy of import tariffs and price support. These measures were designed to keep out cheap foreign produce which could undercut the price of home-grown crops. It could be argued that this was one of the main reasons for both the relative stagnation of French agriculture in the early twentieth century and the maintenance of a very large peasant workforce. In the United Kingdom, by contrast, the policy of free trade meant that farms were forced either to become efficient or to go out of business. As a consequence, thousands of agricultural workers left the land, leaving the United Kingdom with a much more efficient agricultural system than in most of continental Europe.

Several other policies, rather more limited in scope, were introduced before 1960. These included the formation of the Rural Engineering Service in 1918 which mainly provided land drainage and electricity services for rural areas. In 1906, the establishment of an Agricultural Credit Service provided easier and cheaper credit for farmers and farming groups. The *Crédit Agricole* is now one of the largest financial institutions in France. In 1918, the land consolidation legislation, known as *remembrement*, was introduced particularly to assist those areas of northern France affected by trench warfare. The modern legislation on *remembrement* was instituted in 1941, with much more far-reaching effects (see pages 96–8).

Further legislation was introduced in the immediate post-war period. In 1946, the introduction of tenancy legislation gave greater security of tenure to tenants, greater rights to pass on tenancies to heirs and provision for share-croppers to become ordinary tenants. In 1947, the newly-formed National Association for Rural Migration gave grants to farmers and farm workers to encourage out-migration from areas of population pressure. In 1958, the law of land accumulation limited the amount of land which could be farmed (but not owned) by any one person. Although limited in scope and application, these measures provided a legal basis for the removal of some of the severest obstacles in the way of agricultural progress, particularly problems caused by share-cropping, farm fragmentation and the lack of finance for farm improvement.

French national policies after 1960

The farming protests of the late 1950s stimulated the French government into more active and far-reaching policies. Two laws form the basis of the bulk of modern agricultural policy. The *Loi d'Orientation* passed in 1960 was more a statement of principle than an executive measure. It recognized that agriculture should have parity with other sectors of the economy, that price support did not encourage agricultural modernization, and that improvement could come through structural change in farm sizes and the composition of the labour force. Following a further summer of rioting, the *Loi Complémentaire* of 1962 was the bill which put the teeth into the principles of the 1960 act and which established organizations and funds to turn these aims into reality. In particular, the 1962 law, which is the key to French national policy in this period, brought in mechanisms for the structural reform of farming.

Structural reforms hit at the very essence of the old peasant system of subsistence polyculture (see pages 76–9). The policy measures affected three major areas: the labour force, farm size and structure, and farming practices (including the formation of co-operatives and the control of tax and inheritance systems). These national policies principally affect the structure of farming, whereas the EC spends most of its budget on price guarantees and production subsidies. Europe therefore controls prices, export-import and levels of production, but the French national government has a greater influence over the type of farms and the workers who actually produce the goods.

Policies affecting the labour force

In 1960, the *Loi d'Orientation* recognized that farm efficiency could be improved by training the farm labour force and by altering its age structure. The tradition of peasant farming meant that few farmers prolonged their education or went to agricultural college; instead they learned traditional methods from the previous generation of farmers. Furthermore, the difficulties and low returns of agriculture meant that many of the younger and more enterprising people left for other jobs elsewhere, leaving behind an ageing, technologically untrained and conservative workforce.

Retirement premiums: the Indemnité Viagère de Départ

The most extensively adopted of the labour force measures has been the retirement premium or *Indemnité Viagère de Départ* (IVD). The IVD became available in 1963 to full-time farmers aged 65 and over who were prepared to release their land for agricultural use by other farmers (OECD, 1972). Those taking advantage of this premium could keep a plot of no more than one hectare for themselves. The restrictions attached to the pension and its limited financial value meant that the numbers claiming it were not very high for the first two or three years. The premiums became more widely available in 1967, when they could be claimed by farmers aged 55 and over. At the same time, the regulations for transfers of land were relaxed, so that transfers could be made to other members of the same family; additional payments were made to farmers whose land was released for restructuring.

The precise regulations for the award of IVDs have been altered several times. In 1980, the value of the pension was raised and in 1981, pensions became available for farmers aged under 63. These changes in the regulations resulted in an increased annual uptake of the IVD to a peak in 1970, a steady decline in uptake until 1980 and then another smaller increase. In the 1980s, the number of IVDs granted remained steady at about 20 000 per year. This number is likely to be maintained or to decline as the numbers of eligible elderly farmers decrease and as the small numbers born in the depression years of the 1920s and 1930s reach retirement age.

Over 650 000 farmers have received IVDs since 1963 and almost 12 million hectares of land have been transferred (Naylor, 1985). The land transferred has either served to enlarge other existing holdings, or has gone to young farmers setting up in farming. The IVD has therefore not only speeded up the movement of elderly farmers from the land, but has helped to provide opportunities for younger farmers and to enlarge the size of farms. These benefits may sometimes be more apparent than real,

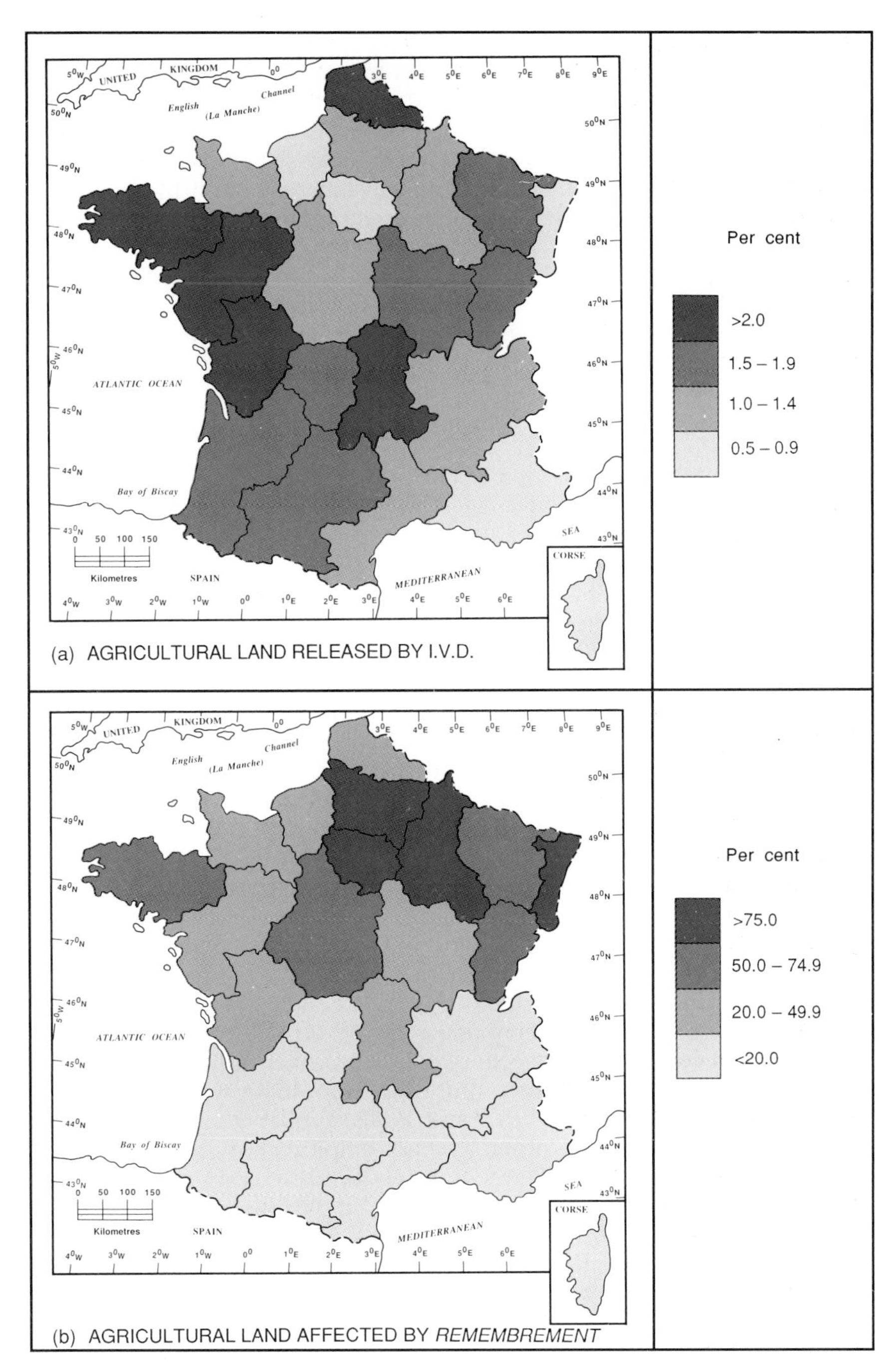

Fig. 3.5 Uptake of agricultural policies from their initiation to 1983 (a) retirement premiums; (b) farm consolidation. (Data from Ministère de l'Agriculture, 1984)

especially in cases where the young farmer is a member of the family and the retiring farmer maintains an indirect interest in affairs. Nonetheless, the scheme has operated on a massive scale and has cost more than FF 1200 million.

Not all eligible farmers have chosen to take an IVD. Many farmers feel that they would be better off both financially and personally by continuing to work. Other farmers have an inbuilt resistance to relinquishing control, stopping work and selling land. A reluctance to part with land is particularly noticeable in areas where the land might later be profitably sold for tourist development or second homes. There is therefore no clear correlation between the areas of need (the areas where there are many elderly farmers) and the areas where IVD uptake has been high (Fig. 3.5a). In the south and east of the country, where elderly farmers are most numerous, IVD uptake is low, whereas there is a much higher rate of uptake in the west and centre of the country where up to half the eligible farmers have taken the IVD.

Assistance to young farmers: training and retraining

The aim of providing assistance to young farmers was to alter the age structure of the farm labour force. Establishment grants to young farmers became available in 1973, originally in the mountain zones and adjacent areas which were experiencing problems of rural depopulation. Since 1977, this finance has been available, although at lower rates, throughout France. The establishment grant, known as the *Dotation aux Jeunes Agriculteurs* (DJA), is available to farmers under 35 who have specific agricultural qualifications. This grant is a very significant aid to young farmers, who otherwise would find it almost impossible to raise the finance to start a farm; its effective value is greater when the DJA and IVD are paid to members of the same family and the land is transferred between them. Since the grants became available, 80 000 young farmers have taken them up, mostly in the west and southwest of the country where family farms are still important. Relative rates of uptake are still significant in the mountain zones.

Since 1962, grants have been made available for the retraining of workers wishing to leave agriculture, and for the improved agricultural training of those wishing to stay in agriculture. The retraining schemes attracted relatively few takers in the 1970s and 1980s, probably as a result of the difficulties of moving into other types of employment at a time of prolonged economic recession. Three-quarters of those receiving grants were improving their agricultural qualifications, in part because the award of the DJA requires a minimum of professional competence. These training and retraining schemes have been followed by some 385 000 people since 1962. They have been particularly important in Bretagne and the west of the country. The emphasis on formal training serves further to break the peasant farming mould and to increase the potential productivity of farms.

These two schemes for old and young farmers have effectively replaced the 1947 scheme which promoted selective rural migration, with a much more comprehensive package of incentives which have so far directly benefited over one million people. These schemes have speeded up the transfer of land from old to young which would have happened naturally by death, retirement and inheritance. Some commentators have estimated that the rate of transfer is perhaps twice that which would have occurred without policy measures being in force. The resistance of some farmers to retirement and land transfer may have kept the pace of change to a rate which is manageable and socially acceptable. Nonetheless, Buller and Lowe (1990: 24) considered that the laws of the 1960s 'sounded the death knell for the traditional peasantry, its way of life and culture'.

Policies affecting farm size and structure

The reorganization of farm holdings was recognized by the 1960 *Loi d'Orientation* as a necessary measure to improve efficiency. Inheritance laws and the peasant farming tradition had combined to make French farms very small and extremely fragmented. Clout (1983: 36) stressed the extreme sub-division of land which occurred not only in areas of subsistence farming, but also in the strips of the open fields of the Paris Basin. In 1908, the land ownership record showed over 13 million land owners, a number inflated by owners holding land in more than one *commune*, but still barely indicative of the extremes of land fragmentation within single *communes*. Farms which themselves were tiny, perhaps five hectares, were generally fragmented into a dozen tiny fields or strips, which could be widely scattered. Legislation concerning farm consolidation dates back to 1918 (see page 91), but policies to increase the size of whole farm units date only from 1962.

Land management: SAFERs

The main mechanism for increasing farm size was the operation of rural settlement companies, *Sociétés d'Amélioration Foncier et d'Etablissement Rurales* (SAFERs). SAFERs were private, non-profit making bodies whose brief was to improve agricultural structures, to increase farm size so that farm units were viable and to help young people set up farms. SAFERs have acted as land management companies with powers to buy up land and to resell it either to enlarge other farms, or in the form of units on which young farmers can become established (usually with the help of a DJA). Much of the land that they buy has come by private agreement from individuals wishing to sell, perhaps as a result of a decision to take the IVD. However, the SAFERs were also given some rights of pre-emption, so they take priority over most other buyers unless the potential purchasers are close relatives, sitting tenants or public authorities. Nevertheless, only 10 to 15 per cent of their purchases have been by pre-emption even though SAFERs buy between 30 to 50 per cent of the land that comes on the market. France has 31 SAFERs which operate throughout almost the entire country. SAFERs have only recently been set up in Corse and in much of the cereal growing region of northern France. Since they started,

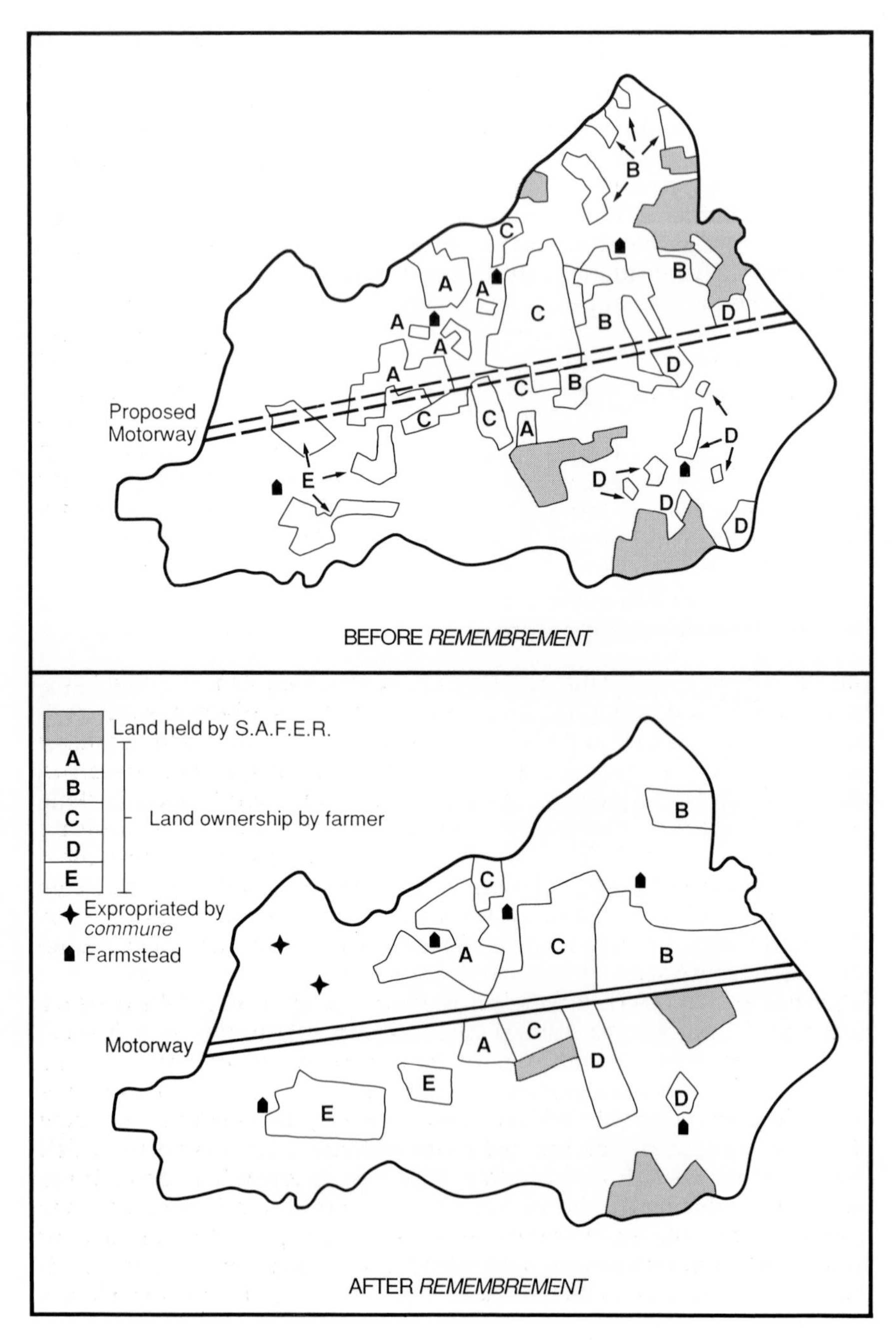

Fig. 3.6 Impact of *remembrement* in a *commune* in Bretagne. (Source: Naylor, 1985)

the SAFERs have bought almost 1 500 000 hectares of land and have resold 1 345 000 hectares. They have therefore dealt with the sale of almost five per cent of France's usable agricultural area and have been a significant stabilizing influence on the land market. Most land acquisition has occurred in the west and centre of France, with relatively small purchases in the commercial north, where land does not often come up for sale.

Land bought by SAFERs can only be held for five years before it must be resold, but usually it is sold more quickly, to maintain cash flow. About 20 per cent of the land resold by the companies has had some improvement, whether this is drainage, soil improvement, clearance of scrub, or the installation of services and equipment such as sealed roads or electricity. The SAFERs are unable to carry out as much improvement as they would wish because of limitations of finance and some administrative delays. Most land sold is used to enlarge existing farms, but an increasing proportion, about 25 per cent, is now used for the establishment of young farmers. Stocks of land may be sold off in connection with *remembrement* schemes (page 98), or used to compensate owners who may be losing land through public schemes such as road widening. An example of this type of scheme is quoted by Naylor (1985) and is shown in Fig. 3.6.

Land consolidation: (remembrement)

Farming of small scattered patches of land was feasible under the traditional peasant farming system. Indeed, a number of advantages existed: fragmented holdings provided farmers with land which varied in quality and capability, the availability of tiny plots meant that farmers could gradually increase their land holdings, and since work was all done manually there was relatively little loss of time or efficiency. Agriculture was adapted to fragmentation and farmers tended to have their most intensive plots close to the house, with more extensively worked fields further away (Baker, 1973). Fragmentation on this scale, however, is unworkable in a mechanized and labour-efficient farming system, particularly one where arable crops are widely grown.

Early legislation concerning *remembrement* of land provided the basis for the consolidation of some 350 000 hectares, mostly in northern France. New laws in 1941 changed the procedure for consolidation from one which relied on the landowners to one which involved consultation between all interested parties. Valuation is a major problem in the exchange of land and the 1941 act outlined a productivity classification by which land could be graded; it also established departmental land committees which could deal with disputes. Disputes are particularly frequent over land with has sentimental value to the farmer (Ardagh, 1982: 216), or over land which may have development potential. Land of the latter sort, usually close to towns or on the coast, is often excluded from consolidation proceedings.

Some 12 million hectares or 40 per cent of the total agricultural area of France has now undergone *remembrement*. The costs of *remembrement* itself are paid by the government, but related activities such as hedge clearance are subsidized at various rates. In Bretagne and Normandie, many of the tiny fields are surrounded by ditches, earth banks and large hedges. This *bocage* landscape is very expensive to consolidate. In these areas, the rate of *remembrement* has slowed because of the additional expenses involved.

The regional distribution of *remembrement* uptake is very concentrated in the commercial cereal-growing areas of northern France (Fig. 3.5b) where the benefits of large open fields are very obvious. Land consolidation is therefore an improvement which has given added advantages to areas which were already reasonably advanced. The effects of consolidation of holdings, after a short period of readjustment, have generally been to increase productivity and to accelerate mechanization and the specialization of agricultural activity. Consolidated areas are also attractive to young farmers and are favoured by organizations such as the *Crédit Agricole* for loans.

Some of the results of land consolidation have not been entirely successful. Consolidation may encourage more people to stay on the land in small farms than is economically desirable. Some of the fields which were consolidated early on are now in need of further consolidation, as field sizes that were economically viable 30 years ago are now too small. Also, in the *bocage* areas, the removal of ditches and hedgerows has had significant ecological and hydrological impact; not only have habitats for wildlife been destroyed, but the fields are now more susceptible to flooding and to erosion (see pages 7–9; 20). Finally, although it arranges the fields more efficiently within each farm, *remembrement* itself does nothing to increase the size of farm unit. Many people are unhappy with the prairie-type landscape which now dominates northern France, as it does in parts of eastern England, since it is monotonous to the eye and almost devoid of wildlife: farming efficiency has its price, and this may be calculated in ways other than the purely financial. Furthermore, *remembrement* is not cheap; it takes a quarter of the national agricultural budget.

Policies affecting farm practice and organization

The policies affecting farm practice and organization consist of a rather disparate set of measures, the most important of which relate to the joint management of farming units and the establishment of co-operative organizations. Laws relating to tenancy and to inheritance also affect the pattern of farm organization. These policies have operated with market forces to produce a pattern of fewer larger farms which increasingly are being managed by firms rather than by families or individuals.

Co-operatives

One possible alternative to farm enlargement is the development of co-operative systems. Co-operatives are of many different types ranging from family groupings of farmers to enormous commercial enterprises concerned with the buying, processing and marketing of agricultural produce. Smaller co-operatives have become very common since 1964 and there are now almost 30 000 agricultural co-operatives of this type, known as *Groupements Agricole en Commun* (GAECs). These cover over two million hectares of land and involve five per cent of the full-time farmers in the country. Two-thirds of the GAECs which have been set up consist only of a farmer with one or more children.

There is a number of advantages for those involved in establishing a GAEC. The inheritance of the farm is easier for the heirs and they are still entitled to claim a DJA (see pages 94–5), subject to the normal age and educational qualifications. There are also financial advantages as GAECs receive priority from the *Crédit Agricole* for farm modernization schemes and loans and grants are easier to come by. For individuals, the split of profits (nominally at least) may reduce tax liability. There are also some social advantages in a formal partnership of this sort, for example the possibility of sharing work at unsocial hours and of taking holidays. Other forms of co-operatives which may or may not be associated with a GAEC include machinery co-operatives, where members buy and use machinery and equipment on a communal basis. It is estimated that 400 000 farmers belong to organizations of this sort (Naylor, 1985).

Larger co-operatives which buy, process and retail produce are particularly common in the north of France, in the Paris Basin and in Bretagne, where agriculture is at its most commercial; they deal with a variety of agricultural produce from cereals and vegetables to milk and meat. Within northern Bretagne, for example, the coastal zone has developed as an area of specialist intensive horticulture and dairying, the produce from which is traded, processed and marketed by major co-operatives. An example is the *Société d'Initiatives et de Coopération Agricole* (SICA) of the small town of St Pol de Léon in north Finistère. This co-operative was formed in 1961 by a few farmers, but has grown to serve over 4000 producers and is the most important dealer in horticultural produce for the whole region. The SICA markets 250 000 tonnes of cauliflowers and 55 000 tonnes of artichokes each year and has specialized sections dealing in cut flowers, carrots, onions, shrubs and flower bulbs, particularly narcissus and iris. The sale of goods no longer requires the carriage of vegetables to market; the goods are sold unseen, but with strict quality control. The bidding is still fast and furious, but is done from computer terminals. The SICA also has its own freezing plant for vegetables, has interests in experimental farms and its own distribution network to other countries of the EC. For example, flower bulbs are traded with the Netherlands and cauliflowers are exported to the United Kingdom (Anon, 1986).

Other SICAs in the same region of intensive farming include the second largest producer of frozen food in France, which trades under the name of *Paysan Breton*, and a dairying co-operative, *Even Dairies*, which serves 2500 producers and processes a range of dairy products, as well as sidelines such as breton pancakes, *crêpes* (Winchester and Ilbery, 1988). The SICAs offer numerous advantages to the small farmers who are members. The farmers benefit from stable prices, an assured market and a share of the profits. The SICAs are able to take advantage of economies of scale, and have vastly improved the turn-round, marketing and packaging of goods. The benefits of the co-operative system have encouraged farmers to produce high-quality goods. The development of co-operatives has therefore indirectly stimulated agricultural modernization and has encouraged the uptake of a variety of government policies.

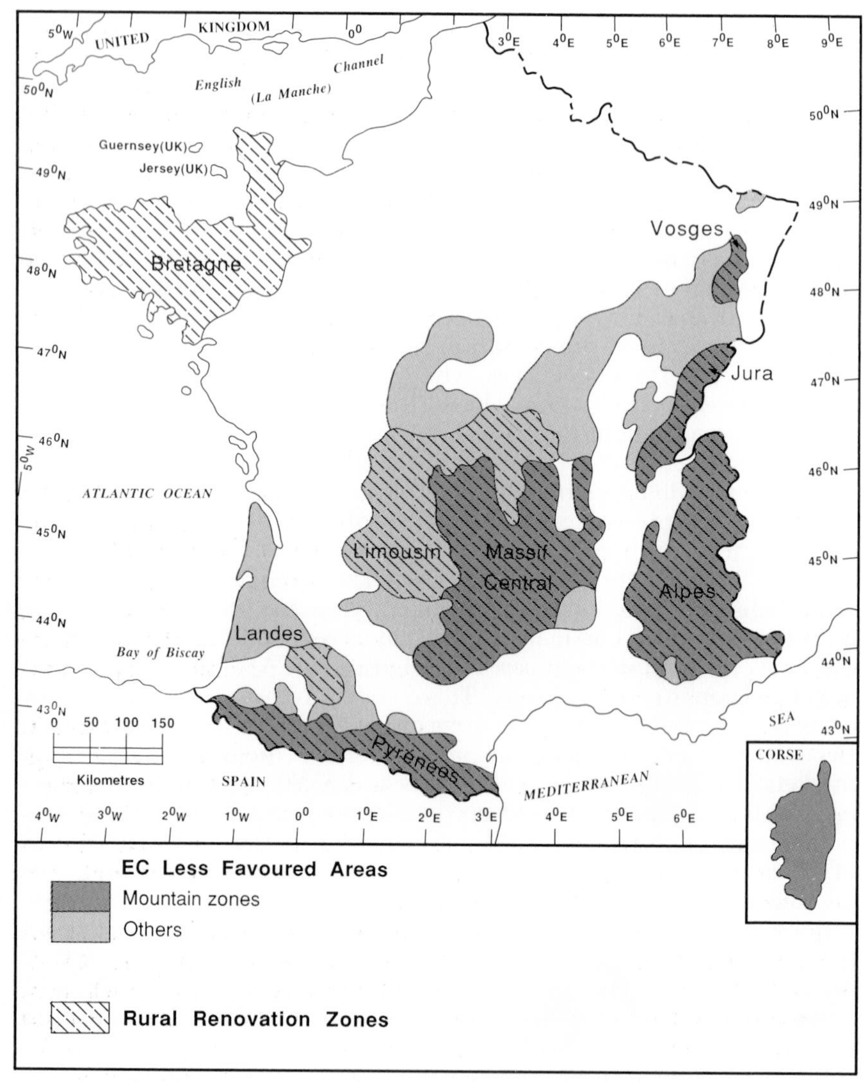

Fig. 3.7 Areas eligible for agricultural assistance, France. (Data from Laborie *et al.*, 1985)

Tenancy and inheritance

Legislation at the end of the Second World War improved the status of share-croppers and introduced longer and more secure leases for tenant farmers. Leases have since been further extended, and the so-called 'life-time lease' of 25 years is not uncommon. Tenants now have more security of tenure than ever before, and since 1975 they have had the option of buying their farm. Over the years, ways of getting round the system of equal inheritance have been devised to avoid further fragmentation of holdings. The heir who chooses to stay in farming has the right to buy out the shares of other siblings, if necessary over a long time period, as well as at a reduced rate because the farmer can claim a wage for working the land. Various co-operative schemes are available to minimize tax liability, and as is the case in Britain, a certain amount of land may be inherited without tax and some gifts made in the parents' lifetime are tax-free.

The policies affecting farm organization provide an environment in which entrepreneurial individuals can prosper without the fruits of their hard work being lost on their death or retirement. In the success or failure of modern farming, the legal context and the balance between taxes and grants are often factors of more day-to-day significance than the quality of the soil or even the choice of farming enterprise.

Regional aid to agriculture

Most French agricultural policy measures are available in any part of the country, but their uptake has varied according to the type of farm or type of farmer to which they applied. It is noticeable from inspection of Fig. 3.5 a and 3.5 b, however, that there are areas of the country which have not gained significant benefits from many of the policies which have been introduced. This is especially true of those parts of the country which are most backward, most deprived and most in need, particularly the mountain and Mediterranean zones.

Official recognition of the regional aspect of farming problems came in 1967 when *Zones de Rénovation Rurale*, rural renovation zones, were designated covering some 27 per cent of the country (Fig. 3.7). In these zones, priority investment programmes were created in any sector of the economy. This investment, as with the DJA, was originally only available in these problem regions, but the French government rapidly succumbed to pressure to make it available everywhere.

In 1975, the EC established policies in favour of Less Favoured Areas (LFAs) which particularly include the mountain zones (see pages 103–4; Fig. 3.7). Policies in the mountains and LFAs are now seen as much more of a coherent whole, with agriculture viewed as part of a wider economic system. In these regions, therefore, grants are not only available for agricultural improvement, sometimes at higher rates than elsewhere or more highly subsidized by the EC, but grants are also available for the development of other sectors of the economy, particularly tourism, craft industry, forestry and infrastructure. In these areas of integrated programmes, the national policies work alongside the full gamut of the EC policies, which include finance not only from the CAP, but also from the European Regional Development Fund and the Social Fund. An integrated programme has covered the *département* of Lozère since 1979.

A particularly good example of this type of integrated approach may be seen in the Mediterranean regions, which were singled out for special assistance in 1986 after the accession of Spain and Portugal to the EC. Many of the Mediterranean regions were poorly developed as a result of isolation and a lack of modern industry. They were handicapped both by their physical conditions and by their backward marketing and transport systems.

Table 3.3 Integrated Mediterranean Programmes: policy measures and levels of European assistance to Mediterranean regions of France

Measures	Present level of assistance (%)	New level of assistance under IMP (%)	Cost to the Programme (million ECU)	Type of measure
Agricultural				
Physical improvements				
Reparcelling	–	50	28	New
Irrigation	50	50	176	Broader coverage
Land improvement	–	50	32	New
Socio-structural directives				
Cessation of farming	25	50	25	Increased EC contribution
Compensatory allowance	25	50	49	
Measures for inland areas				
Livestock farming	–	50	27	New
Crops	–	50	20	New
Measures for lowland areas				
Fodder and protein crops	–	50	25	New
Irrigated crops	–	50	16	New
Unirrigated crops	35–50	50	83	New and broader
Non-agricultural				
Forestry	50	50	75	Supplementary
Fisheries	25–40	50	60	measures
Processing and marketing	25–35	45	60	More assistance
Support for producer groups	25	50	3	More assistance
Craft industries	0–70	50–70	261	New and broader
Rural tourism	50	50	38	Broader coverage
Renewable energy sources	30–70	30–70	36	Broader coverage
Infrastructure	40	50	50	More assistance
Advisory and training services	0–50	50	61	New and broader
Research	0–50	50	9	New and broader
Total IMP cost for the French regions:			1134	
Total IMP cost for the Italian regions:			2951	
Total IMP cost for the Greek regions:			2542	

The EC recognized a number of related problems. Agricultural problems were seen to be only part of a complex of economic problems being experienced by peripheral regions of Europe. The industrial sector was seen as backward, with a large number of small firms operating in traditional and declining sectors, and relatively few firms producing high-technology goods. The tertiary sectors, while underdeveloped, often showed an over-dependence on tourism and consumer services which are both seasonal and unreliable, while the rapidly-growing area of producer services was under-represented. Moreover, many Mediterranean areas were over-dependent on the fishing industry which suffers severely from inefficiency and obsolescence. In general, the economic environment of these areas was considered to be sluggish, badly affected by economic recession (shown particularly by above-average unemployment figures) and unable to take full advantage of the economic unity of the Community. In some cases, it is recognized that regional disparities between the Mediterranean regions and the core of Europe have grown wider in recent years. Furthermore, these regions have been poorly treated by the EC itself. The finance that was made available to these regions between 1973 and 1983 amounted to only six per cent of Community expenditure, an amount insufficient to meet the scale of the problem.

Programmes for mountain and Mediterranean areas

Since its expansion to include Spain and Portugal, the EC has introduced special programmes aimed at the most deprived rural areas of the Mediterranean. These Integrated Mediterranean Programmes (IMPs) have two basic aims, first, to raise income levels, and secondly, to improve the number and range of employment opportunities. The measures are different for France, Italy and Greece, with France paying a greater contribution than the other two countries in recognition of its greater wealth and less severe problems. The programmes are intended to supplement existing EC policies for agriculture, fishing and regional development. In France, the programmes apply to the five most southerly planning regions, with the exception of the major cities and areas of coastal tourist development (see Fig. 7.2).

The types of measure being introduced or supplemented are shown in Table 3.3, which also includes an estimate of the costs involved. Incentives which are additional to those available in the rest of the country include finance for the related work associated with land consolidation, for example, levelling and ditching; higher levels of income support and retirement premiums paid from the Community rather than the nation; and increased assistance for irrigation.

Policies for the mountain zones were instituted by the EC in 1975 under the LFA programme. These were supplemented by the French government in 1976 and are applicable in the mountain massifs (see Fig. 3.7), defined by altitude, slope, agricultural productivity and demographic change. Farms in these areas are disadvantaged agriculturally as a result of altitude and isolation (Plate 3.2). A special subsidy, known as the *Indemnité Spéciale Montagne* (ISM), is payable to offset these disadvantages. Farmers aged under 65 who farm a minimum of three hectares are eligible for this. Analysis of ISM payments shows that they are most frequently taken

Plate 3.2 The only road to the village of Oulles, Isère. Mountain areas suffer from a combination of problems including isolation and steep relief. (Photo: S J Gale)

up in the more commercial farming areas of the Jura and the Massif Central, where the farming population is younger than in the other mountain zones. The subsidy itself is generally spent on agricultural expenses, rather than on capital developments, investments in machinery or domestic improvements. In this way, by subsidizing the costs of farming in marginal areas, the ISM may actually help preserve the small unproductive farm units that other farm policies are attempting to eradicate. However, it is recognized that in many of the Less Favoured Areas there are few employment opportunities outside agriculture, and that in such areas farmers have an important role in conserving the landscape.

An evaluation of national agricultural policies

French national agricultural policies have been wide-reaching since 1960. They were further extended in 1980 by a new *Loi d'Orientation* which stipulated that each *département* should draw up an Agricultural Structure Plan, thereby furthering the regional emphasis of recent policy. The policies form a coherent socio-structural programme with an increasing regional emphasis.

It is clear that the policies have had measurable effects: 40 per cent of land consolidated, five per cent of land bought and sold by SAFERs, 650 000 farmers paid off with retirement premiums, 350 000 farmers trained or retrained, 80 000 young farmers helped to set up in business. It is also clear that the policies for land have stimulated and speeded the increase in farm size and the reduction in farm parcellation, and as a by-product have stabilized a potentially volatile land market. Similarly, the policies for labour have accelerated the reduction of the labour force, improved agricultural professional training and introduced more young farmers into the farming community. The development of co-operatives has increased the availability of machinery, improved working conditions for farmers and encouraged the modernization of processing and marketing.

It is less easy to disentangle the results of market forces from the results of government intervention. It is impossible to estimate the rate and direction of agricultural change had the policies not been in operation. However, it is clear that state policy, in working with market forces rather than against them, has achieved results which have transformed the traditional face of French agriculture.

It is very difficult to operate an agricultural strategy in isolation and therefore some of the results of government intervention have not always been quite as intended. Thus, separate policies for land and for labour may each have unintended effects on the other, as both are related parts of the same agricultural system. For example, schemes directed at the agricultural labour force, which help old farmers retire and young ones get established, may indirectly maintain the presence of relatively small and marginal holdings, especially when both policies operate together on the same farm. Similarly, land improvements or farm consolidation may provide better job prospects for the same workforce which labour policies are trying to reduce. However, in a number of cases, the organization of the SAFER in manipulating the land market has provided a bridge between different aspects of policy.

The other implication of agricultural policies in contemporary France, and one which has been squarely faced since the 1980s, is the spatial and regional one. It is apparent that many of the improvements and benefits have accrued to those areas which were already well advanced, and at a smaller scale have particularly favoured those farmers who were already wealthy. Many of the poorest and least efficient farmers, including those who work part-time, may not be entitled to claim benefits, or are least able to cope with the paperwork and bureaucracy that is required. These farmers are concentrated in the poorest agricultural regions in the south-east and centre of the country, and new measures are now being taken to cope with these problem regions.

The relationship between national and European policies

Most French socio-structural policies are subsidized by the EC, but the policies instituted by the EC itself as part of the Common Agricultural Policy (CAP) fulfil a rather different role. The CAP is the cornerstone of EC policy and the major item in the European budget. The 1991–2 farm budget is likely to cost 32.5 billion European currency units, about 67 per cent of the total EC budget. The CAP was designed to provide a stable farming environment in which productivity could be increased, farm incomes could be assured and supplies would be readily available to consumers at reasonable prices (for further discussion, see, for example, the texts by Hill, 1984 and Bowler, 1985). Since 1962, when the Agricultural Guidance and Guarantee Fund (FEOGA) started, the CAP has spent most of its budget on guaranteed prices and export subsidies and relatively little on structural improvement. In the late 1980s, the proposed spending was weighted 75:25 in favour of guarantees over structural policies, whereas the actual expenditure was 96:4. The high expenditure on guarantees has helped maintain European prices at a very high level and production has soared. In a number of sectors, this increase in productivity has brought about over-production and has necessitated the introduction of severe structural measures to cope with the problem, as with the introduction of quotas for dairy produce in the mid-1980s. Furthermore, the system of intervention buying has caused a major escalation in the costs of agricultural subsidy and the price of goods to the consumer. Paradoxically, the high costs of intervention buying and storage have meant that the huge expenditure on farming has not been reflected in increased farm incomes.

The policy of guaranteed prices has undoubtedly been to the benefit of large farmers with a capacity for large production and to the benefit of rich modernized agricultural areas. The policy of high prices has also tended to keep some marginal farmers in production who otherwise could not have made ends meet. The operation of the CAP has therefore worked predominantly to the benefit of commercial northern France rather than the peasant south. In particular, the Mediterranean regions, the very poorest, have suffered considerably as most of their produce has not had the advantage of guaranteed prices. Two serious implications arise from this discussion. One is that in concentrating almost exclusively on guarantees, the EC has maintained the small inefficient farmers. These are the marginal farmers that national structural policies are trying to eliminate. There is therefore some conflict, often unintended or indirect, between national and supranational policies. Secondly, the national and supranational policies have combined to benefit the advanced temperate regions of France to the detriment of the mountain and Mediterranean zones. Both groups of policies, although not explicitly regional until a few years ago, have had profound regional impact, and it is only in the last 10 years or so that nation and EC have combined to tackle the regional problem and the regional disparities within agriculture.

In the 1990s, the EC sees its agricultural role becoming more diverse. In areas adjacent to large conurbations, the main concern will be the protection of the rural environment and the limiting of ecological damage from intensive agriculture. In remoter areas, continued emphasis will be placed on economic development, by stimulating indigenous development potential in all sectors. Farmers will be encouraged to respond to the consumer demand for greater variety and quality in foodstuffs, including the production of organic foods. The EC is also actively funding research and development in biotechnology to make agricultural products more competitive and to widen their range of uses from foodstuffs to industrial raw materials.

CHAPTER 4

The French industrial miracle

Industrial, a French word, said to mean mechanical: lately adopted by the English newspapers.

Vesey (1841: 82)

Phases of industrial development in France

Controversy exists over whether or not France was industrially underdeveloped prior to 1945. Many authors have argued that French industrial development before the Second World War was retarded (Landes, 1969). From a British perspective, France never experienced a 'real' industrial revolution and the French economy before 1945 was seen as a 'still life' composed of 'miserly landlords, hesitant entrepreneurs, careless politicians . . . (and) . . . infertile couples' (Aldrich, 1987: 89). Recently, Aldrich and others have emphasized that French industrial productivity was proportional to slow demographic growth, just as British industrial productivity was proportional to faster demographic growth. Furthermore, French industrial productivity progressed steadily from the eighteenth century, without the marked break of the Industrial Revolution (O'Brien and Keyder, 1978). The French path towards industrialization was different from the British, both in the organization of production and in its regional and socio-economic consequences. The organization of production was characterized by small firms, local inputs of technology and capital, family labour and specialized products (Landes, 1966). This small-scale specialized production was reflected in slow urban growth, and in socio-economic and regional inequalities which were less pronounced than those associated with the big industrial cities of northern England.

A major acceleration in the progress of industrialization occurred after the Second World War. A period of extremely rapid economic and industrial growth coincided with massive population growth (see pages 44–5; Fig. 2.2). This post-war period may be divided into three main phases. The first, from 1946 to 1960, was a period of reconstruction with growth in heavy industry fuelling other sectors of manufacturing. Coalfield-based industries boomed, producing regional concentrations of particular industrial sectors. This period, despite population growth, was a period of labour shortage and full employment. The second phase, from 1960 to 1973, was a period of self-sustaining growth, marked by very rapid increases in production and exports from most sectors of industry. This phase was characterized by the development of production-line processes, decentralization from Paris and the growth of new regional specializations. The streamlining of industrial processes permitted continued increases in productivity, while decentralization helped keep land and labour costs low. The benefits from these organizational changes in industrial production coincided with the advantages of entering the enlarged market of the European Community. France, rather later than Britain, entered a period of industrial growth based on powerful monopoly capital interests. During this time, France achieved a take-off of industrial development on an unprecedented scale, which has been termed an 'economic miracle'.

The 1973 oil crisis ushered in a period of industrial restructuring in European industry. Much of French industry did not suffer severe readjustment until the 1980s, in part because of its late development. The oil crisis precipitated industrial changes, but the underlying causes of these have also been related to the end of one of the Kondratieff long waves and a change in the technological base of western society (Dicken, 1986). The deep-seated trends which became apparent after 1973 included the shift of industrial production from the family firm to the transnational corporation, the internationalization of production and competition from newly industrializing countries, and the decline in the power of the state to regulate production and trade. Individual industries adapted to increased competition and reduced demand by the restructuring of work practices and streamlining of operations. European industry as a whole was diagnosed as suffering from 'Eurosclerosis', an inflexibility brought about by high wage levels, low wage differentials and strong unions which hindered labour mobility and structural adjustment (Grahl and Teague, 1989). In France, the regional consequences of restructuring were felt most strongly in the coalfield regions of heavy industry, where closures brought about job losses and industrial dereliction. Government policies were introduced to soften the impact of restructuring by providing new jobs in growth sectors as well as redundancy and retraining packages. The 1980s were therefore a decade of stagnation and retrenchment in many sectors of manufacturing and a period of rising unemployment and inflation. The transition from monopoly capital to global capital was accompanied by marked political changes in the state's regulation and deregulation of industry. By the early 1990s, however, the massive political changes in eastern Europe were being seen as great opportunities, opening up vast potential markets for a revitalized manufacturing sector.

Early industrialization

Proto-industrialization is seen as the first phase of the industrialization process and as part of the transition from feudalism to capitalism (Mendels, 1972; Kriedte, 1981). This early industrialization occurred predominantly in rural regions. Agricultural workers participated in industrial production, often on a seasonal basis. The goods produced were highly specialized and were sold outside the region; proto-industrialization may therefore be distinguished from the petty industry established in many parts of rural France to supply purely local needs (Coleman, 1983). The development of proto-industry was intimately associated with the commercialization of agriculture, which provided a food surplus, and with the growth of towns where marketing and trading functions were located. This early phase of industrialization thus provided the impetus for population growth and capital accumulation, which enabled further industrial development to occur (Braudel, 1982).

There is considerable evidence throughout France for widespread rural industry in the early modern period, from about 1740. The most widespread of rural industries were the textile and metallurgical industries which produced goods both for home consumption and for wider national and international markets. Eighteenth-century workshops-cum-factories, not necessarily using mechanical power, produced munitions for war and consumer goods for the rich (Weber, 1977). Although almost every small town had its own domestic textile industry (Pounds, 1979: 220–36), regions of more specialized production served wider markets. In particular, the area around St Etienne and Lyon on the eastern fringe of the Massif Central was the hub of silk production destined mainly for the British market; sub-districts produced regional specialities such as ribbons, gloves, hosiery and brocade (Sheridan, 1979; Tilly, 1979). The rural silk industry in the Ardèche in the 1830s utilized mainly young female labour which was both cheap and docile. This rural manufacture did not always lead to large-scale industrialization in the late nineteenth century; in the Ardèche the demise of proto-industry was part of a sustained period of demographic and economic decline. However, the textile industry around St Etienne was transformed to factory production with the development of the coalfields, although the handloom weavers of Lyon resisted mechanization until late in the nineteenth century (Sheridan, 1979).

The traditional view of French industrialization is that it was hampered by low demand from a peasantry living at subsistence levels: Landes (1969: 128) commented on the 'circular link between poverty . . . (and) . . . the absence of industry'. Industrial performance was also seen to be inhibited by political instability in the early nineteenth century, by excessive or inappropriate state intervention and by protectionism. Sheriff (1980) argued that industry was impeded by restrictive social attitudes, a preference by the elite for land and property rather than for manufacture and trade, and by the conservatism and passivity of family firms. It is difficult to evaluate the relative contributions of these factors against the general lack of raw materials, the limited development of national transport networks and the slow growth of population emphasized by recent authors. However, the products of proto-industrialization were destined for a wider international market and were not affected by a lack of raw materials or transport. Britain, which is often taken as the model of industrial development, was also affected by political upheaval and rigid class attitudes. Slow industrial development in France may best be viewed as an appropriate response to the nation's particular demographic circumstances rather than as a poorer and slower version of the industrial transformation of Britain.

The modernization of industrial production required its organization into larger factory-based units using new sources of power and new technology. However, modernization was no sudden transformation from a 'closed, traditional, immobile set of social worlds' (Tilly, 1979: 21), as France at the start of the nineteenth century was 'connected, mobile and even, in its way, industrial' (Tilly, 1979: 37). Early factory-based manufacturing, as opposed to proto-industry, developed in the nineteenth century in specific sectors, mainly textiles and iron. These two key sectors were widely dispersed throughout France, reflecting the pattern of previous rural industry. Other more specialized types of manufacturing production were more restricted to specific locations because of raw material availability, such as glass-making at St-Gobain in Picardie and sugar-refining and soap-making at ports (Pounds, 1979).

The textile industry developed from both a rural craft base, which supplied local needs, and a proto-industrial base supplying specialized products to wider markets. Between 1840 and 1880, the textile factories and associated supply industries around St Etienne transformed the city and region into one of the busiest and most rapidly developing industrial areas of Europe (Hanagan, 1989). The woollen industry became factory-based in the northern cities of Reims, Roubaix, St Quentin and Elbeuf. Some of the provincial centres became uncompetitive, while others such as Carcassonne and Louviers survived by producing high-quality goods. The introduction of mechanization was inhibited in many places by the variable quality of the raw wool, the best of which had to be imported from Spain.

In the nineteenth century, France was the major producer of cotton textiles in continental Europe, although with an industry only a quarter the size of Britain's (Landes, 1969). Cotton textiles were spun and woven predominantly north of the Loire; they were found in Normandie around Rouen, in the Nord around Lille and Roubaix-Tourcoing, in Alsace and the Vosges, and initially also in Paris itself and other scattered centres. Most cotton manufacturers used cheap water power and cheap rural labour. The cotton industry in Alsace was more technologically advanced than that of the other centres because of its association with other industries, which provided the local manufacture of machinery and of advanced chemical dyes. The Norman industry had the best locational advantages, close to the ports which imported cotton from America, close to Parisian and overseas markets, and rich in water power and cheap labour. These locational advantages produced an inertia in the industry which combined with state protection from British competition to inhibit the changes required for successful adaptation to the new economic circumstances of the twentieth century.

The iron industry underwent a similar process of concentration during the nineteenth century. Many hundreds of small ironworks closed down: Weber (1977: 214) cited the closure of over a hundred ironworks in the Dordogne between 1830 and 1865. This rural iron industry had used charcoal from the forests and local ores. The factory production of iron accompanied the shift to a coal-based industry, although many new plants continued to use pre-existing techniques. The iron industry expanded rapidly from the 1840s around the coalfields of St Etienne, located at Charenton, Le Creusot and Vienne. The focus of the iron industry shifted to Lorraine in the late nineteenth century, where technical advances made it possible to use the local phosphoric *minette* ore for steelmaking. Iron and metalworking also developed on the coalfields of the Nord and Pas-de-Calais. The iron industry of these areas complemented and reinforced the pre-existing concentration of industry based on textiles. Since then, deposits of iron ore at Briey in Lorraine and of coal in Lorraine were developed respectively in the late nineteenth and early twentieth centuries. This industrial development has further reinforced the basic pattern of heavy industry in France, which is highly concentrated on the coalfields of St Etienne, the Nord and Alsace–Lorraine.

The regional distribution of nineteenth-century heavy industry changed little during the first half of the twentieth century. This period was one of difficulty for industrial development, disrupted by two World Wars, global depression and inflation. A number of new industries developed, stimulated by the demands of war, particularly communications, engineering, armaments and other strategic industries. One of the most important of the new industries was the manufacture of motor vehicles. In France, this was dominated by Renault, Citroën and Peugeot; their plants were concentrated in Paris and between them they provided 75 per cent of French vehicle sales by 1938. Other industries, which brought significant discoveries into commercial use, also thrived. These included the chemicals industry which produced fertilizers, lubricants and pharmaceuticals, and the electrical and electricity supply industries. To a lesser extent, the production of consumer durables also grew in response to increasing consumer demand, caused by higher wages. Important consumer goods included bicycles and radios, although the more widespread use of consumer durables occurred after the Second World War. At the same time, other traditional industries such as textiles and metalworking fared less well and overall industrial growth was slow. Between 1929 and 1949, the growth of industrial production averaged a mere one per cent annually. However, demographic growth was also very slow at this time (see pages 39–44, Fig. 2.2), only increasing after 1942–3. The relative stagnation of industry, coupled with the effects of wartime occupation, left France, with its anachronistic industrial structure and undeveloped rural areas, in a weak position to face the economic competition of post-war Europe.

The post-war industrial miracle

Despite the inauspicious circumstances at the end of the Second World War, France has experienced an industrial transformation which has elevated the country to its current status as the second largest industrial nation in Europe. In the immediate post-war years, overall reconstruction was slow, based on the restructuring of the heavy industries of coal and steel. Reconstruction was a national priority and was assisted by American aid. The reconstruction of heavy industry laid the basis for unprecedented growth in other sectors, particularly chemicals and metal goods. By the 1960s, expansion was occurring in light industry, construction and the production of capital goods such as vehicles and electronics. However, by the 1960s, the beginnings of decline were evident in heavy industry from changes in consumer demand and foreign competition. Nonetheless, during the 1960s, industrial output increased by an annual average of more than six per cent and GNP increased annually by 5.8 per cent, a rate higher than that of any other country of the EC. Between 1958 and 1972, exports increased between five and six times and over half of these exports were manufactured goods. By 1972, the volume of goods exported was sufficient to rank France as the fourth exporter in the world.

Post-war industrial reconstruction

The reasons for the dramatic turnaround in France's industrial fortunes are complex. State intervention and foreign aid provided the initial impetus to industrial redevelopment. Between 1947 and 1952, France received about US $ five million from the USA, at least half of which was used directly in industry. France's First National Plan of the post-war period emphasized the need for the modernization of industry and increased production, and presaged the nationalization of the key industries of coal, electricity and the railways. International aid and state intervention thus facilitated capital growth.

The industrial 'miracle' was also related to a fundamental restructuring of French manufacturing. French industrial development in the twentieth century changed from a small, family-based industrial structure to a monopolistic structure. Monopoly capitalism allowed new and more productive technology, such as automation, to be introduced. Innovations in production methods were made at different times in different sectors of industry. Some of the earliest industries to adopt automated production-line methods were the vehicle industry and the food industry. By 1974, 14 per cent of employees in the food industry worked on production lines compared with only six per cent in industry as a whole; automation was a factor in severing the linkages between the food industry and the agricultural sector (Nefussi, 1990: 146–7). Decentralization of the vehicle industry from Paris enabled production-line processes to be introduced; these were typically Fordist, in that they involved the standardized production of mass-produced goods (Oberhauser, 1987).

Monopoly control by the state allows similar production advantages to

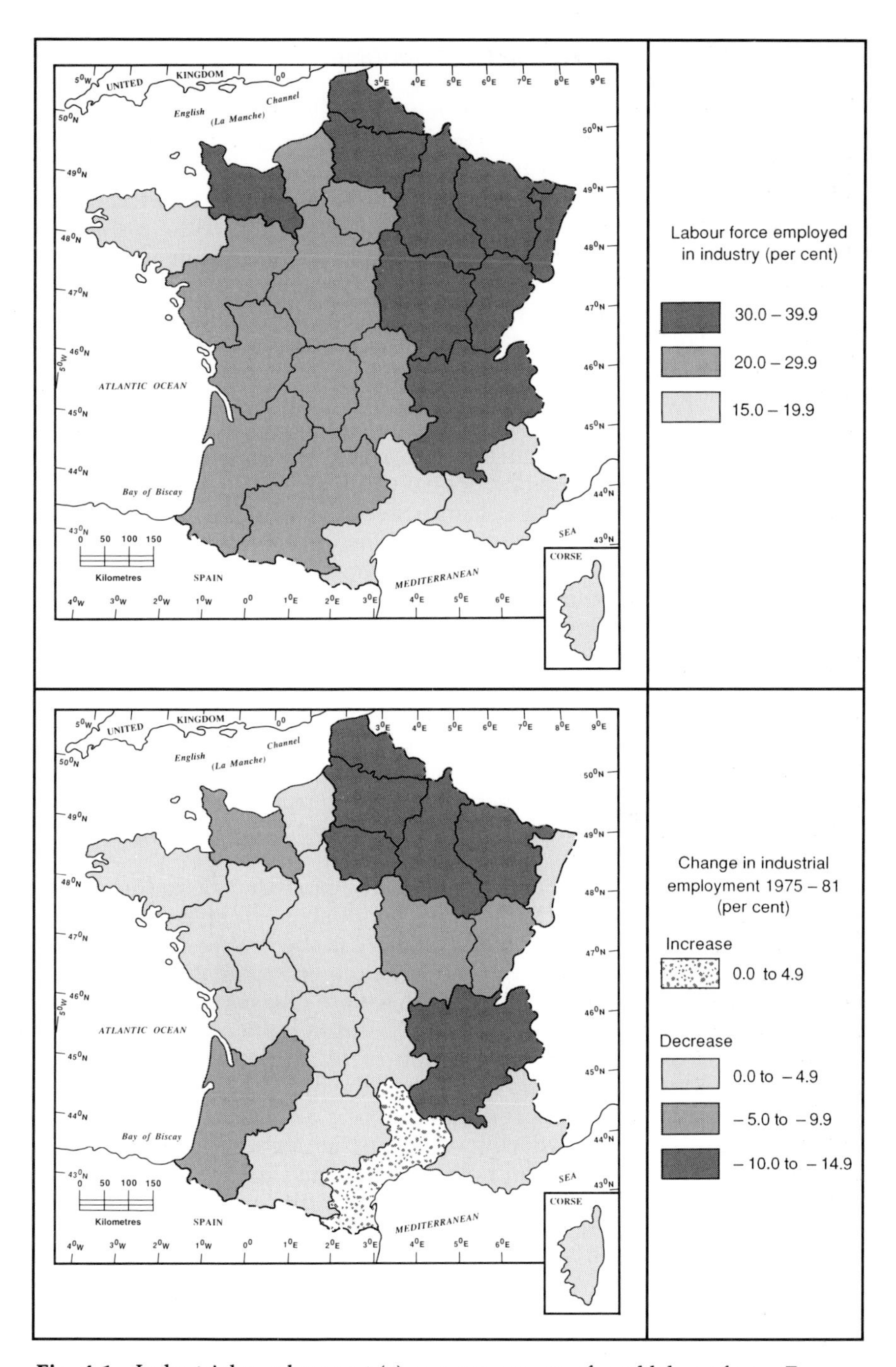

Fig. 4.1 Industrial employment (a) as a percentage of total labour force, France, 1982; (b) percentage change, France, 1975–81. (Source: INSEE, 1982)

those of monopoly control by capital. Moreover, it provides greater control over location and looser requirements for immediate profits. Changes in the structure of industrial control occurred contemporaneously with changes in the relationship between capital and labour. As the role of the state in industry has increased, so business leaders have made strenuous efforts to improve their image, and labour has gradually improved its bargaining position. A period of labour shortage after the war forced up wages; these were recycled into greater consumption of the industrial goods produced. French industrial production was caught in an upward spiral of production, wages, demand and consumption. The demand, especially for consumer goods, was stimulated by a rapid change from a rural peasant society to an urban industrial economy and by rapid population growth, itself stimulated by state policies of pronatalism (pages 46–9). The continued demand for consumption goods reduced the impact of the recession of the mid-1970s in France (Boyer and Mistral, 1978).

Post-war industrial restructuring occurred in a changed international environment. The reconstruction of heavy industry was facilitated by the formation of the European Coal and Steel Community in 1952. This organization performed two functions: to incorporate Germany into the European economy after the war and to rationalize and expand the European coal and steel industries, which were both damaged and outmoded. Other moves towards international co-operation, such as the formation of the World Bank, helped create a climate of opinion conducive to industrial expansion. The most significant change in the international environment occurred with the formation of the European Economic Community, dating effectively from 1960. This union of major European countries expanded the market for agricultural and manufactured goods. The restructuring of French industry at this time exposed it to competition but also allowed the French to take advantage of greater marketing opportunities.

The industrial development of the early post-war period induced changes in the locational pattern of French industry. The strengthening of the capital goods sector occurred particularly in the Rhône-Alpes region, while continued concentration in the Paris region stimulated policies of decentralization. The consequences of this were that manufacturing employment became more evenly distributed through the *régions* of eastern France (Fig. 4.1a). The decentralization of state monopoly industries helped distribute key economic sectors into the periphery of France, such as aviation and aerospace to Toulouse in the southwest. This period of prolonged economic growth facilitated industrial dispersal from the existing areas of concentration and helped to narrow regional inequalities, although the west continued to offer fewer industrial employment opportunities than the east (Fig. 4.1a).

Industrial restructuring

In the early 1970s, the sudden increase in oil prices precipitated a decade of economic crisis for those western industrialized countries dependent on imported oil. In France, the oil crisis had less dramatic effects, although it did cause an increase in the price of manufactured goods and some realignment of trading patterns and reassessment of consumer demand. The oil crisis was in many ways a trigger for change, hastening underlying

Table 4.1 Classification of French industries, by rates of growth

	Agriculture & food industries	Intermediate goods	Equipment industries	Consumer goods
Growth industries	Animal feed; non-alcoholic drinks	Glass/plastics; semi-finished non-ferrous metals	Aeronautics; telephone, television & radio equipment; armaments	Sports equipment; pleasure boats; records/cassettes; kitchen equipment
		Heavy chemicals	Electronics & information equipment; medical & surgical equipment	Pharmaceuticals Photographic goods
Stagnating industries	Food processing	Tyres & paints; phosphate & nitrate fertilizers; light metal goods	Naval & service vehicles; cars; bicycles; electrical equipment	Perfumery; toys; clothing; printing
		Construction materials: bricks, tiles, cement, prefabricated buildings; paper, pulp & cartons	Clocks, optical equipment & precision engineering; household equipment	
Declining industries	Flour milling; alcoholic drinks, especially aperitifs; vegetable preserving	Iron and steel; non-ferrous metals; artificial textiles; other fertilizers	Industrial machinery; machine tools; rolling stock & heavy plant	Wooden goods & furniture; natural textiles; shoes & leather; lingerie & hosiery

Source: Dutailly and Hannoun, 1980.

structural transformation. The major difference for French industry was the increased internationalization of production and trade, shown in the control of large sectors of industry by transnational corporations. These corporations have dramatically influenced the global distribution of industrial production by the transfer of routine production to places with lower factor costs. Accordingly, French industry has suffered competition from newly industrializing countries and has been forced to undergo rationalization and labour loss. A related international influence has been the shift to a new form of capitalism, that of flexible accumulation, whereby international producers shift their interests between locations and sectors in search of profit. Industrial plants, now often branches or subsidiaries of international concerns, have to be flexible to meet changing demands. This flexibility has necessitated further changes to both the location of industry and to the industrial workforce.

For many reasons, industry since the 1970s has undergone a prolonged and sometimes painful period of rationalization. Over that period, the contribution of industry to the formation of financial capital has fallen from 27 per cent to 22 per cent, three quarters of a million of industrial jobs have been lost (Fig. 4.1b), the unemployment rate has risen to over 10 per cent, and previously buoyant sectors of the economy have stagnated or declined. In the middle of this period of restructuring, Dutailly and Hannoun (1980) classified French industries by their rate of growth on the basis of production levels, import resistance, export penetration and overall commercial performance. The growth sectors were predominantly the high-technology equipment industries and some specialized consumer goods, such as sporting and photographic equipment (Table 4.1). However, there was no growth in the production of most intermediate goods, that is those manufactured goods used in further manufacture or construction. Industries which were the backbone of growth in the 1960s, such as motor vehicles and precision engineering, were also stagnating. Changing patterns of consumer demand and expenditure contributed to a decline in the production of many traditional industries, including textiles and iron and steel (Table 4.1).

The restructuring of French industry has had a variable impact on the regions. Three effects have been very clear. First, the traditional industrial regions of the north and northeast which developed in the nineteenth century have suffered severe decline of their heavy industries. Secondly, the boom areas of the post-war period, notably the Rhône-Alpes, have been unable to sustain their industrial momentum which was based chiefly on the production of cars, electrical goods and engineering. Thirdly, the industrial periphery has undergone very variable industrial development. One area with little comparative advantage is the southwest; unlike the west, it is too far from Paris to have benefited significantly from industrial decentralization and unlike the southeast, it lacks the climatic and perceived advantages of the Mediterranean for tourism and footloose industry.

Concentration of industrial firms

A major aspect of the process of industrial restructuring has been the significant increase in the average size of firms and a concomitant reduction in the number of firms. This reorganization reflects the need for higher productivity in a competitive international economy. The number of industrial mergers and takeovers has increased markedly since 1977, with a further increase after 1988. In 1985, there were 251 industrial mergers in France, while in 1988 there were 526. The most recent wave of mergers has been most evident in the capital goods sector and has been stimulated by the prospect of a European Single Market from 1992 (OECD, 1991: 89–90). The process of internal expansion, as with Renault, or of company mergers, typical of the brewing industry, has enabled large firms to take advantage of economies of scale. Mergers have occurred not only in heavy industry (for example, Usinor and Sacilor in the steel industry), but also in other sectors of the economy (Di Méo, 1984). The French government and the EC have implemented close controls on mergers and their monopoly effects since 1990.

The structure of French firms has changed dramatically from the many small and medium-sized firms which existed before the Second World War. In 1980, four per cent of French firms employed 60 per cent of the workforce, while accounting for 69 per cent of turnover and 79 per cent of total investment. The 15 largest French companies employed a million and a half people and generated the majority of their sales outside France. In times of crisis these large firms have been seen as the major agents of investment and the spearhead of competition (Bellon and Chevalier, 1983), but on the other hand some of these large firms such as Usinor and Sacilor have also been subject to major job losses as a result of restructuring. Small industries have been able to create relatively more employment and have taken the opportunity to supply marketing niches that large firms were too inflexible to fill. Since 1974, the growth of large industrial alliances has been counterbalanced by a revitalization of small industries. Between 1980 and 1989, the percentage of workers employed by large firms (with 200 employees or more) fell from 32 to 25 per cent. During the same period, the proportion of workers in firms employing fewer than 50 people increased from 46 to 53 per cent (OECD, 1991: 130).

The process of merger and concentration is exemplified by the motor vehicle industry. France has two major vehicle producers, Renault and Peugeot. Peugeot has grown by a process of merger, with Citroën in 1974 and with Chrysler-Europe (Talbot) in 1978. Renault has taken over the firms of Saviem and Berliet and has incorporated them in its industrial vehicles division (Tuppen, 1983). Vehicles are still produced under their brand names but the firms producing them are no longer autonomous. The concentration of the vehicle industry has had both negative and positive effects. The negative effects have been the severe local impact of recession and rationalization and the greater susceptibility to state regulation (both of which may have occurred without merger). The main

Plate 4.1 One of the breweries of northern France dominating a traditional urban-industrial landscape. (Photo: S J Gale)

positive effect is the increased international competitiveness of the French vehicle industry, although both firms predominantly supply the national market.

Merger and concentration have also affected the brewing industry. The French beer industry, although much less well known than the wine industry, nonetheless supplies important markets, especially in the north of the country: half of the nation's breweries are located in the *région* of the Nord (Jackson, 1977) (Plate 4.1). Mergers between brewing companies have resulted in a domination of the market by the 'big six' and the elimination of hundreds of small local breweries. The largest company holds half the French market and trades under names including La Meuse and Kronenbourg. The trend to niche marketing and the renewed interest in real ale and boutique beers has enabled a few of the more commercial small breweries to maintain or even increase their production.

State intervention: nationalization

The nationalization of many French industries may be seen as part of the process of concentration, and also as an indicator of state intervention. French industry has experienced successive waves of nationalization. The first of these occurred after the First World War, when strategic industries confiscated from Germany (chemicals, fertilizers and petroleum) were put under state control. In the 1930s, again for strategic reasons, the aeronautics industry was nationalized. A second wave of nationalizations took place after the Second World War, as a means of revitalizing key economic sectors. Nationalization was also used as a sanction against firms which were seen to have collaborated overtly with the occupying German forces, a fate which befell Renault, which became the national corporation of Renault factories, the *Régie Nationale des Usines Renault* (Di Méo, 1984). The major strategic sector affected by post-war nationalization was energy. Coal, electricity, gas and nuclear energy were all put under the control of national boards; *Charbonnages de France* (CDF) (coal), *Electricité de France* (EDF) (electricity), *Gaz de France* (GDF)(gas) and the *Commissariat à l'Energie Atomique* (CEA) (nuclear power). There was also some public intervention in the petroleum industry, with the formation of three companies, which later amalgamated to become Elf-Aquitaine.

The most far-reaching wave of nationalizations was set in motion by the socialist government under President Mitterrand in 1982 (Holton, 1986). Nationalization was adopted to reduce the monopoly control of economic sectors by a narrow range of private interests, to continue the programme of restructuring, and to gain control of strategic industries which were relatively autonomous because they had become part of transnational corporations. Four groups of nationalizations were announced, together with all banking services (Table 4.2). As a result of these nationalizations, France became the most state-controlled capitalist country in the world. In 1980, French state industries had employed 10 per cent of the industrial labour force; by 1984 this proportion had increased to 22 per cent, compared to only 12 per cent in Italy and the United Kingdom. The French government took upon itself the direct responsibility for many sectors of industry and acquired much greater control over the location and direction of industrial development.

Table 4.2 Industrial nationalizations, 1982

	Reasons for nationalization	Firms and their major specialisms
1.	Strategic military activity	Dassault-Bréguet; Matra (armaments)
2.	Strategic activity; subsidiaries of foreign-based multinationals	CII–Honeywell–Bull; ITT France (information industries); Rousell-Uclaf; Hoechst AG (chemicals)
3.	Major industrial groups exercising quasi-monopoly in their sectors	Compagnie Générale d'Electricité; Thomson-Brand-CSF (electrical and electronics industries); Rhône-Poulenc (chemicals); Péchiney-Ugine-Kühlmann (electro-metallurgy); Saint Gobain–Pont à Mousson (construction materials)
4.	Restructuring of iron and steel	Usinor; Sacilor

The nationalizations were a form of protectionism, and restructuring continued, albeit subsidized by the public purse. Subsidies from the state to the nationalized groups rose from FF five thousand million in 1982 to FF 13 thousand million in 1984. Austerity measures introduced in 1984 limited further borrowings and imposed a requirement for state industries to balance their books. In the case of the steel, coal and shipbuilding industries, improvements in their financial position were brought about by further job losses. Other groups brought about profitability by the sale of profit-making parts of their enterprises, for example the sale of the *Compagnie Electro Financière*, a successful finance company which was a subsidiary of the *Compagnie Générale d'Electricité*. By rationalizing labour, reselling parts of the nationalized groups and by closure of superfluous plants, companies such as Thomson, Rhône-Poulenc and Péchiney reduced their losses from 1982 to 1986, and by 1986 only one of the nationalized companies was 'in the red'. However, the austerity of the rationalizations reduced political support for state intervention, and in 1986 during the brief period of '*cohabitation*' between the socialist President and a right-wing government, a policy of privatization reversed the trend of nationalization (Tuppen, 1991). Industries were again to be exposed to the competition of the market place after being revitalized by state intervention.

Internationalization of French industry

The merger and consolidation of firms is linked to the increasing control of French industry by transnational companies, most based outside France. On the other hand, the process of industrial nationalization may be viewed as a reaction against outside control. Foreign-based transnational corporations have not only penetrated the French domestic market, but they have invested heavily in productive capacity within France. France has the highest proportion of externally-controlled industry in Europe, with foreign companies controlling 28 per cent of manufacturing production and 19 per cent of manufacturing employment (Dicken, 1986: 64). The level of foreign control varies between industries, but is particularly marked within the information equipment sector, where 80 per cent of investment is dominated by IBM France. External control is also significant in civil engineering, where the foreign companies of Caterpillar and Komatsu provide 44 per cent of investment. Approximately 40 per cent of foreign-controlled jobs originate from USA-based corporations. The French react chauvinistically to this perceived invasion of their economic space. This is partly because the process is very one-sided; only 6.5 per cent of French enterprises are located outside metropolitan France.

Since 1970, some French industries have made global investment decisions, and a shift of some French labour-intensive productive capacity to parts of the third world has occurred. Examples of this trend include a Thomson television assembly plant in Singapore and a knitwear factory in Tunisia under the auspices of a Roubaix woollen company, *Lainière de Roubaix*. The use of cheaper foreign labour for assembly-line work is part of the new international division of labour, whereby large firms decentralize production processes to the cheaper periphery. This process may make the company more internationally competitive, but is obviously detrimental to French industrial employment in the short term. Other major French companies are becoming significant players on the international scene: Péchiney is a major international aluminium company with manufacturing plants in Canada and Australia, while Pernod-Ricard has recently acquired established Australian vineyards in the Hunter Valley of New South Wales. French producer services are also internationalizing rapidly: the French reinsurance market, while lagging well behind more traditional providers, has ranked fifth in the world since 1985 (Wolkowitsch, 1989: 653).

Regional changes in industry

Changes in French industrial structure, control and rationalization have had significant effects on regional industrial distribution. At the end of the Second World War, the major industrial regions were the coalfields of Nord and Lorraine on which heavy industries were concentrated. Paris was a major industrial centre, the focus for a great range of consumer products such as newspapers and fashion goods (Plate 4.2), and for other manufactured items, as well as the major centre for service industries and

Plate 4.2 The garment industry is still concentrated in the centre of Paris. This labour-intensive industry is disproportionately staffed by migrant workers. (Photo: S J Gale)

business headquarters. Industry was also well-established in parts of the Rhône-Alpes where metalworking and other diverse industries were based on water power as well as coal. This region also produced a host of traditional manufactures from gloves and ribbons to pottery and guns. Industrial activity was therefore concentrated in a broad arc of northern and eastern France, although each of the four industrial regions possessed a distinctive industrial character. By contrast, with the exception of particular regional urban centres, the west, centre and south were comparatively underindustrialized. Within these regions, much of the industry was labour-intensive, technically out-dated and catered for static or declining consumer demand. These industries were the remnants of local rural manufacturing using textiles, wood, leather and other local materials. These industries have suffered competition from mass-produced goods and have been in decline since the early twentieth century.

During the period of post-war industrial expansion, three major changes became evident in the regional distribution of industry. First, the traditional regions of heavy industry, the Nord and Lorraine, experienced a marked decline in employment as heavy industry passed its peak of productivity in about 1960. These regions also suffered from a decaying infrastructure and urban environment, with significant areas of polluted and derelict land. Secondly, the Paris region experienced deconcentration, both as a result of diseconomies of scale in the agglomeration and as a consequence of a series of government incentives. Thirdly, the growth of technical industries during the 1960s and early 1970s was particularly marked in the west and centre of France, including Bretagne, mainly as a result of decentralization to areas not too distant from Paris.

The regional dimension to government industrial strategy was introduced in a modest manner in the 1950s. State industrial policies in the immediate post-war period were concerned with the reconstruction of industry rather than its regional distribution. The regional policy was introduced as part of a wider commitment to reduce the over-concentration of economic activity in and around Paris and to spread job opportunities to the less developed regions. Government initiatives consisted of a combination of 'carrot' and 'stick' policies. In 1955, restrictions were placed on the creation or extension of industrial premises within Paris over a threshold size of 500 m^2. Government approval, known as an *agrément*, became necessary in these circumstances. Further control over Parisian expansion was imposed by a new development tax, the *redévance*, in 1960, but these policies were gradually liberalized from 1967 onwards (House, 1978). Conversely, grants became payable for companies which elected to move to the areas of declining heavy industry.

The inducements for industries to relocate to the provinces were probably the most influential of the locational policies in determining industrial location in the post-war period. The relocation grants available have fluctuated in their amounts and their spatial coverage. In 1955, conversion zones '*zones critiques*' were designated; these were limited areas undergoing industrial decline and suffering from high rates of unemployment. Grants of 20 per cent of investment costs were payable to firms relocating in those zones. In 1960, regional aid was extended to rural areas suffering from underdevelopment and underemployment. House (1978) saw this extension as a break point in regional policy, as this was the first time that the national economic space was viewed as a whole. Throughout the 1970s and 1980s, regional grants were available at differing levels for development zones and conversion zones.

The effectiveness of such policies is always difficult to assess. Between 1954 and 1975, over a million jobs were created within the secondary sector. Of these, 40 per cent were located in the Paris Basin and 30 per cent in the industrially underdeveloped western regions of Bretagne, Pays de la Loire and Poitou-Charentes. Two major reservations about the effectiveness of this industrial development in western France should be noted. First, the creation of industrial jobs failed to match the decline in agricultural employment; hence these regions experienced a net loss of jobs. Secondly, decentralization occurred relatively rapidly and easily in the early 1960s, but the rate of movement had declined well before the oil crisis year of 1973. Furthermore, the peripheral areas gained least, as most firms chose to locate as close as possible to Paris. Nonetheless, a number of significant industries were located away from Paris, bringing a more equitable distribution of jobs and resources for a number of years.

Since the onset of recession in 1973, decentralization to the regions has slowed. As three quarters of a million jobs have been lost from industry, there have been relatively few jobs available for relocation. Continued job loss was greatest from the traditional industrial areas, the Nord, Lorraine and Paris. Losses were also experienced around Paris as branch plants

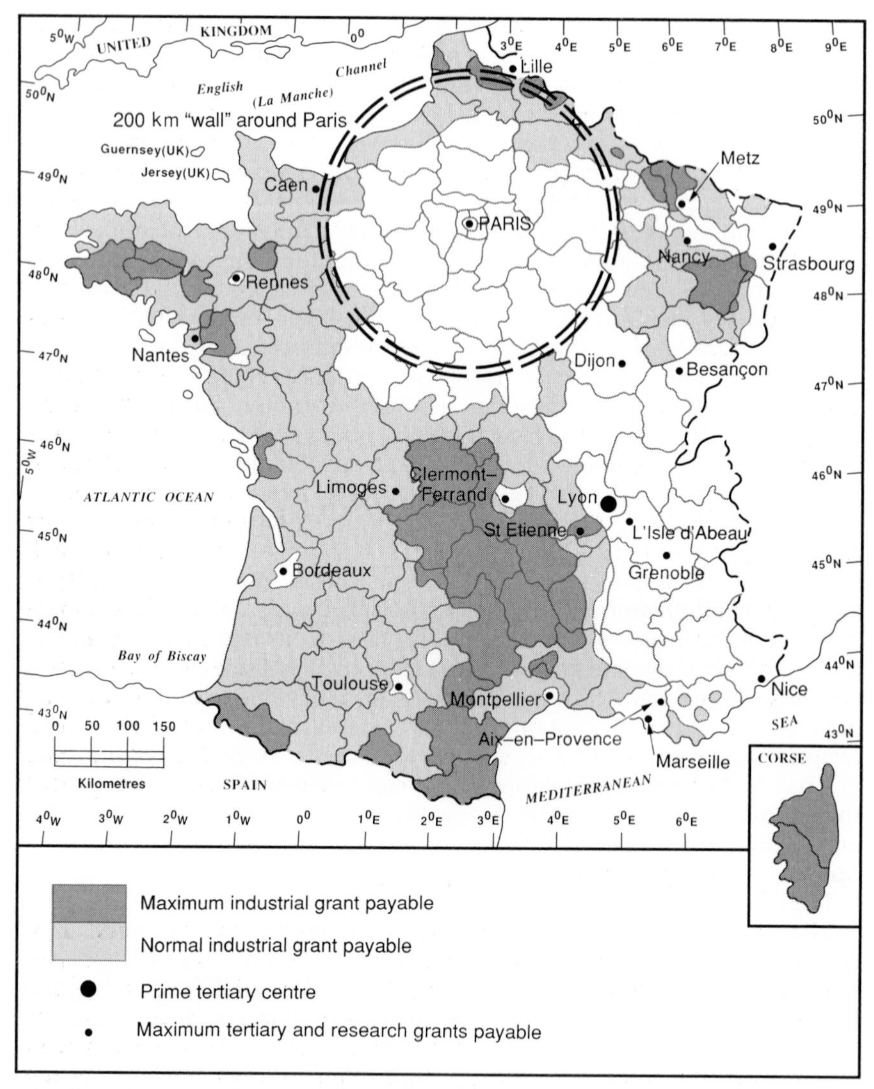

Fig. 4.2 Areas eligible for industrial and tertiary development grants, France. (Data from Laborie *et al.*, 1985)

closed and in the Rhône-Alpes as previously booming industries stagnated. In response to economic changes, the system of regional grants was modified in 1982, when two main subventions became payable. The amount of the grant is dependent on the number of jobs created, the level of investment and the region in which they are to be established. Maximum payment of the regional development grant, the *Prime d'Aménagement du Territoire*, is payable in central France and in areas of special difficulty (Fig. 4.2), notably the Massif Central, Corse, parts of Bretagne, the Nord and Lorraine. Maximum grants are therefore payable in both development and conversion areas. The other type of grant available, the small business grant or *Prime Régionale à l'Emploi*, is a more flexible and local form of aid. It is granted by local authorities and allows them responsibility for allocating priorities in local regional development.

Despite the grants available, the locational pattern of industry in the 1980s differs only slightly from that of 1954. In 1982, the four major industrial areas of the Nord, Lorraine, the Paris Basin and the Rhône-Alpes still accounted for half the country's industrial activities as they had in 1954, containing respectively nine, six, 22 and 12 per cent of industrial jobs. However, substantial job losses have occurred from the Nord and Lorraine. In the Nord, most industrial activities reduced their labour forces between 25 and 40 per cent between 1973 and 1983; the rate of job loss was double the national average, causing high unemployment and out-migration (Zukin, 1985: 360). Lorraine, with its heavy reliance on steel, lost more jobs than its neighbour, Alsace, which had a more diversified industrial base (Nonn, 1989). Although the Paris Basin has lost industrial employment (only 21 per cent of its inhabitants are employed in industry), it has maintained a control over business organization. The rural periphery has shown some internal differentiation, although much of it remains relatively unindustrialized and has few locational advantages for industry. However, there is some evidence of the development of industry in many small towns. Almost a third of the industrial workforce is located in small towns and rural areas, notably in Basse-Normandie and Poitou-Charentes. This new locational pattern may be related to the phenomenon of counterurbanization (see pages 53–5; 58–60). The general decline in industrial employment not only makes decentralization more difficult, but also brings its appropriateness as a stimulus to development into question.

Changes in the labour force

The changing structure and location of employment opportunities cannot be viewed in isolation from the changing composition of the labour force. The size of the labour force was virtually static in the first half of the twentieth century, but has expanded in the post-war period as a result of both natural increase and immigration. This increase in numbers has been accompanied by significant changes in the age and sex structure of the labour force, by an increase in the general level of educational attainment, by changes in the country of origin of migrants (see pages 60–3) and by continued high levels of unemployment since about 1984.

Between 1962 and 1987, the proportion of the labour force employed in each major sector of the economy has altered dramatically. Employment in the primary sector (predominantly agriculture) has declined both absolutely and relatively (see pages 81–4), while employment in the secondary sector grew until 1975 and then declined. By 1987, the numbers employed in industry, construction and public works were lower than in 1962 although the total labour force had grown by 25 per cent; manufacturing workers represented only 22 per cent of the labour force. By contrast, the tertiary sector has undergone marked expansion both in absolute and relative terms. These major trends of continued decline in employment in the primary and secondary sectors, and growth in the tertiary sector, were evident throughout the 1980s (Table 4.3).

The growth of the labour force helped stimulate the consumer demand which fuelled the industrial miracle of the post-war years. The reconstruction period immediately after the war was a period of labour shortage, partly filled by immigration. From the 1960s, the size of the indigenous labour force increased and by the 1980s there was evidence of labour surplus. The rapid increase in the labour force since 1962 is attributable to three factors. First, the large birth cohorts of the baby boom reached the age of employment, while simultaneously many of the cohorts reaching retirement were relatively small. Secondly, migratory movements brought back numerous ex-colonials, including *pieds-noirs* from north Africa, as well as labour migrants from former French colonies and from Mediterranean Europe. Finally, the proportion of women in paid work outside the home rose markedly. Increases caused by these factors have been partially offset by an increase in the number of years spent in full-time education and by retirement at earlier ages.

Unemployment: a problem of the 1980s and 1990s

Unemployment has become a structural aspect of the French labour force. The boom years of the 1960s were times of full employment, with unemployment rates of less than two per cent. One of the results of changed economic circumstances and policy has been a continued high level of unemployment around the 10 per cent mark since 1984; the unemployment of over two and a half million workers is a persistent political and economic problem. This significant unemployment has become the target of government policy. The privatizations of industry in the late 1980s were accompanied by a broad economic programme to lift price controls, introduce a shorter working week and stimulate youth training. More recently, legislation has been adopted in relation to redundancy provisions, early retirement and contract working. Many of these provisions aim to reduce the size of the workforce and hence to reduce unemployment. Since 1986, France has increasingly adopted a market-driven economic approach, with individuals, branch plants and firms exposed to competition and to international influence.

State policies directed towards the labour force have also encouraged labour flexibility, deregulation of employment conditions and a new workplace culture. These provisions have been less far-reaching than those of the United Kingdom. Legislation on contract working and flexible working practices was also introduced to make industry more competitive by reducing rigid working practices; however, many industrial sectors in France were only weakly unionized and so had preserved more labour flexibility than industry in some other parts of Europe (Grahl and Teague, 1989). It is likely that the introduction of deregulated labour markets and increasingly flexible working practices may lead not only to higher unemployment but also to greater inequalities in wages and working conditions, more 'atypical' forms of employment (Grahl and Teague, 1989: 101) and the growth of an unemployed underclass. Indications of these changes are the persistence of long-term unemployment, a reduced vacancy rate and marked increases in part-time and contract working (OECD, 1991: 53).

The sex structure of the labour force

The rate of female participation in the paid labour force has risen markedly from a steady level around 35 per cent between 1954 and 1968, to 37 per cent in 1975 and 41 per cent in 1982. This increase was caused by a complex of changes in the status and role of women, a reduction in marriage and fertility rates and changes in family lifestyles. Figure 4.3 shows the increase in paid activity rates for women of various ages since 1962. For example, 40 per cent of women aged 30 were defined as economically active in 1962; this proportion had increased to 58 per cent by 1975 and 69 per cent by 1982. The rates for particular birth cohorts can also be traced (Fig. 4.3); 48 per cent of women born in 1940 were defined as economically

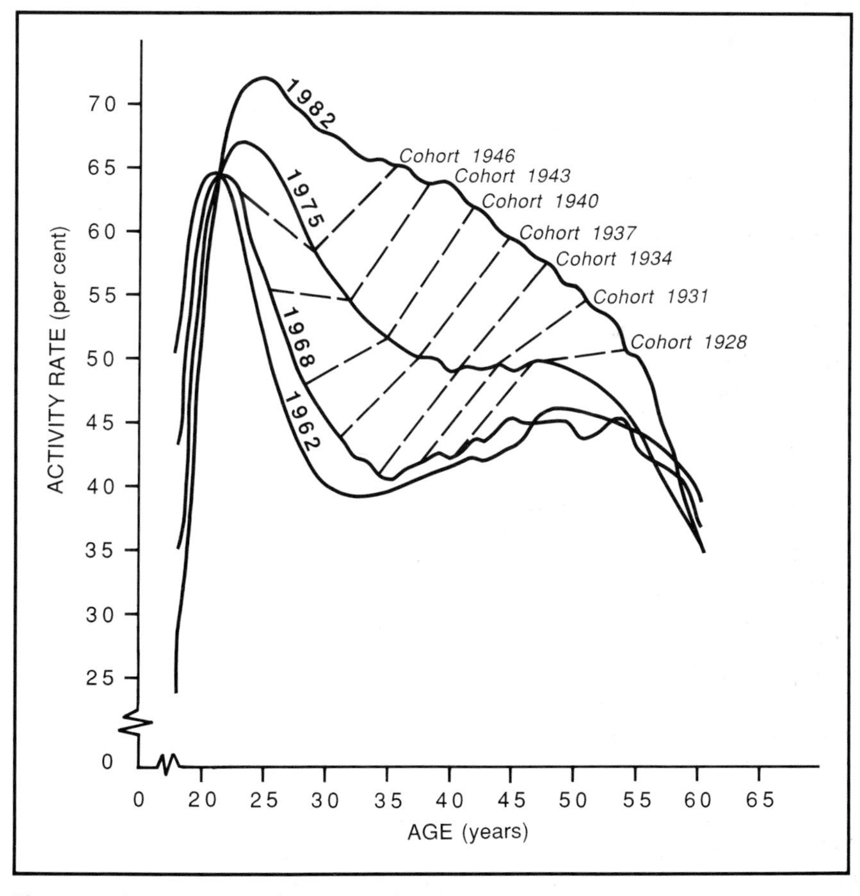

Fig. 4.3 Activity rates, by ages, of women in the paid labour force, France, 1962–82. (Source: INSEE, 1982)

active in 1968 when they were 28 years old; these same women became increasingly engaged in work outside the home as they got older. In 1975, 52 per cent of them were 'economically active' when 35 years old and 62 per cent 'economically active' in 1982 when 42 years old. This change reflects not only the move to the paid labour force after bringing up children, but also a great change in the attitudes to women working outside the home, by women themselves and by society in general.

The women who have moved into paid employment have taken up jobs in the developing tertiary sector, rather than in agriculture or industry where job opportunities have been declining. Women's participation in the paid labour force is highest in central Paris where fertility rates are lowest (Fagnani, 1987). Such women, who choose to postpone or curtail child-bearing, have been termed *Malthusiennes* (Fagnani, 1988) (see pages 46–9). In the tertiary sector, women outnumber men as clerical workers by two to one, whereas men are over-represented both as manual workers and as managers and professionals. Thus the structure of socio-economic groups has a distinctive breakdown by sex. Furthermore, the numbers of women unemployed and their rates of unemployment are consistently higher than those of men; women also tend to have a higher rate of undeclared unemployment. Unemployment rates are particularly high for young women aged under 24 and for foreign women. In 1987, 29 per cent of women aged under 24 years were unemployed compared with 24 per cent of men; while 20 per cent of foreign-born women were unemployed compared with 12 per cent of foreign-born men (INSEE, 1987). These statistics reflect a labour market segregated by sex, class and race.

The age structure of the labour force

The age structure of the labour force has become increasingly compressed since 1962, as a result of earlier retirement and an increase in the length of full-time education. The decline in the aged labour force has been precipitated by economic recession. During the recession, there has been a very marked decline in activity rates for the over-55 age group as a whole. In 1962, 33 per cent of men still worked at the age of 70 and 50 per cent at the age of 62, whereas in 1982 barely five per cent and 33 per cent worked at those ages respectively. The trend towards early retirement has been accentuated since 1983, when the normal age for the receipt of retirement benefits was lowered to 60 years.

At the other end of the spectrum of working ages, the proportion of teenagers working has declined substantially as the length of full-time education has been extended. In 1962, about 33 per cent of 18-year olds were engaged in full-time education, but by 1982 this proportion had risen to 60 per cent for men and 70 per cent for women. The highly qualified labour force fulfils a demand for skilled professional and managerial jobs especially in service occupations and in high technology industries. Increasing educational attainments allow the indigenous labour force to move up the socio-economic ladder, the lower rungs of which have been filled by immigrant labour.

France has become an increasingly technocratic and meritocratic society; in particular, business managers in the public sector and civil service are drawn from the ranks of the ex-students of the *grandes écoles* and especially the national business school, the *Ecole Nationale d'Administration* (ENA). The graduates of the ENA are seen as an elitist, Paris-based and

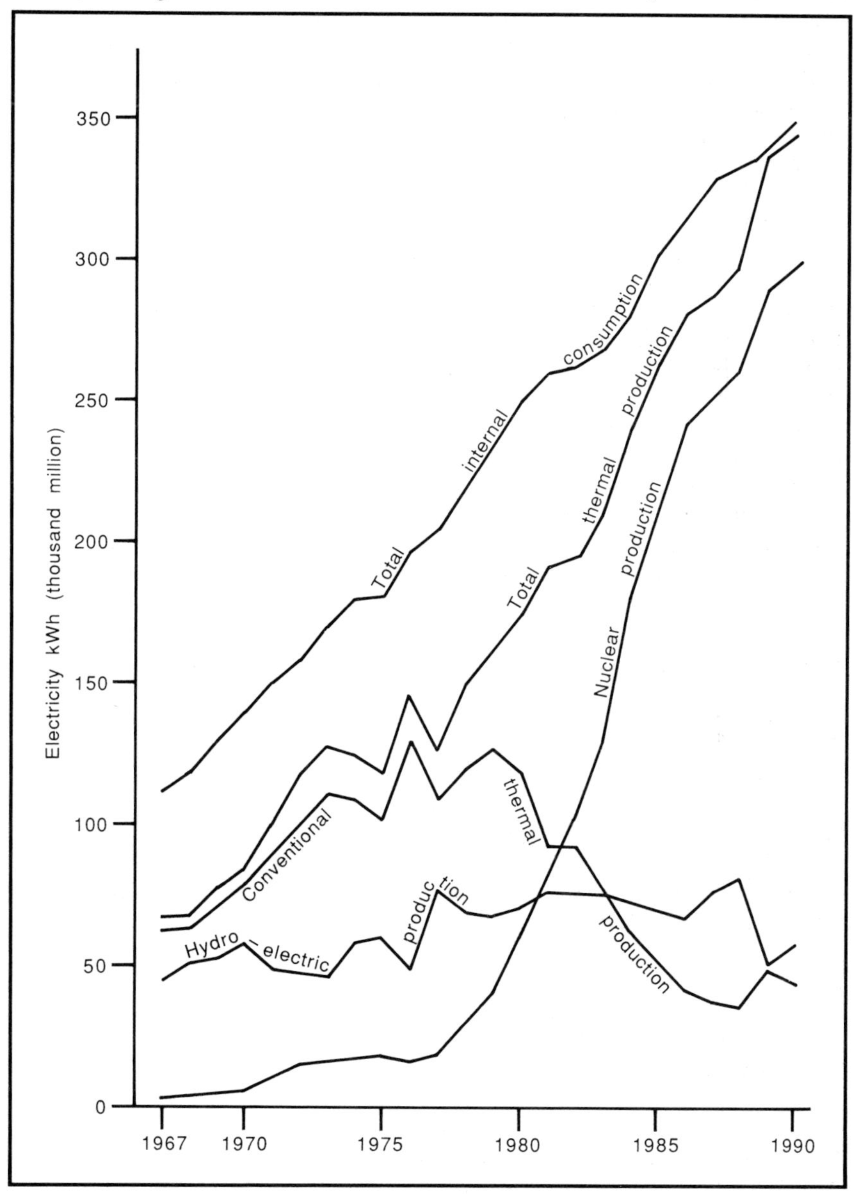

Fig. 4.4 Energy sources used for electricity production, France, 1967–90. Conventional thermal production includes coal, oil and gas. (Source: Electricité de France, 1991)

highly competent administrative class, often dubbed the *énarchie*. This group of technocrats has had a profound influence on government and business management in the post-war industrial economy.

Major industries

The changing energy industry

The French energy industry, as with the industrial sector in general, experienced post-war expansion followed by a period of rationalization and contraction after 1973. Demand for power increased during the 1950s and 1960s, caused by population and urban growth, industrial expansion and higher living standards. Domestic energy sources (pages 28–9) were stretched to capacity to meet demand. Most of the energy industry, except petroleum, was nationalized immediately after the war in recognition of its critical importance in industrial reconstruction.

Since the Second World War, there have been major changes in the sources of energy used in France. At the end of the 1940s, the majority (86 per cent) of France's energy requirements was supplied by coal. The other energy sources were hydro-electric schemes and imported petroleum. Coal production increased to meet demand throughout the 1950s. However, oil rapidly supplanted coal as the major energy source because of its low price, ease of use, general substitutability and widespread availability as an import. Gas and oil rose steadily from less than 20 per cent of all fuel consumed in 1950 to 75 per cent in 1973. By this time, French industry had become excessively dependent on imported oil, while the domestic coal industry was in decline. In 1970, the French government launched a bold policy to develop oil refining capacity in the expectation that continued substantial imports of oil would occur. These refineries were located principally at the lower Seine and Marseille-Fos, areas which were themselves energy-deficient yet with excellent port facilities. Unfortunately the total refining capacity of 170 million tonnes per year was only 65 per cent utilized in 1975, leaving France with a problem of massive over-capacity.

The oil crisis of 1973 precipitated a re-evaluation of the national energy strategy and in 1975 a national energy plan emphasized the need to develop domestic resources, rehabilitate the coal industry, reduce consumption of foreign fuel and undertake a massive nuclear power programme. Since the early 1970s, considerable diversification has taken place in the sources of energy used. Since 1981, petroleum has supplied less than 50 per cent of national energy requirements, while nuclear power has increased dramatically in importance (Fig. 4.4). Since 1982, national energy policy has focused even more strongly on reducing energy consumption, by investigating ways of using waste energy and by developing renewable energy resources including solar and geothermal energy (Agence Française pour la Maîtrise de l'Energie, 1985).

The development of nuclear energy has become a major tenet of French energy policy. By 1989, there were 30 nuclear power stations in France, widely located but with concentrations on the Loire and Rhône rivers (Fig.

4.5). They supplied 65 per cent of the nation's energy requirements and the vast majority of its electricity (Fig. 4.4). By this time, France had become the second largest producer of nuclear energy in the world after the USA (Electricité de France, n.d.(a)). The initial enthusiasm for its production has dissipated because of the powerful anti-nuclear lobby, growing environmental concern after Chernobyl, and economic recession. Furthermore, nuclear energy is primarily suited to the production of base-load electricity as it is inflexible in operation; this structural characteristic is likely to limit the extent of its further adoption. The opponents of

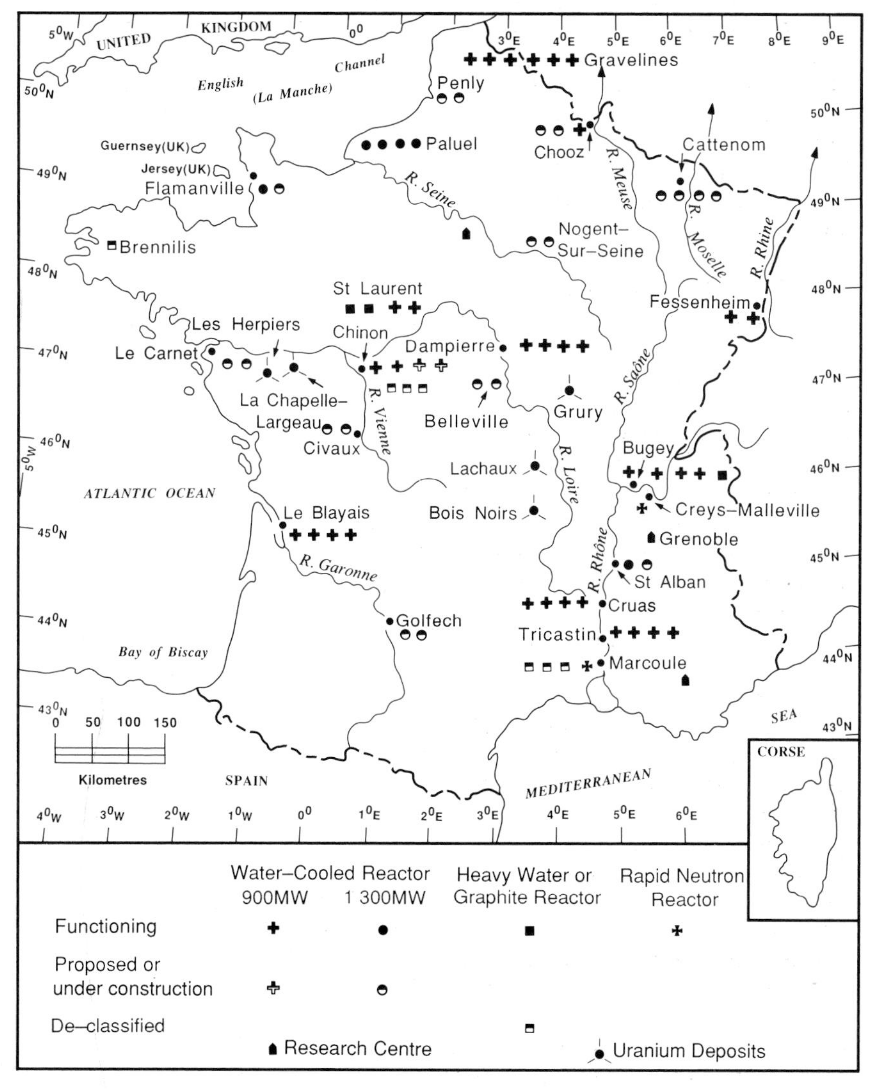

Fig. 4.5 Location of the nuclear power industry, France, 1990. (Source: Electricité de France, n.d., (a))

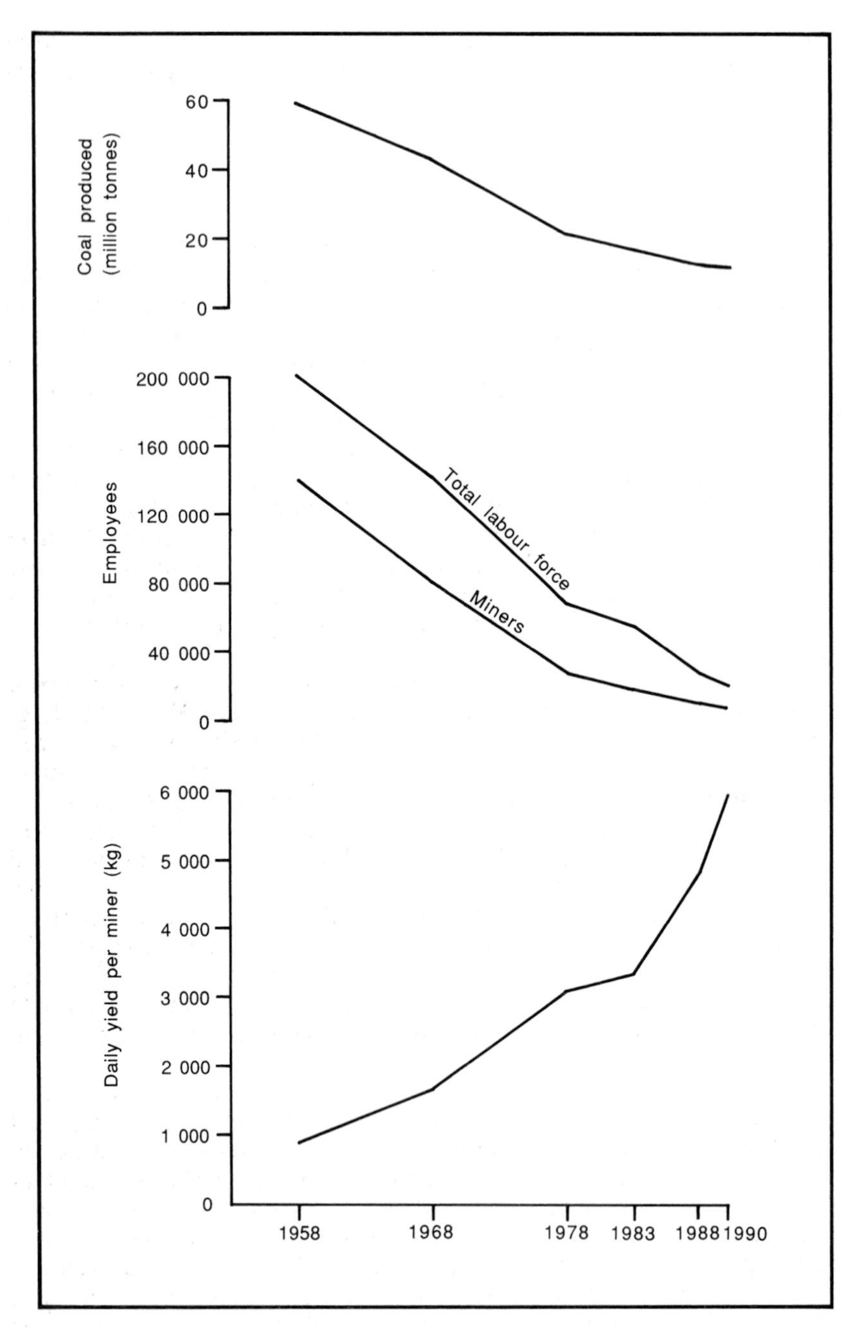

Fig. 4.6 The coal industry: production, labour force and daily yields, France, 1958–90. (Data from Charbonnages de France, 1991)

nuclear energy argue that, apart from the environmental and health issues, the reliance on nuclear sources merely shifts the dependence from one type of imported fuel to another.

The coal industry

The French coal industry is still the major domestic source of energy and is economically significant because of its regional concentration, environmental impact and relationship with heavy industry. Coal production reached a peak of almost 60 million tonnes per year in the late 1950s. Its decline since that date is a result not only of competition from oil, but also is because of the fragmented nature and excessive depth of the coal reserves (page 29). The period of adjustment and decline since the 1960s has therefore been particularly severe in France (Scargill, 1990).

The oil crisis, although temporarily making coal cost-competitive, did not assist the coal industry in the long term. French coal production continued to decline sharply (Fig. 4.6). The decline in production is related to the loss of markets for coal; the railways, domestic consumers, the steel and the electricity industries all reduced demand. The decline in domestic production has increased reliance on imported coal, the consumption of which, at 25 million tonnes per year, greatly exceeds the domestic production of less than 15 million tonnes.

The reduced production of coal has inevitably been accompanied by a massive decline in the labour force, from a peak of 200 000 in 1958 (Fig. 4.6). The rationalization and concentration of the industry had reduced the labour force to less than 40 000 by 1990. The austerity measures introduced in 1984 to curtail financial losses and borrowings by the nationalized coal industry had the dual effect of reducing production targets and of precipitating further job losses. Figure 4.6 shows the changes in coal output, employment in the coal industry and yield per miner. The massive restructuring of the industry, which has resulted in employment contraction and improved productivity, has had a severe regional impact. By 1988, Lorraine was the only productive coalfield of real significance, although scattered Alpine coalfields retained a limited production (Plate 4.3), while open-cast mines in the south of the country were increasing in relative importance.

The effective closure of the Nord–Pas-de-Calais coalfield has been dramatic (Scargill, 1990). In 1946, 110 mining centres were in operation, all of which had closed by 1988. The loss of jobs in coal and other heavy industry has prompted government action to create new jobs in the region. Over 50 000 jobs were moved to the region between 1967 and 1977, about half of which were in the vehicle and vehicle component industries. Most of these jobs were located in the coal basin around Lens, Béthune and Douai; for example, a Renault plant was established at Douai which employed 7000 in the 1970s. Critics of the scheme argue that the number of jobs has been insufficient to stop migration loss, that dependence has merely shifted from one industry to another and that there are very few jobs for women (see also page 137).

The Lorraine coalfield was developed mainly after the Second World

Plate 4.3 One of the smaller coalfields in the Alps, now declining. (Photo: S J Gale)

War, reaching peak production of 15 million tonnes in 1958. In absolute terms, production has declined but relatively it has become much more significant, from 30 per cent of the national output in the 1960s to over 90 per cent in the 1990s. The shedding of jobs has been less severe than in the Nord and similar policies of job replacement have been pursued. Tuppen (1983) considered that the Moselle Valley benefited more than the coalfield areas from job creation schemes, but that workers have also been able to find jobs by increasing their commuting range, even across the German border.

The steel industry: heavy industry in decline

Heavy industry boomed in the 1950s and has been declining since the 1960s. In the post-war period, iron and steel production was spatially concentrated, with 66 per cent of steel output being produced in the eastern region (Lorraine) and a further 20 per cent in Nord–Pas-de-Calais. Steel production increased steadily from 1952 when annual production was 10 million tonnes, to 1974 when annual production reached 27 million tonnes. At this time, the demand for steel from other manufacturing industries was buoyant and a combination of government and private initiatives resulted in investment in new production technologies and in new integrated steel works. The new works at Marseille-Fos and at Dunkerque came on stream just as demand faltered because of recession and restructuring. Output remained steady at 22–3 million tonnes in the late 1970s, falling sharply to 19 million tonnes in 1984 and to 18 million tonnes in 1987. By the early 1980s, therefore, the French steel industry was working at about only 80 per cent of its capacity; it has, however, shown signs of stability and even a slight upturn since 1988.

The steel industry provides an excellent example of the processes of concentration and state intervention outlined on pages 117–20. A first wave of amalgamations occurred in the 1950s, forming three major groupings. These were the Union of Northern Steelmakers, the *Union Sidérurgique du Nord*, known as Usinor; the Union of Lorraine Steelmakers, the *Union Sidérurgique de Lorraine*, known as Sidelor; and Lorraine-Escaut, a firm with regional interests in both the Nord and Lorraine. In the 1960s, smaller companies were formed with regional interests in the Loire Valley and the Moselle. At this time, each major group had a productive capacity of 3–4 million tonnes per year, each rather smaller than their European competitors. The requirements of international competition led to a second wave of amalgamations; in 1966 Usinor absorbed Lorraine-Escaut and Sidelor combined with the de Wendel group to form Sacilor. Similarly, in the special steels sector, Creusot-Loire was formed in 1970 from three other groups to rival Ugine-Kuhlmann as a major producer. In 1978, 80 per cent of French steel-making capacity was owned by Usinor and Sacilor, which were effectively government-owned from that date and officially nationalized in 1982. Between 1982 and 1986 state involvement in steel was virtually total, as both Creusot-Loire and Péchiney-Ugine-Kuhlmann were nationalized, although Creusot-Loire was taken into receivership in 1984.

Despite huge state intervention and investment, the French steel industry has undergone dramatic decline, because of severe international competition and a world glut of steel. Demand for steel has fallen since 1974 because of the downturn in construction, vehicle manufacture, and shipbuilding and allied engineering industries. Other major producers such as Spain and some of the newly-industrializing countries have also increased their market share of world steel at the expense of traditional producers such as France and the United Kingdom.

The rationalization of steel production and of the labour force employed in steel has been severe and prolonged. The 1977 Davignon plan for European steel introduced voluntary production quotas and aimed to provide an integrated European plan for the rationalization of steel-making capacity and the gradual reduction of national government support. The French reduced their steel-making capacity by just over five million tonnes in the early 1980s by the shutting down of smaller and less efficient plants. These closures accounted for 20 per cent of the total EC planned reductions in capacity. Plant rationalization and reduced production have resulted in severe job losses. In 1974, 156 000 were employed in the industry, but this had shrunk to 96 000 by 1982 and 60 000 by 1987.

Closures announced in 1978 in Lorraine resulted in strikes, civil disobedience and violence which constituted the biggest breakdown in public order since the student riots of 1968 in Paris (Hudson and Sadler, 1989: 89). As a compensation for the loss of employment, the government announced retraining schemes for redundant workers, the movement of branch plants of nationalized industries to affected areas and new infrastructure projects including motorways (Hudson and Sadler, 1989: 126–8). The affected areas also benefited from loans from the European Community for industrial conversion. The restructuring of the steel industry has had a lasting impact on the relationships between capital, labour and the state: 'the new directors have transformed staff management methods . . . the trade unions have shifted from a hard-line ideology of class struggle to greater realism . . . the State has extended some of the arrangements arrived at in the steel industry to all wage earners . . .' (Piganiol, 1989: 630).

Restructuring has had a profound spatial impact on the steel industry. Within Lorraine, production has become progressively more concentrated, especially at the major works at Gandrange and Seremange, each of which has a capacity of over three million tonnes. Job losses have been particularly severe at Longwy and at Neuves-Maisons and at older plants which have a poorer environment and less space for renewal. Within the Nord–Pas-de-Calais region, the major centre of the industry has moved from the Sambre and the Escaut valleys to the coast at Dunkerque which has access to a deep-water port. The other major integrated works at Marseille-Fos, although working at less than capacity, benefits from similar locational and technological advantages.

The motor vehicle industry: product of the post-war boom

The motor vehicle industry in France has become a major sector of French industrial production since the Second World War. By 1979, annual production of vehicles was over 3.5 million, half of which were exported, while the industry directly employed 530 000 people and indirectly supported a similar number in ancillary industries. In the early 1980s, production slumped to around 2.7 million vehicles, but this had improved again to 3.2 million by 1988. However, in the intervening decade, the industry had shed 200 000 workers and had become more efficient.

The production of motor vehicles is dominated by two major groups, Renault and Peugeot, both of which have absorbed other manufacturers (page 117). The industry has therefore undergone the processes of concentration characteristic of recent industrial change (see pages 117–19). The vehicle industry is partly nationalized (page 119) and has become increasingly international in operation. Renault now has production links with Volvo of Sweden and has some productive capacity in Taiwan. Both major French manufacturers are international in their penetration of external markets, while they have maintained the lion's share of the domestic market.

The vehicle industry's expansion during the boom years was accompanied by a significant locational change. In 1968, 63 per cent of car workers were located in the Paris region, the major exceptions to this being the Peugeot factory at Sochaux in eastern France and the Berliet factory at Lyon. By 1982, fewer than 30 per cent of car workers were located in the Paris region, as the industry decentralized to underdeveloped and conversion zones. Decentralization has occurred in response to market forces and to government intervention. Diseconomies of scale within Paris, especially rising costs and pressures on space, made expansion *in situ* difficult and expensive. Government intervention in the location of the industry resulted from the high level of state control, its rapid expansion and its perceived role as a leading sector in regional economic growth. Furthermore, the major labour requirement for unskilled or semi-skilled workers could be met almost anywhere in France.

Early decentralization was directed towards the under-industrialized regions of western France and a string of car plants was established from Paris to Rennes and in the wider Paris Basin (Oberhauser, 1987). A second wave of decentralization has directed car plants to the declining areas of heavy industry in the Nord and Alsace-Lorraine. Decentralization of this type, for example of Peugeot to Valenciennes, was heavily subsidized by the state (Oberhauser, 1990: 64). The company's employment policy favoured workers without industrial experience; relatively few ex-steelworkers were employed and the few skilled workers required were brought in from outside. This employment policy allowed the company easier negotiation of new terms and conditions of work. Oberhauser (1990: 66) concluded that the decentralization of this plant did little to help regional transformation, but assisted Peugeot, France's largest private company, to compete in international markets. The subsidized decentralization also permitted the introduction of automated assembly techniques.

The value of the motor vehicle industry as a tool of decentralization policies changed significantly in the 1980s. The industry suffered from falling demand and itself was forced to make redundancies; for example, in Alsace, a wave of redundancies in 1980–1 was followed by the loss of 1370 further jobs in 1984–5 (Nonn, 1989: 147). Peugeot, in attempting to rationalize, announced 2900 redundancies at the Talbot plant at Poissy, northern Paris, in 1983. This announcement triggered industrial unrest and a prolonged strike; the dispute had racist overtones as many of the workers threatened with redundancy were foreign migrants. The rationalization of the labour force occurred at the same time as new robotized and just-in-time production systems, which allow more flexible production, were introduced. The traditional production line has been replaced by short-run production lines staffed by autonomous teams.

The vehicle industry shows clear links between growth rates, state intervention, spatial distribution and the production system itself. Industrial expansion through the 1960s and 1970s was accompanied by government-sponsored spatial decentralization and by the introduction of Fordist (assembly-line) production methods using cheap unskilled labour. The period of recession and restructuring in the 1980s has resulted in contraction, accompanied by new forms of capital-intensive production, including just-in-time systems and modular computerized systems which require a flexible and multi-skilled labour force (Womack *et al.*, 1990). The reorganization of the French motor vehicle industry has occurred not only in response to the changing circumstances of the 1980s but also in anticipation of the European Single Market of the 1990s.

High-technology industry: growth sector

High-technology industries are those industries which require high levels of scientific skills, usually in the sectors of micro-electronics, telecommunications and aerospace. The definition of high-technology is sometimes extended to include biotechnology or aspects of design, research and development in almost any sector of manufacturing. Research and development accounts for a steadily increasing proportion of Gross Domestic Product, from 1.8 per cent in 1980 to 2.3 per cent in 1989 (OECD, 1991). Many of the scientists engaged in research and development are located in the Paris Basin (Brocard, 1981), as are many of the most highly skilled segments of the French labour force (Pottier, 1987). Nonetheless, the concentration of high-technology industry in Paris has dispersed a little since the early 1960s. Between 1963 and 1975, the proportion of scientists located in Paris fell from 65 per cent of the nation's total to 55 per cent, as a result of the decentralization of numerous research centres during this period (Brocard, 1981: 62). The decentralization of research and development is facilitated by state control of the process: in France 41 per cent of research is carried out by state organizations and just over 50 per cent is financed by the state, proportions much higher than in Germany (27 and 37 per cent respectively) (Brunat and Reverdy, 1989: 284).

The high-technology sectors are particularly concentrated in Paris and in major towns of the west and southwest of France. These sectors are relatively undeveloped in the conversion zones of the Nord and Lorraine, which is in part a problem of regional image (Gachelin, 1987). The reasons for location in the provinces vary between industrial sectors. Micro-electronics firms are located in the provinces for two main reasons. First, American firms which established plants in France in the 1960s use these locations as part of the international division of labour: headquarters functions are maintained in the USA, production-line processes are in the newly industrializing countries, and France is used as an intermediate location for research of a limited scope or for specialized production. Examples of this approach include the siting of Motorola in Toulouse and Texas Instruments in Nice. Both these cities have particular advantages, including international airports, universities and attractive physical environments (Pottier, 1987: 206–7). Secondly, French micro-electronics firms have also relocated to the provinces in search of cheaper labour. Examples of this are the movements of Thomson and RTC to the university cities of Caen and Grenoble.

Other state-run industries have also decentralized since the 1960s. Early strategic decentralization of the aerospace industry occurred to Toulouse, a process which was started even before the Second World War. Similar decentralization to some of the major cities in the urban hierarchy included Bordeaux, Marseille and Nantes-St Nazaire. A prime example of state decentralization is that of the telephone industry to Bretagne; in 1982, the state controlled both major manufacturers, Thomson and the CGE group, and about 9000 jobs were created in a relatively unindustrialized region (Pottier, 1987: 209). On the other hand, the relatively new sector of robotics is mainly in private ownership and is more footloose and widely dispersed, although robotics component manufacturers appear to cluster because of their complementary needs.

The experience of high-technology industry in Silicon Valley, California, was that it thrived on a combination of university research, innovative production and the pleasant living environments desired by high-earning professionals (Saxenian, 1984). Many countries have tried to emulate the conjunction of circumstances which occurred there. France has attempted to develop a number of technopoles as triggers for regional economic development; in 1989 there were 30 throughout the country (Benko, 1991). The technology parks have sites in a variety of locations: in new towns, near airports and in historic satellite towns (Dézert, 1989: 28). Cergy-Pontoise is a Parisian new town which brings together electronics, informatics and robotics industries. Paris-Nord, five kilometres from the Roissy-Charles de Gaulle airport and adjacent to the exhibition centre, is a prestige location benefiting from excellent transport links; while the high-status towns of Versailles and St Germain-en-Laye have also attracted high-technology industries because of their perceived prestige and their high proportion of skilled residents. Further expansion of these centres of technological excellence is planned in the 1990s (see pages 193–4).

The earliest technology parks were created in 1983–4 adjacent to large regional centres: Sophia-Antipolis was developed by an individual promoter, while Grenoble and Nancy were developed by local municipalities. The Angers-technopole, started in 1984, now has two centres of excellence in biotechnology and electronics/computing, with developing specializations in medicine and fashion/textiles. The technopole was stimulated by the creation of higher education institutions, the *écoles supérieures* and the universities, and by the relocation or formation of research centres. In 1989, Angers contained 26 research centres and five *grandes écoles*. The technopole, located in the science park of Belle-Beille, brings together research, teaching, industrial production and industrial incubator units

Fig. 4.7 Angers-technopole, major land uses, 1989. (Source: Jeanneau, 1989: 92)

on a 300 ha site with excellent communications (Fig. 4.7). The potential regional advantages of such centres may be diluted by their increasing number: for example, Angers' neighbouring cities of Le Mans and Nantes also have plans for technopoles which may be competitive rather than complementary (Jeanneau, 1989: 93).

A recent trend in industrial development is the establishment of industrial incubators, *pépinières d'enterprise* (Fig. 4.8). In 1983, there were only two, but by 1988, France had 71 industrial incubators, the majority financed by the state and established in association with industry, universities and research and development. Some are established and financed by local authorities, for example outside Pau a small incubator unit has been established in association with the university and the council to support new agricultural and artisanal industries. Most are relatively small, with an average of 24 firms employing 112 people in total (Benko, 1989: 641). Their distribution shows a concentration in Paris and the Rhône-Alpes, but they are also more widely distributed throughout the

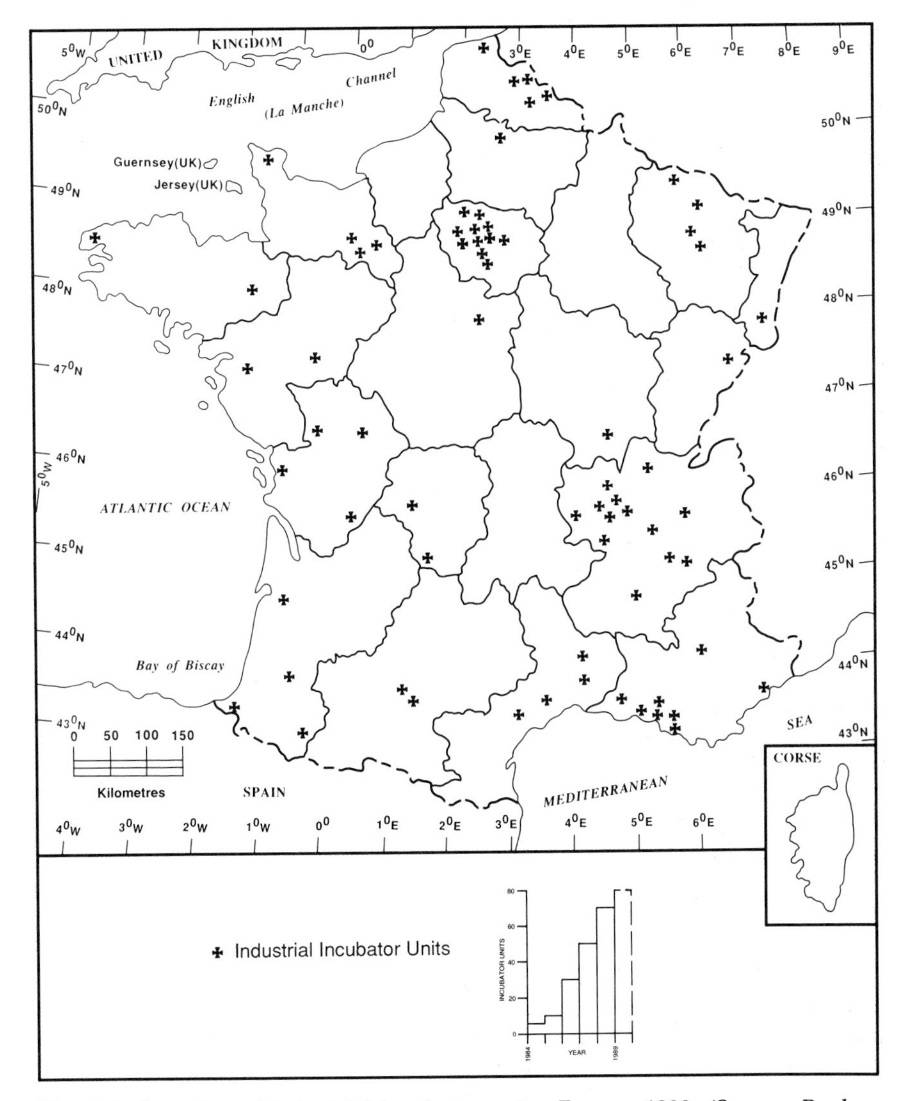

Fig. 4.8 Location of industrial incubator units, France, 1988. (Source: Benko, 1989: 640)

regions (Fig. 4.8). The role of incubator units in regional development is as yet limited, but Benko (1989: 644) considered that their formation and their usual association with high-technology industry may be indicative of the dynamism, prestige and security which are intangible factors which promote industrial and regional development.

Table 4.3 Employment by sector in France, 1980–7

	1980 total employed (thousands)	**% labour force***	**1987 total employed (thousands)**	**% labour force***	**% change 1980–7**
Primary sector:					
Agriculture	1853.7	8.6	1488.7	7.0	−19.7
Mining	140.1	0.6	102.9	0.5	−26.6
Total employed in primary sector	1993.8	9.2	1591.6	7.5	−20.2
Secondary sector:					
Manufacturing	5494.8	25.4	4636.8	21.8	−15.6
Gas, water, electricity	187.3	0.9	209.2	1.0	+11.7
Construction	1841.5	8.5	1522.3	7.2	−17.3
Total employed in secondary sector	7523.6	34.8	6368.3	29.9	−15.4
Tertiary sector:					
Trade	3460.7	16.0	3538.0	16.6	+2.2
Transport	1333.9	6.2	1368.9	6.4	+2.6
Producer services	1553.0	7.2	1804.6	8.5	+16.2
Community services	5773.1	26.7	6608.6	31.0	+14.5
Total employed in tertiary sector	12 120.7	56.0	13 320.1	62.6	+9.9
% of tertiary sector employment in government non-traded services		21.9		25.4	
TOTAL LABOUR FORCE	21 638.1		21 280.0		

Source: INSEE, 1989.

*Note that column percentages may not sum to column totals because of the effects of rounding

Service industries: growth sector

The decline in industrial and agricultural jobs has been compensated for by a substantial increase in service sector employment. The service sector has grown by approximately two per cent per year since 1960 and accounts for over 60 per cent of employment and value-added (Guillauchon, 1986) and 66 per cent of production (OECD, 1991). The rise of the service industries has been associated with a transition from an industrial to a post-industrial society (Bell, 1973). As industry has become more efficient, so jobs have been created in fields which are not directly productive, such as research and development, and finance and legal services. An increasing proportion of these jobs requires tertiary education and high levels of expertise.

The growth in the tertiary sector is by no means uniform (Table 4.3). A substantial growth has been recorded in the public sector, in hospitals, education, community services and national and local government. However, there has also been a marked expansion in some types of private sector services. This expansion has been greatest in producer services, such as financial and legal services to other industry and government. From 1980 to 1987, the number of employees in the producer services sector, which includes finance, insurance and real estate, rose very sharply (Table 4.3). The expansion of producer services has occurred mainly in Paris and in major urban centres where industrial headquarters and government offices are located. A recent study of insurance in France (Wolkowitsch, 1989) reported a doubling of employment in this sector between 1960 and 1984, a preponderance of female employment (62 per cent) and a remarkable concentration of this industry in Paris.

Other types of consumer services have shown variable growth. Flockton and Kofman (1989: 56) emphasized the growth of some major French firms in retail trade and tourism: firms such as Carrefour and Club Méditerranée have shown dramatic penetration into overseas markets, especially in the United Kingdom. On the other hand, transport and communications have undergone many of the restructuring processes experienced by the secondary sector. Despite increased investment in infrastructure, such as for the TGV (the high speed trains), employment growth has been relatively slow. In general though, the consumer services such as wholesale and retail trade, and tourism and hospitality have grown rather more slowly than producer services. They tend to be more widely dispersed than producer services and many rural towns depend for their major employment on public and community services, tourism and trade.

CHAPTER 5

The urban system

Look on thy country, look on fertile France,
And see the cities and the towns . . .

Shakespeare, *I Henry VI*, III, iii, 45–6

Phases of urban growth in France

France has many ancient and long-established cities, some dating back at least to Roman times. The foundations of the present urban system were laid in the Middle Ages and many cities bear the imprint of their early origins. The urban hierarchy was therefore well established by the time of the French Revolution in 1789. Although France had experienced urbanization by that time, the cities were relatively small. City growth throughout the nineteenth century was slow and steady, but reached more spectacular rates after the Second World War. Post-war urban growth was related both to the population baby boom (see pages 44–6) and to the industrial resurgence of that period (see pages 107–8; 112–14). Since the 1970s, overall urban growth has stabilized; many large cities and city centres have experienced population decline, whereas smaller urban centres have grown. The recent growth of small towns is related both to suburbanization and counterurbanization.

The slow growth of French cities until the Second World War may be related not only to limited population growth and industrialization, but also to the overwhelming dominance of Paris in the economic and political life of the country. The national government was concentrated in Paris and the rest of the nation's affairs were controlled from 90 *préfectures*, the capitals of the *départements* (see Frontispiece). Tertiary functions were therefore widely spread among a large number of relatively small towns. The geographical spread of services and competition with Paris stunted the development of large regional centres. The French urban hierarchy therefore became rather imbalanced and Paris, the primate city, is many times larger than the next city (see page 150; Fig. 5.3).

The rapid post-war growth of French cities has stimulated urban planning both within individual cities and throughout the system as a whole. The planning of the urban system has been one of the main mechanisms of regional planning. In the late 1960s, eight of the largest cities were designated as counter-magnets, *métropoles d'équilibre*, to balance the dominance of Paris. From the 1970s, policies were designed to stimulate regional and local centres of opportunity at lower levels in the urban hierarchy, particularly in the towns around the edge of the Paris Basin and in a number of medium-sized towns, the *villes moyennes*. New towns have also been used as elements in France's regional planning strategy and as foci for channelling growth. The five new towns of the Paris Basin are a significant element in the planned development of the city and region (see pages 197–200), while the other four new towns are all associated with other large provincial cities (see pages 157–9).

Within cities, urban planning and policy developed relatively late. The structure and morphology of French cities have been greatly influenced by their early origins, by the construction and removal of defensive ramparts and by nineteenth-century Haussmannization (see pages 159–60). Their subsequent post-war growth has followed a general pattern of inner-city decline and renewal on the one hand and peripheral expansion on the other. This post-war pattern has been increasingly channelled by land-use regulations. Since the 1960s, the focus of urban policy has been directed towards the conservation of historic sites and the rehabilitation of degraded city centres (see pages 162–3). These physical changes have often been accompanied by social changes, notably the gentrification of inner cities.

The early urban system

The present urban system has been greatly influenced by its historical development. Service and administrative functions were gradually complemented by industrial functions. By the sixteenth century, a system of urban central places was well established throughout France. These cities grew very slowly, often responding to cyclical changes of growth and decline. They were constrained by limited resources and technology and by inhibiting socio-economic and political institutions (Hohenberg and Lees, 1985). Proto-industrialization stimulated some urban growth, but much industry remained basically rural (see pages 109–11). Silk, for example, was produced in the rural hinterland of Lyon in areas such as the Cévennes. Nevertheless, spinning and weaving eventually became urban functions, with industrial control vested in the regional centre of Lyon. Even before the nineteenth century, many large towns, such as Rouen, Roubaix and Nîmes, gained additional industrial functions which complemented their regional marketing and service activities.

Large-scale industrialization in the nineteenth century added new elements to the existing hierarchy of central places (Fig. 5.1). The prime example of the specialized industrial city was St Etienne (Merriman, 1991), which underwent very rapid expansion from the 1840s as a result of textile

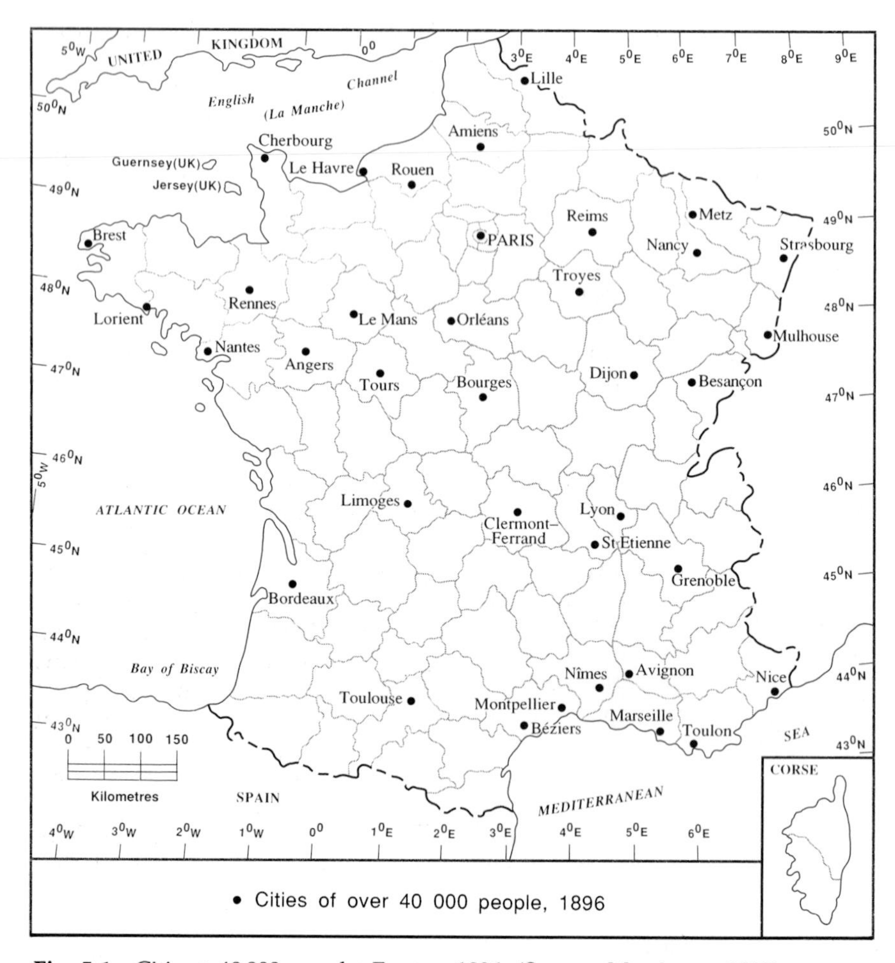

Fig. 5.1 Cities >40 000 people, France, 1896. (Source: Merriman, 1982)

and metallurgical manufacturing. Other cities, particularly on the coalfields of the Nord and Lorraine, experienced specialized industrial development. Often this development took place not in the older-established towns with central place functions, but in new towns which developed close to their resource base. In this way, a network of cities with complementary roles became established. Furthermore, this industrial urbanization was regionally differentiated. North and northeast France experienced a pattern of urbanization and industrialization which was not very different from that of the industrial towns of northern England. Growth occurred in new industrial centres, while long-established non-industrial cities, such as Orléans and Dijon, failed to grow. Urbanization around Paris itself was inhibited by the dominance of the capital. In south and west France, only the regional capitals and some major ports grew rapidly; cities such as Nice, Marseille and Bordeaux combined these two functions.

In 1800, fewer than 10 per cent of France's population lived in urban areas, a lower proportion than England or Italy, but more than Germany, Portugal or Scandinavia (De Vries, 1984: 39). The development of industrialization brought people to the cities during the nineteenth century; in 1850, 25 per cent of the population had become urban and by 1900, 40 per cent (Fig. 5.2). During the nineteenth century, the larger cities, of 40 000 or more (Fig. 5.1), grew most rapidly, but by 1911 these still contained less than 10 per cent of the French population. Paris maintained its status as the second city of Europe from 1750 through to 1950, but the other large cities of Lyon, Bordeaux, Marseille and Rouen, although all in the top 30 in 1750, had dropped from the European rankings by 1950. The rate of urbanization was much more gradual than in Britain, as a consequence of the different patterns of population and industrial growth (see pages 37–44; 107–8).

Historical influences on the shape of the French city

Many French cities reveal their early origins in their morphology, streetscape and street pattern. Roman remains are still evident in a number of southern towns, such as Orange, Vienne and Béziers. More commonly though, the oldest remaining buildings and street patterns in city centres date from the Middle Ages. Churches, castles, market squares, narrow winding streets and timber-framed buildings are to be found in many centres, from major cities such as Rouen and Bordeaux to small country towns such as Carcassonne and Quimper. Many of these old areas are now designated as conservation zones and pedestrian precincts.

Braudel (1981) has suggested that the physical and social structure of the pre-industrial cities of Europe was greatly influenced by their controlling

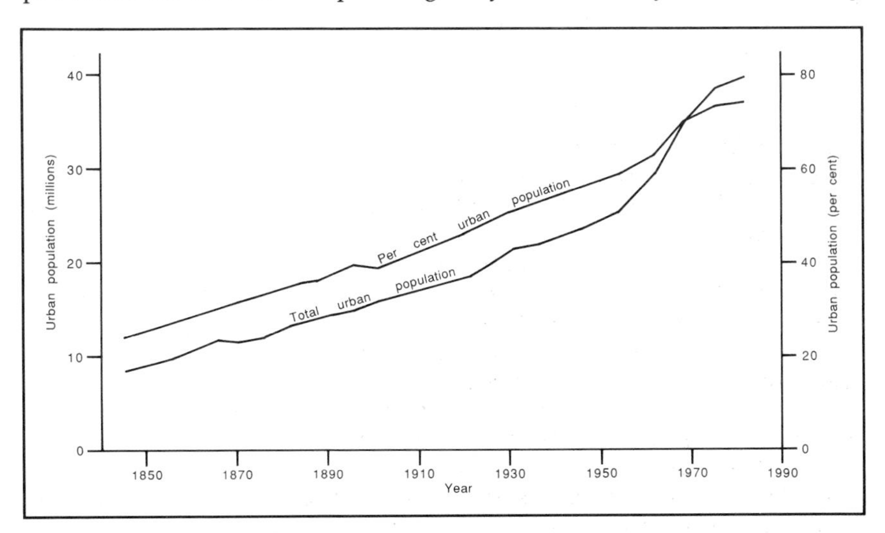

Fig. 5.2 Urban growth, France, 1846–1990. (Data from INSEE, 1982; OECD, 1991)

elites. White (1984) classified these towns as feudal, mercantilist or absolutist; all three types are represented in France. The feudal city was one controlled by a centrally-located elite group, which dominated the core of the city. The peripheral zone of such cities contained the working population, with outcast groups living literally beyond the walls. This type of city, which corresponds most closely to Sjoberg's (1960) idea of the pre-industrial city, was common in southern Europe. It is exemplified in France by the city of Bordeaux, which had a centrally-located elite and, in medieval times at least, a Jewish ghetto located beyond the city ramparts. The mercantilist cities were dominated by more varied interest groups which were more widely distributed throughout the city. Many important northern cities were controlled by powerful mercantile interests through the craft guilds, each of which regulated activities in a certain trade. Cities such as Lille came to contain a number of functional zones, each the focus of a particular craft; within each zone, however, there was a mixture of land uses including workshops, residences and stores. The absolutist city was subjugated under the rule of one powerful leader, such as a king or bishop, as in Nancy which was dominated by the rulers of Lorraine. In absolutist cities, the very structure of the urban area reflected the power structures within society. Typically the city would be dominated by a palace or castle on an elevated site, with avenues leading towards it; together the vista and monument created a form of prestige space, constructed by and for the ruler. The lower and peripheral spatial location of the masses accurately reflected their lower status. Cities of this type were more likely to be controlled by early planning regulations, such as on the heights of buildings. In many instances, the impact of such regulations and of the early structure of the city are still evident in the morphology of the central zone. The conservation of some such zones is considered on pages 162–5.

City morphology has been profoundly influenced not only by medieval controlling interests, but also by successive phases of fortification (White, 1984: 22–7). City walls and gates were useful not only for defence but also for taxation purposes and for control of the often-turbulent suburban populations. Walls were initially simple and could be rebuilt when necessary, forming a 'corset' which could easily accommodate expansion. From about the sixteenth century, as firearms became more widely available and as the firing range of weapons increased, city defences became more complex and costly to rebuild. An essential feature of such fortifications was a *glacis*, a clear zone free of buildings to provide a line of fire from the walls on to potential attackers. From the sixteenth century onwards, the corset constrained city expansion and urban growth resulted in an increasing density of people and dwellings within the walls. The physical impact of fortifications can still be seen. In some cases, remains of walls still exist, but in many cities they have been replaced by wide boulevards. In Grenoble, for example, at least five sets of walls dating from Roman times to the 1830s can be traced as physical remnants, as street lines and as *quartier* boundaries. Frequently, the line of walls may be detected by changes in land use, for example, where industry had been located outside the city walls, or by a change in the age and density of buildings. In several frontier cities, many strong and distinctive fortresses dating from the late seventeenth century remain as historic sites, museums and tourist attractions. Each of the cities was constructed to a similar pattern under the supervision of Vauban, military engineer to Louis XIV. Examples of these fortified cities include Strasbourg, Brest, Antibes and Briançon.

Nineteenth century growth caused expansion beyond the walls in many cities, as the development of industry attracted large numbers of migrants. These populations beyond the walls were also beyond city jurisdiction; they were marginal both spatially and socially (Merriman,

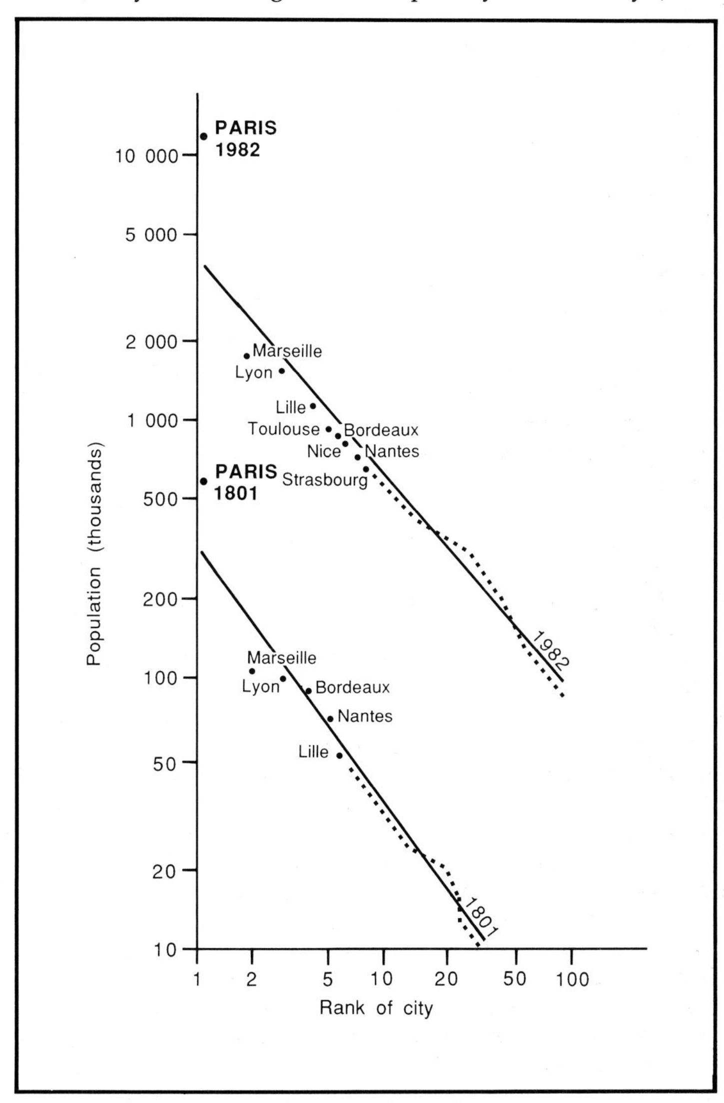

Fig. 5.3 Rank-size of cities, France, 1801 and 1982. (Sources: INSEE, 1982; Noin, 1987)

1991). A response to the threat of the industrial masses was to enlarge the legal urban area to incorporate the suburbs. Another method of control was by direct removal. Many slum districts, especially in the central areas of cities, were cleared and replaced by wide boulevards, allowing the free flow of vehicular traffic. This was a process that came to be known as Haussmannization, a form of early urban planning originating in Paris (see page 176) which was emulated in many cities (White, 1984: 28). These physical changes were often accompanied by the introduction of height and building regulations and by statutes concerning public health and hygiene.

Twentieth century urban growth

The urban hierarchy

Urban growth was rapid in the twentieth century (Fig. 5.2). Half of the French population lived in cities by 1926 and 70 per cent by 1968. The most dramatic period of urban growth occurred between 1954 and 1968, when the population of urban areas increased by 10 million people. The pace of urban growth slowed abruptly from the 1970s, and by 1982 the percentage of the population in urban areas was little more than in 1968 (73 per cent). In the early post-war period the big cities continued their faster growth, while from 1968 the small cities and towns have grown most rapidly. Most growth has occurred at the city edges, in previously rural *communes* which have become part of an expanded urban area.

Twentieth century urban growth has affected the urban hierarchy less than might be expected. In the late twentieth century, the hierarchy is still overwhelmingly dominated by the primate city of Paris (Fig. 5.3). This excessive dominance is clearly related to the long period of political centralization and to the role of Paris as a major European city and capital of a nation and an empire. A notable feature of the rank-size distribution of French cities from 1801 to the late twentieth century is the relatively small size of the next rank of urban centres, such as Lyon, Marseille and Bordeaux (Noin, 1987). However, the smaller cities formed an almost perfectly logarithmic rank-size distribution both in 1801 and 1982 (Fig. 5.3).

Changing city structure

The massive early post-war growth of urban population necessitated an emergency housing response; this took the form of the construction of large apartment blocks of public housing on vacant land at the edge of cities. These *grands ensembles* were often poorly built and inadequately serviced by transport, shops, schools and other facilities. They have often subsequently become problem areas, housing a disproportionate number of the unemployed, the disadvantaged and recent migrants. Further peripheral growth has occurred with the sprawl of private individual housing, known as *pavillons*, into suburban areas, especially since 1968.

Growth on the urban periphery was matched by decline in the city centres, often accompanied by clearance and renewal. A number of French cities, such as the Breton port city of Brest, suffered war damage and were almost totally reconstructed. Many others faced the problem of industrial dereliction and the filtering-down of old housing in city centres. The response to inner city change has shifted from one of clearance and renewal in the early post-war period to conservation and rehabilitation from the 1970s. Traditional areas of mixed land use in the city centres are becoming increasingly subject to single land-use zoning and to a variety of urban planning measures.

The characteristic phasing of city development outlined above has enabled White (1984: 188) to produce a spatial model of the west European city which is applicable to many French cities. It is essentially a model of physical historical elements in the city and their present social use (Fig. 5.4). The historic city centre contains the landmarks built by former authorities and is occupied by the social elite, who have maintained their preference for inner-city living. A line of former city walls demarcates the centre from the zone-in-transition, often an area of nineteenth-century industry, containing a mixture of workshops, warehouses and the old and

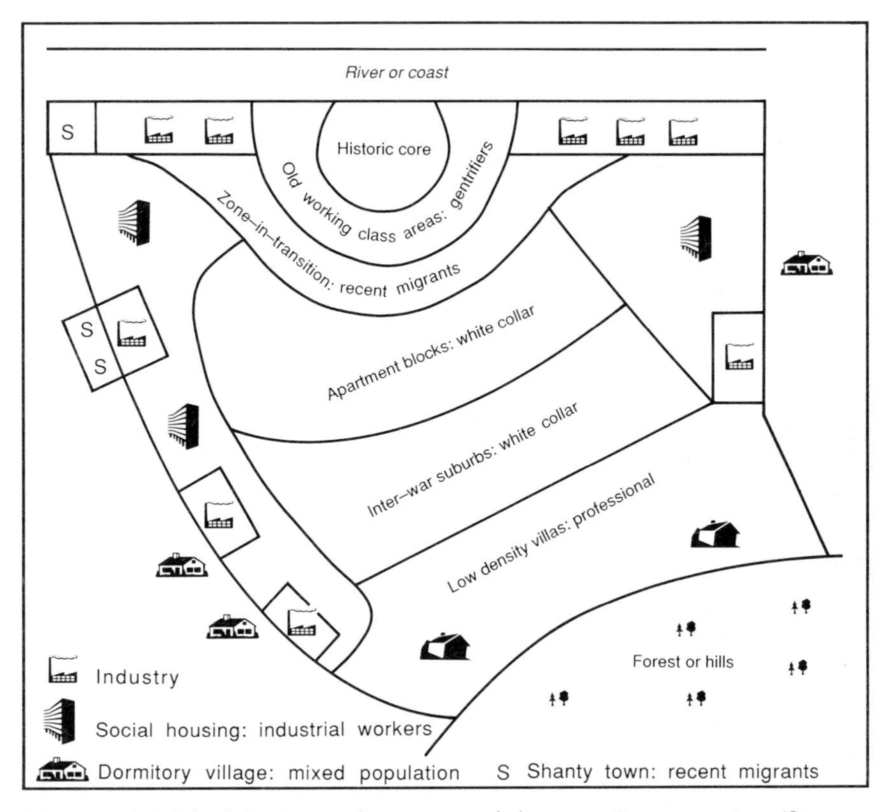

Fig. 5.4 Model of the internal structure of the west European city. (Source: White, 1984:188)

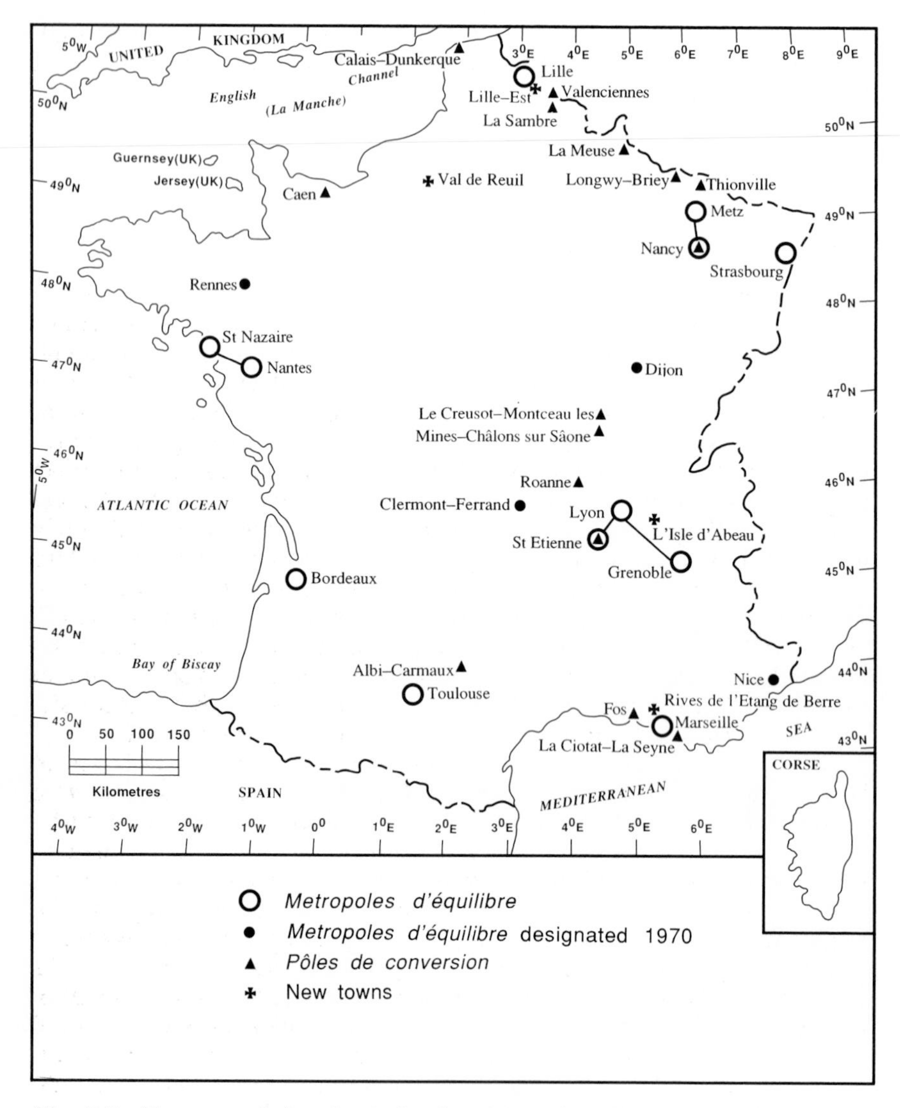

Fig. 5.5 Elements of the planned urban hierarchy, France, 1982. (Data from Laborie *et al.*, 1985)

new working classes. Newer housing occurs at increasing distances from the city centre, with post-war social housing on the periphery and new individual housing in the more attractive areas. Although this model does not have the elegance of the traditional models of Burgess and Hoyt and is inductive rather than deductive, it does reveal some enduring features of French city structure. These include at a general level the influential legacy of history, the incompatibility of certain land uses and the dominance of high-status groups in determining the social structure and built form of the city. More specifically, the model pinpoints the complexity of urban structure brought about by the continuance of mixed land uses and high densities in west European cities and the characteristic preference of Europeans for city living.

Planning the urban system

The large cities: métropoles d'équilibre

In the Fifth National Plan (1966–70), the strategy of spatial planning through the urban system was introduced. Eight counter-magnet cities, *métropoles d'équilibre,* were selected to be generators of regional development. They were to receive activities decentralized from Paris and would function as high-order self-sufficient centres. The designation of the eight cities was influenced by Perroux's (1970) theory of growth poles, whereby particular economic functions were seen to have a spread effect into the wider economic system. The cities selected were the largest centres of tertiary activity at that time. The eight centres were not all single cities but sometimes consisted of groups of adjacent and complementary towns (Fig. 5.5). In 1970, four other towns were assimilated into the programme: Rennes, Clermont-Ferrand, Dijon and Nice (Laborie *et al.*, 1985). At the time of designation, it was expected that urban populations would double by the year 2000, and that the large cities would maintain their attraction for migrants. Subsequently, of course, these extrapolations of the trends of the 1960s have proved to be over-optimistic.

The initial stages of the scheme necessitated the public acquisition of land for future development. In the 1970s, 600 000 hectares of land were acquired and detailed planning strategies, the *Schémas directeurs d'aménagement et d'urbanisme* (SDAUs), were published. In many cases, the planning of these large urban centres necessitated the formation of planning groups from a number of *communes* within the metropolitan region. The large-scale renewal required was facilitated by changes in urban planning procedures (see pages 161; 217). The significant results of the *métropoles d'équilibre* policy include the renewal of city centres, the establishment of new urban transport networks and the implantation of new industrial activities, industrial zones, and more recently, new tertiary employment. Furthermore, three new towns have been designated in association with three of the centres, Lille, Lyon and Marseille (Fig. 5.5).

Large-scale clearance and redevelopment has been most significant in Lyon, where the Part-Dieu district consists of a totally new and dramatically designed retailing and office complex (Plate 5.1). The multi-level shopping centre, which opened in 1975, is adjacent to a planned office centre, containing the offices of the city and regional administration, of public bodies and of private firms, including those of multinationals (Bateman, 1988). Similar renewal has occurred in central Bordeaux in the Mériadeck district. This was an area of marshland drained in the nineteenth century to provide working-class housing. The dwellings were sub-standard when they were built and by the mid-twentieth century formed a grim reservoir of over 2000 old and decaying dwellings, and

Plate 5.1 The Part-Dieu in Lyon: part of the new commercial centre. (Photo: S J Gale)

more than 500 small workshops and commercial activities, including scrap metal dealers, rag and bone merchants and brothels. The district was completely cleared in 1963 and has been replaced by an air-conditioned shopping centre and extensive new offices including the city and *département* administrative headquarters. Other similar major renewal schemes have been undertaken at St Sauveur in Lille, in the Ste Barbe district of Marseille and the Ile Beaulieu in Nantes.

New transport networks have overcome some of the problems of intra-urban commuting, particularly the new *métro* systems of Marseille, Lyon and Lille. The Lyon *métro* links the Part-Dieu to the old city centre and provides fast frequent access to both from a large number of suburbs; the Part-Dieu is also connected to the rest of France by a new railway station served by the high-speed trains, *trains à grande vitesse* (TGV). The Lille *métro* links the new developments at the university and the new town at Lille-Est by a system which is both efficient and architecturally striking. Lyon has also benefited from the development of the new airport at Satolas to the east of the city, adjacent to the new town of L'Isle d'Abeau.

New industries have also been stimulated, both by decentralization from Paris (see pages 123–5) and also by the designation of new industrial zones, as at Plaine de l'Ain in Lyon or in the port-industrial zone, *zone industrialo-portuaire*, at Fos near Marseille. While the decentralization of high-technology industries was successful in cities such as Toulouse and Grenoble, the planning for growth in oil refining, petro-chemicals and steel in centres such as Fos was unfortunately timed during recession. Subsequently, the emphasis has shifted away from manufacturing industry to the tertiary sector and the designated cities all receive grants for the development of tertiary employment. The most significant development is at Lyon where there has been marked growth in financial and producer services; it has the most important of the regional stock exchanges, dealing with over 70 per cent of provincial transactions (Labasse, 1985: 178–9).

The projects started under the *métropoles d'équilibre* policy have continued into the 1990s, although the policy itself was effectively replaced after 1974 (Laborie *et al.*, 1985). Pumain and St Julien (1984), in an analysis of French urban change since the Second World War, considered that the *métropoles d'équilibre* have become more functionally similar over time. This is partly a result of industrial decentralization from Paris to cities which had formerly not been industrial, and partly a result of increased tertiary activities in those that were industrial, especially Lille and Lyon.

The impact of designation and large-scale renewal has been profound. The redevelopment of the centres of major cities altered much of the traditional mixed land use of these areas and many of the long-standing small industries and inhabitants were displaced. However, the *métropoles d'équilibre* policy was increasingly seen to be over-ambitious and inappropriate and the proposed tertiary redevelopment in Marseille was reduced in scope. Furthermore, the emphasis on the major cities left relatively little financial assistance available for the smaller towns, which were the parts of the urban hierarchy growing in population. From 1974, there was a shift in government policy towards smaller-scale projects more sensitive to local needs: these projects had the added benefits of being dispersed through the regions, more highly visible to the electorate and relatively cheap.

Planning for smaller towns

The strategy for the medium-sized towns, the *villes moyennes*, was contained in the Sixth National Plan (1971–5). These towns were perceived to be capable of rapid growth but also of maintaining a sense of community; the overriding concern of the policy was for a human scale of development which would enhance quality of life and allow the active conservation of the urban heritage (House, 1978: 376–8). The policy was designed to prevent both inappropriate development, such as *grands ensembles* at the edges of small towns, and the insensitive urban renewal of their historic centres. It was expected that such policy would enhance the attractiveness of these towns for residents and so would help retain population in the provinces. The policy was active from 1973 to 1979, when towns of this size were in any case benefiting from changed patterns of migration and urban living (see pages 51–5).

The initiative which enabled the *villes moyennes* to obtain national funding had to come from the participating towns themselves. Most of the 73 towns which established contracts with the government contained between 20 000 and 100 000 inhabitants. The contracts established a plan of urban improvements to be carried out over a three-year period. Most of the contracts were concerned with rehabilitation and beautification projects; pedestrianization of central areas featured highly on the list of urban improvements, as did the provision of cultural facilities, tourist attractions and a limited amount of housing upgrading. A highly significant feature of the policy, foreshadowing later decentralization of power, was the level of local input to the schemes, in terms of both design and finance.

The policy was extended on a regional basis in 1975 to the towns at the edge of the Paris Basin which encircled Paris and hence were dubbed 'crown towns'. These towns, such as Orléans and Reims, formed the upper end of the urban hierarchy within their regions, but were relatively unindustrialized, isolated and lacking in facilities. The policy has particularly encouraged new and innovative uses of old buildings, the modernization of infrastructure and the provision of local and community facilities. It failed, however, to enhance the range of employment opportunities in any significant way and the policy came under criticism before the 1977 election. Nonetheless, a number of the 'crown towns' benefited from their designation; for example, Reims experienced reasonable industrial growth in the early 1970s as the appointed centre of the 'north Champagne support zone' and has been undergoing renewed development in the late 1980s as part of a growth sector incorporating Châlons and Troyes (Bazin, 1990).

From 1975, the smallest towns of the urban hierarchy became eligible for government assistance. Small towns in the population range 5000 to 20 000 were able to apply for three-year contractual arrangements with the government, similar to those in operation for the *villes moyennes*. These contracts were designed to support economic development, reinforce local services and improve quality of life; their scope ranged from the provision of open spaces to the protection of historic sites. The contracts have proved popular, with over 300 small towns, mainly in the west of France, entering into them (Laborie *et al.*, 1985). It is arguable that many of the small towns most in need of redevelopment have lacked the initiative or resources to apply for this assistance. House (1978) considered the policy to have a number of internal contradictions which were difficult to balance, particularly between growth and economic development on the one hand, and the conservation of heritage and the preservation of quality of life on the other.

The planning of the whole urban system therefore reached a peak in the mid-1970s. This elegant and technocratic policy was a product of a centralized system which was undergoing rapid economic growth. It has subsequently been profoundly shaken by a period of demographic and industrial recession and by political decentralization. Nonetheless, the concerted effort to manage the urban network was seen by House (1978) as an outstanding achievement; it left many concrete improvements in towns of all sizes. Furthermore, it left a series of planning instruments which, in various forms, have continued to be utilized in the planning of the individual towns which make up the urban system.

Urban restructuring: pôles de conversion

France in the 1980s felt the belated impact of the oil crisis and economic recession. The impact of industrial restructuring (see pages 114–16) was most marked in single-industry towns. In 1984, 14 towns or groups of towns were designated as foci for restructuring, *pôles de conversion*, in order to address deep-seated problems of declining employment and damaged environments (Fig. 5.5). Most of these towns were located in the old industrial areas of north and east France, but it is notable that they also included the former growth centres of Fos and the more industrial parts of the *métropoles d'équilibre*, such as Nancy and St Etienne. The aid available is for new industrial enterprises and some retraining schemes. Some assistance is also given to *communes* which have suffered financially from the loss of local industrial taxes, for example as a result of the closure of power stations (Bruyelle, 1987).

The French new towns

The French new towns were conceived in the 1960s as a solution to the pressing problem of rapid urban growth. The first new towns were designated in the Paris region as a means of absorbing population growth and relieving some of the pressures felt in Paris itself. Eight Parisian new towns were proposed in the Paris SDAU of 1965; it was predicted that each of them could house between 300 000 and a million inhabitants by the year 2000. A slowing of population growth after 1965 resulted in a sharp downward revision of population estimates; the number of proposed new towns was reduced from eight to five and their target populations lowered to between 150 000 and 300 000 (see pages 197–200; Fig. 6.7). The new towns were seen as key elements in solving the problems caused by population growth in Greater Paris, particularly traffic congestion, lack of open space, land-price inflation and poor housing. They were also designed to provide employment and service foci for the haphazardly-developed suburbs.

In the 1970s, several provincial new towns were designated. Three of these were closely associated with *métropoles d'équilibre*, and were designed to absorb some of their expected growth, while avoiding problems of further congestion in the large centres. These towns were Rives de l'Etang de Berre (usually known as Etang de Berre), adjacent to Marseille-Fos; L'Isle d'Abeau near Lyon (see pages 153–4) and Lille-Est (Villeneuve-d'Ascq) on the outskirts of Lille (Fig. 5.5). A smaller new town of Val de Reuil (formerly Le Vaudreuil) was designated as a growth centre close to the city of Rouen in the lower Seine Valley. Other new towns or new suburbs have also been created based on high-technology industry, such as Sophia-Antipolis near Nice, or as part of privately developed commercial centres and residential complexes, such as Parly 2 in the Paris suburbs. These are new towns in all but name, but are not designated as new towns under formal planning syndicates.

The administrative organization of the new towns proper is complex, with the planning and construction carried out by a mixture of private and public enterprise. This administrative structure was established in 1970 by the *Loi Boscher* (Allen, 1984). Generally, each of the *communes* in the designated area continues to operate as a local government body and receives residential taxes, but they may also form a syndicate of *communes* for planning purposes. Most of the planning and infrastructure provision is undertaken by the technical arm of each syndicate, the *établissement public d'aménagement* (EPA). In Marne-la-Vallée, there is one syndicate of six *communes* known as Val Maubuée, one large separate *commune*, Noisy-le-Grand, and one probable future syndicate: there is also an EPA for land pre-emption, management and the provision of infrastructure, but the boundaries in which the *communes*, syndicates and EPAs operate are not necessarily coincident (see also page 197). Lille-Est had an unusually simple administrative structure, which was totally controlled by a single *commune*, Villeneuve d'Ascq. All the EPAs were established between 1969 and 1973 and their activities, general policies and finances are co-ordinated by a national body.

The construction of new towns required land, which was acquired either by government or by the development agencies of the new towns. New planning legislation permitted these bodies to purchase land for future 'deferred' development. By 1985, the Parisian agencies had acquired 22 000 hectares of land and had kept the price stable: the policy of land acquisition has therefore been considered successful (Flandrin, 1987). One aspect which was initially less successful was the sale of land to private developers under a leasehold system; this had to be changed to a freehold system because of investor suspicion. The interest shown by private developers in the sites has varied, affected by location and by economic circumstances.

The provincial new towns

The provincial new towns were all associated with large cities and were intended to channel urban growth. The smallest was Val de Reuil, with a target population at the end of the century of around 100 000, while the largest was Rives de l'Etang de Berre, which had an enormous population target of 750 000. Lille-Est and Val de Reuil, the smaller towns, are of traditional design and have a high proportion of collective housing (70 per cent and 63 per cent respectively). The other towns, L'Isle d'Abeau and Rives de l'Etang de Berre, are larger and more diffuse in layout; half their housing is individual housing rather than collective units. The rate of growth of the provincial new towns is similar to that of the Parisian ones (see pages 197–200), although the growth of L'Isle d'Abeau has been slower than average.

The most ambitious new town scheme was that of Rives de l'Etang de Berre, which was designed as a growth pole for marine industrial development adjacent to Fos (see pages 154; 157). It was expected that the new town would absorb half the region's anticipated population growth and that it would provide a complementary residential centre to the industry at Fos and the services at Marseille. The town has grown more quickly than the other new towns, but its success in attracting industrial and office jobs has been severely limited. Despite its actual and projected growth, Rives de l'Etang de Berre did not benefit from new transport connections, and it has suffered severely from the economic recession which affected Fos and resulted in its designation as a *pôle de conversion*.

Despite their slower than anticipated growth, the provincial new towns have been successful as growth foci, as centres of technological innovation (an example of which is the automated *métro* at Lille-Est) and in their urban design. The design of L'Isle d'Abeau, for example, incorporates green open space into housing areas and makes sensitive use of topographic features and colour to provide an attractive residential environment very different from the older industrial suburbs of Lyon.

The planning of cities

Early urban planning

Planning of cities before the Second World War was limited in scope, yet it affected the structure and morphology of many of the larger French cities. The impact was seen in two main ways: in street layout and building regulations, and in the construction of wide boulevards during the nineteenth century. At the same time, a general lack of planning also left its mark on many cities, particularly those with large industrial suburbs which often grew up outside the existing city limits. These were characterized by a chaotic mixture of land uses and by housing which was often sub-standard.

The planning of French cities has been heavily influenced by events in Paris. In the pre-Revolutionary era, planning proceeded by royal edict which affected Paris almost exclusively. A series of regulations on building lines, street widths and building heights in the period 1607 to 1783 formed 'the most comprehensive building code in Europe' (Sutcliffe, 1981: 128). This code, revised in 1807 after the Revolution, was adopted in the late nineteenth century by over 200 towns. Further regulations on matters such as ventilation, sewerage and water supplies were introduced in response to increasing public concern over public health and contagious disease, especially tuberculosis. However, the planning of French cities lagged behind that of other countries of western Europe, as there were no physical planning mechanisms in place to provide industrial zones, parks, suburban transport or slum clearance. Even where city councils were informed and enthusiastic, they had little scope for physical planning even within the city boundaries and none whatsoever within suburban areas. For example, in the city of Lyon, until the city boundaries were extended, chaotic development occurred in the neighbouring suburban *commune* of Villeurbanne. The council could bring in ordinances about public health, but it could not extend the city boundaries, could not appropriate land and could not provide housing. As a consequence, most cities during the nineteenth century continued to be laid out haphazardly by individual landowners and developers.

Many French cities did, however, undergo a period of modernization in the mid-nineteenth century, emulating the Haussmannization of Paris (see page 176). The impact of this was twofold: slum areas were removed from city centres and wide boulevards were provided for wheeled traffic. The patterns of boulevards imitated those of Paris, with crescents and radial streets and some newly-laid out parks (often accompanied by provision of piped water and sewerage). Numerous large cities adopted this style of planning: Lyon, Marseille, Toulon, Toulouse, Brest, Bordeaux and Rouen among others bear the imprint of Haussmannization. One of the most admired cities was Le Havre, where Jules Siegfried established the first public hygiene service in France. He also instigated the municipal construction of new streets and boulevards, undertook the provision of a piped sewerage system, and devised slum clearance and building regulations for new suburbs.

The grand Haussmannization schemes were completed by the end of the 1860s and there was little further progress towards planning until the early twentieth century. This lack of progress may be attributed to the slow rate of urban growth, the under-representation of urban interests in government, the relative isolation of the French from planning trends abroad and a sense of satisfaction with the progress of the 1850s and 1860s. As a consequence, although a number of laws appeared on the statute books, there was much less interest in urban planning than in Britain or Germany (Sutcliffe, 1981). France did not develop the same traditions of philanthropic housing, municipal socialism or concerns over public health, but neither did it experience problems of the same magnitude as those in the industrial cities of Britain and Germany.

Post-war urban planning

Modern urban planning, *urbanisme*, effectively dates from 1943, when an administrative, planning and building code was established. The post-war priority was the construction of urban housing, a need brought about by population and urban growth and exacerbated by an unimproved and ageing housing stock (Chaline, 1984: 328). The creation of housing units, mostly *grands ensembles* at the urban periphery, required a mechanism for the expropriation of land. Zones for future urbanization were designated and acquired, invoking a new procedure which has been widely used in a variety of similar forms. The zones are known by a series of acronyms: among others, there are *zones à urbaniser à priorité* (ZUPs), priority urbanization zones; *zones d'aménagement concertée* (ZACs), joint development zones; and *zones d'aménagement deferrée* (ZADs), deferred development zones. The differences between the type of zone used depends largely on the required speed of development; for example, the ZAD has been widely used in the new towns where long-term planning is required. There are also administrative differences, in particular, the ZAC is normally controlled by a mixture of public and private enterprise (see House, 1978 for a fuller discussion).

The new planning mechanisms of the post-war period were used until the late 1960s mainly for the construction of large housing projects and for large-scale clearance and renewal in the major cities: Chaline (1984: 329) has referred to the period 1945–65 as the 'age of renewal'. However, both the *grands ensembles* and the wholesale clearance of degraded but historic central areas (see pages 162–3) aroused considerable opposition. From the late 1960s, French urban planning has been more focused on small-scale projects, on conservation and rehabilitation and on local and community participation. The need for rehabilitation of obsolescent urban areas was expressed in the Seventh National Plan; this change in focus and attitude in city planning parallels the change in approach to the planning of the urban hierarchy (pages 155–6).

Land-use planning

Urban land-use planning is based on local land-use plans, drawn up by *communes* or groups of *communes*, possibly with assistance from higher levels of government. There are two sorts of land-use plan, the *plan d'occupation des sols* (POS), which is a local land-use plan, and a *schéma directeur* (SD), a long-term planning document for major urban areas. Each *commune* with a population of 10 000 or more, or groups of *communes*, may draw up a POS, which consists essentially of a zoning map for the area, together with an explanatory report and the regulations governing future changes to land-use. The POS has to conform to national legislation, for example in relation to national highways, national parks or coastal development (see pages 31–4). In theory, *communes* are not required to produce a POS, but in practice, as development decisions would then pass to the *département*, it is in the local interest for a POS to be produced (Braid Wilson, 1983). Indeed, Braid Wilson (1983: 164) has suggested that local interests are all too frequently to the fore, with local politicians protecting their own property interests, and elected mayors 'unashamedly' pursuing projects such as 'swimming pools and stadiums which they imagine will be popular with the electorate'. The POS is open to public scrutiny and objection, but public participation has been relatively limited. Once the POS has been agreed with the *préfet* of the *département*, it forms a legally binding document, which may be inflexible and difficult to adapt to changing circumstances (Braid Wilson, 1983).

In 1982, city councils were required to produce not only the local POSs but also the SD, a strategic document outlining phases of development. The SD is an updated version of the SDAU, over 160 of which were produced between 1967 and 1980 (see page 153). The SD has a clear legal status, which the SDAU had not; for example, in Grenoble, the SDAU was accepted by a majority of the 113 *communes* in the agglomeration, but could only function given the political will to work together. The SDs are produced by representatives of the *communes* concerned, with assistance from the *département* and the *région*. The decentralization legislation, brought in by the *Loi Deferre* of 1982, encourages such collaboration even at the *commune* level. Small *communes* often lack the personnel and the expertise to engage in all aspects of urban planning and so are increasingly obliged to use the services of higher level authorities. The decentralization of planning powers may in theory enhance local control over local issues, but the increased emphasis on joint planning may in practice bring about greater centralization of control and conflicts of interests.

Plate 5.2 Preservation of half-timbered façades in the *secteur sauvegardé* of Rouen. (Photo: S J Gale)

Urban conservation

France, despite its tardy adoption of general urban planning, has been in the forefront of urban heritage planning. As early as 1913, a procedure for the listing and protection of historic buildings was instituted. However, the basis of contemporary procedures for urban conservation was instituted by the *Loi Malraux* of 1962, which introduced conservation zones, *secteurs sauvegardés* (Kain, 1982). The aim of the legislation was to encourage positive conservation, so that historic sites have a functioning economic role rather than merely being preserved as relics of the past. The law was introduced partly as a reaction against the massive urban renewal that was going on at that time and has proved to be a successful model which has been emulated by other European countries (Kain, 1982).

Over sixty *secteurs sauvegardés* were established between 1962 and 1976, in a great variety of sites, including the historic centres of a number of cities, such as the royal squares of Nancy, the bourgeois merchants' houses around the central square of Lille and the medieval centres of Bordeaux, Rouen (Plate 5.2) and Chartres. In many cases, such as Bordeaux and Rouen, the designation has been accompanied by the pedestrianization of streets, the replacement of workshops by boutiques and restaurants catering for visitors rather than local people, and the displacement of working class populations by incoming gentrifiers. Nevertheless there are still many valuable historic sites which have not been designated as *secteurs sauvegardés*. Some of the notable omissions include the Roman cities of Nîmes and Orange, and the perimeter of many historic sites such as the streets surrounding the Grand Place at Lille (Kain, 1982).

Furthermore, the sites designated date predominantly from the *Ancien Régime*, while buildings from periods such as the nineteenth century are under-represented, largely because many are still considered to be eyesores.

While the *secteurs sauvegardés* have been used for the restoration of major sites, the rehabilitation of more modest buildings and dwellings has been undertaken since 1977 by the designation of housing improvement projects, *opérations programmées d'amélioration de l'habitat* (OPAHs). OPAHs are schemes put together by local initiative; owners, tenants and the local authority may join together to apply for funds to upgrade housing amenities and to provide some open space and public services in areas of need. They are usually relatively small-scale projects affecting fewer than 300 houses; the improvements have to be completed within three years. The flexibility and local control of the scheme has made the OPAH a successful formula; 379 OPAHs were in operation by 1980. These were well distributed throughout the regions and the urban hierarchy and were very well represented in the west and southwest, a distribution attributed by Kain (1982: 414) to the elevated regional consciousness of those areas.

The OPAHs have been generally welcomed for their responsiveness to local needs: in particular, they have been successful in retaining the existing population (Heugas-Darraspen, 1985). Nevertheless, Chaline (1984) cited a number of objections to their operation. One has been that the OPAHs have been monopolized by private developers, as has been the case in Bordeaux (Clout, 1984), rather than encouraging participation by individual tenants and owner-occupiers. Another complaint is that the OPAHs can deal with only a fraction of the upgrading needed, and that the schemes can tackle less than five per cent of the urban rehabilitation which takes place each year. On the other hand, the schemes have been considered significant enough to be condemned as electoral bribes and vote-catchers. On balance, they appear to be relatively small-scale and pragmatic solutions to some urban problems, which work well in particular areas depending on the circumstances and the individuals concerned.

Béziers: an example of urban planning

The town of Béziers is the second largest urban centre in the *département* of Hérault in Languedoc with a population in 1990 of 76 300. It has never undergone major urban renewal, although many of the urban planning mechanisms described above have been employed in the town. The town council has used the POS, the OPAH and its own municipal powers in an imaginative and far-reaching programme of urban rehabilitation. The old heart of the city, built within ramparts above the River Orb and crowned by the fortified medieval cathedral of St Nazaire (Fig. 5.6), had become a problem area by the mid-1970s. Successful viticulture in the hinterland of the town had caused pressures for urban expansion during the nineteenth century. As a result, the city ramparts were demolished and new *quartiers* were constructed to the east away from the river, effectively separated from the old city by wide new boulevards built in 1894. Throughout the twentieth century, the old centre, the *quartiers* of St Jacques, St Nazaire

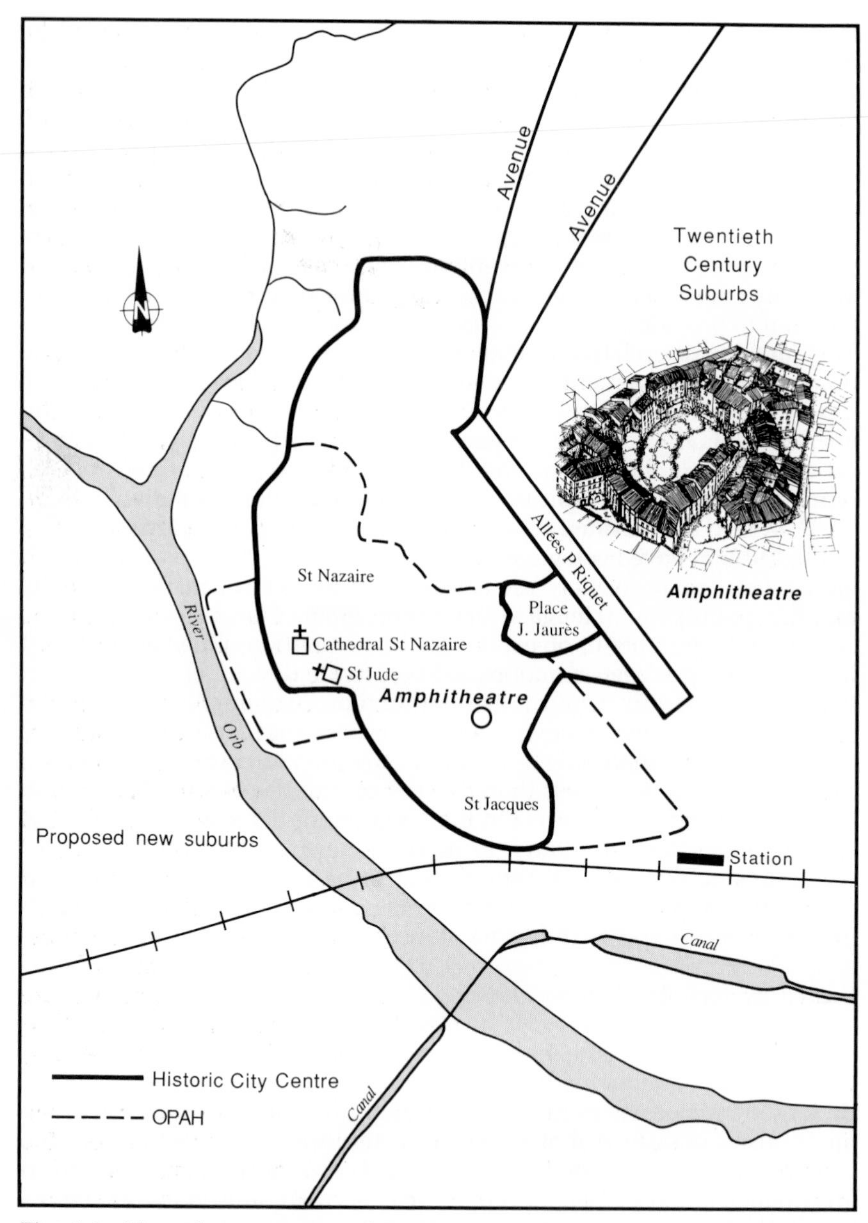

Fig. 5.6 The urban structure of Béziers, Hérault: planning for conservation. (After Merlin, 1986)

and St Jude, became gradually more dilapidated and insalubrious. The centre suffered from selective depopulation as families moved to the suburbs and were replaced by a marginal population composed mainly of foreign workers, the elderly and other single-person households. At the same time, traditional industry and commerce in the old centre declined.

In 1975–6, the town commenced a strategy for the rehabilitation of the old centre. The first step was the signing of a *contrat ville moyenne* (see pages 155–6) which delimited the historic centre and established pedestrian zones within its narrow streets. In 1980, the POS was revised to shift the city's centre of gravity back to its original site, by proposing new western suburbs beyond the Orb to balance its twentieth century eastern growth (Fig. 5.6). Both the *contrat ville moyenne* and the POS served to refocus attention on the historic centre and to facilitate access, commerce and the preservation of monuments and churches within the area, but did little to improve the poor housing conditions which prevailed (Merlin, 1986).

The improvement of housing conditions was tackled by the designation of two OPAHs. The first, in the *quartier* of St Nazaire, contained 240 dwellings which were renovated. The original population was retained by the provision of rental housing and HLMs. In 1984, a further OPAH was established in the most degraded *quartier* of St Jacques. In this area, houses of two or three storeys existed, tightly packed on narrow blocks, often hemmed in on all sides by other housing and makeshift lean-to sheds and workshops. The improvement of this area has included the provision of basic amenities for houses, the clearance of accretionary structures and the installation of skylights and terraces to improve light and ventilation to the houses; however, the housing itself has remained modest, which has helped maintain the existing population in place.

Within the *quartier* of St Jacques, at the *îlot des anciennes arènes*, lie the ruins of a Roman amphitheatre. Only the western and southern sides of the feature still remain, the rest having been used as a source of building materials over the centuries. The site of the amphitheatre was densely built over (Fig. 5.6). Excavations have revealed some archaeological remains, but others are known to exist several metres below the surface (Chappal and Martinez, 1986). Since 1975, the town council has purchased a large number of dwellings in the *îlot*, either as they fell vacant or by pre-emption. The purpose of this is twofold: first, to enable further archaeological excavation to occur and secondly, as a means of facilitating housing improvement. The rehabilitation of this area has required more sweeping powers than elsewhere in the city centre in order to allow integrated development of the site to take place (Commune de Béziers, 1987).

Since 1985, the city council has instituted a wide-ranging plan for urban development in Béziers. The plan contains a range of projects from the construction of a new bridge over the Orb and the provision of underground car parks in the city centre to the establishment of museums, archaeological site interpretation and tourist facilities. As such, the urban planning has gone beyond issues of heritage conservation and housing improvement to a consideration of fundamental issues of transport, accessibility and economic development. The aim of the municipal project is to help restore business confidence in the city and to attract private investment to improve job opportunities and services in the centre (Commune de Béziers, 1987).

Rehabilitation of the grands ensembles

The *grands ensembles*, constructed mainly in the 1950s and 1960s at the edges of many of the larger cities of France, provided accommodation for over 10 per cent of the French population. The housing shortage stimulated construction, much of it of public housing for low and middle income groups in HLM accommodation, *habitations à loyer modéré*. The state support for public housing at this time contrasts markedly with the emphasis on private renting before the Second World War and owner-occupation after 1965. It was, in essence, an emergency response to population and urban growth.

Much of this peripheral public housing was of low-cost construction and of prefabricated design, often in monumental tower blocks such as those of the Haut du Lièvre estate outside Nancy, where two blocks 15 storeys high and 400 metres in length each house 10 000 people. Such estates were usually badly built and badly serviced and as housing supply has improved, many families have moved out to newer individual housing. By the 1980s, the *grands ensembles* were suffering from a vicious cycle of out-migration, physical deterioration, an accumulation of disadvantaged groups, vandalism and social problems. Ironically, the physical deterioration of the buildings became critical at the same time that many of the peripheral estates began to acquire a normal level of urban services (Tuppen and Mingret, 1986).

From 1977 to 1983, a rehabilitation programme for these estates was established known as *habitat, vie sociale*, the housing and social life programme. Fifty separate reconstruction projects were undertaken in the worst estates. In the Ninth National Plan, the problem of the *grands ensembles* was viewed as a priority, given added impetus by urban social movements for reform in Paris (Castells, 1983) and by riots in Vénissieux, Lyon in 1981 and 1983. In 1983, the estates project, *projet des quartiers*, initiated 21 physical improvement schemes, including one at Vénissieux. The innovative aspect of the programme is the belated introduction of local social, economic and political activity organized by the local authority, ranging from educational programmes to social services and training courses for unemployed teenagers, as well as greater attempts to encourage public participation in local government.

The problem of suburban disaffection worsened during the economic recession of the 1980s. Riots occurred in October 1990 in Vaulx-en-Velin, a suburb of Lyon, following the death of a motorcyclist in a collision with a police car. In the riots, a shopping centre was burned down, riot police were called in and 30 people were injured. Similar riots occurred in Narbonne in June 1991. The incidents were underlain by high levels of youth unemployment, long-standing racial tension and allegations of persistent police harassment of young people. In June 1991, a further financial package was announced to provide infrastructure and training programmes in these estates (see pages 187–8). In some *grands ensembles*, however, the only solution has been demolition. Many others will require enormous subsidies to make them viable, especially as vacancy rates are high and as much of the public housing has been let to low-income groups

(Flockton and Kofman, 1989). The *grands ensembles* now contain a high proportion of immigrant groups, single parents, the unemployed and others who cannot afford private housing. These estates have become a residual housing sector with attendant socio-economic problems, requiring action at a structural level as well as physical rehabilitation.

Urban classification

A study of the major functions of French cities, based on a cluster analysis of employment in 1968 and 1975 was undertaken by Pumain and St Julien in 1978 and up-dated in 1984. Their studies noted the great stability of city functions after 1954 and possibly even from the late nineteenth century (Pumain and St Julien, 1978; 1984). Two main types of cities were identified: industrial cities in which over half the employed population worked in industry, and tertiary cities in which less than 40 per cent of the employed population worked in industry. The most specialized cities were the textile, mining and metallurgical centres of the Nord, Lorraine and the eastern Massif Central (Fig. 5.7). The tertiary cities also contained some specialized sub-groups, notably the defence centres such as Brest and Toulon; the transport centres, including the Channel ports, such as Le Havre and Dieppe; and the resort towns such as Nice and Cannes. The largest class of town, where administration was over-represented, included 30 cities of which 29 were *préfectures*.

In a principal components analysis of socio-economic, demographic and welfare indicators for the largest 88 agglomerations, two major components of variation were extracted (Pumain and St Julien, 1984) (Fig. 5.8). The first component was defined as image, *image de marque*, which consisted of the major economic activity of the city related to wealth and recent demographic change. The explanation of this component is clarified by examining cities with extreme component scores. At one extreme were the growing, attractive, wealthy, service-oriented cities populated by the social elite, such as Cannes and Nice. At the other end of the spectrum were those cities characterized by ageing industrial structures, working class populations, low standards of living and demographic stagnation or out-migration. Examples of this type included Denain, Douai and Montceau-les-Mines.

The second major component was termed a modernity component, *axe de modernité*, which was not evident in studies of the 1950s, but reflected post-war employment change. The most rapidly modernizing cities showed new industrial growth and high wage levels and contained significant proportions of their workforces in professional and administrative occupations. These towns included Lyon, Annecy and Montbéliard. The most traditional cities were those with a high proportion of small retail businesses, low wage levels and relative stagnation, such as Agen, Périgeux and Perpignan.

All of the 88 cities could be categorized in relation to these two axes (Fig. 5.8), and seven groups of cities were distinguished (Pumain and St Julien, 1984). Very few cities were deemed to be both traditional and to have a

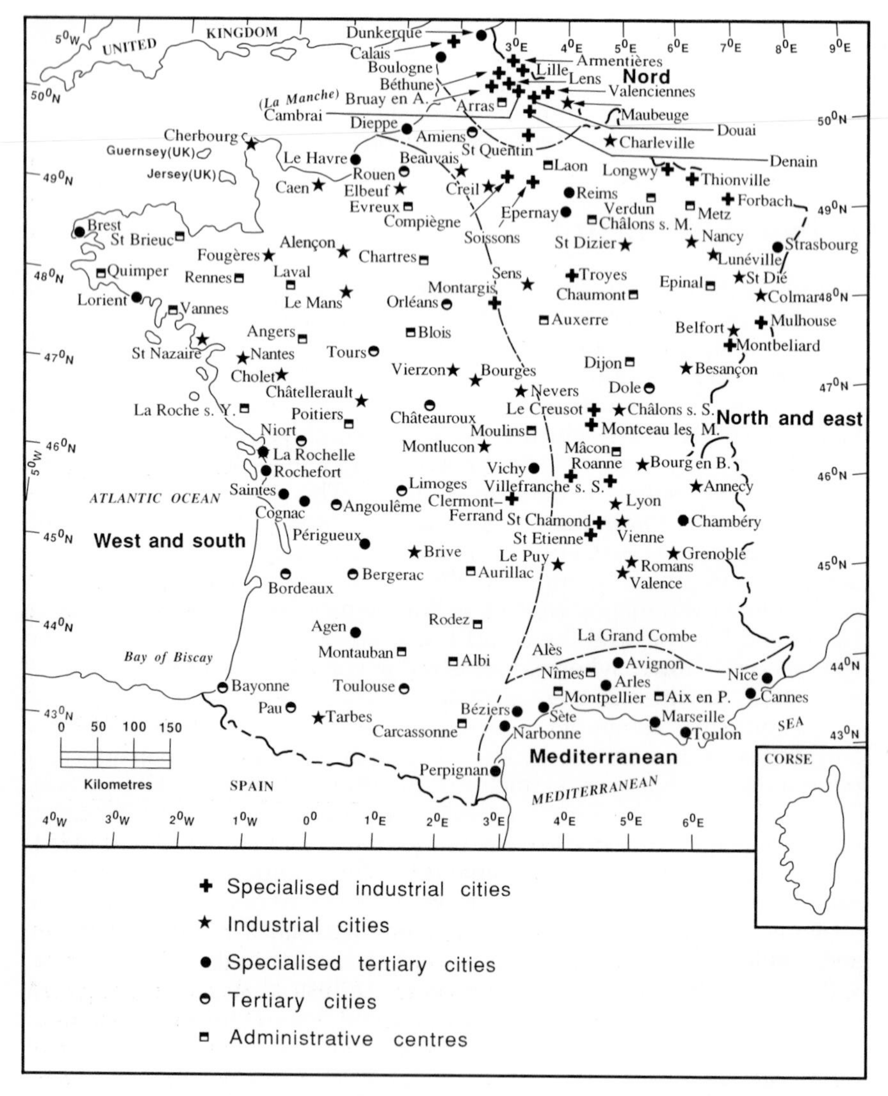

Fig. 5.7 Functional classification of cities, France, 1982. (After Pumain and St Julien, 1984)

poor image, although Calais lay firmly within this class (Fig. 5.8). The most successful group of cities rated highly on both components and were therefore modern with a favourable image; such cities included Strasbourg and Grenoble. The various groups of cities were regionally differentiated, with the more traditional cities located predominantly in the west and southwest and the more advantaged cities located in the Rhône-Alpes, the east of France generally and the Paris Basin. This regional distribution reflects the patterns and processes of industrial urban growth evident in the nineteenth century (see pages 110–11).

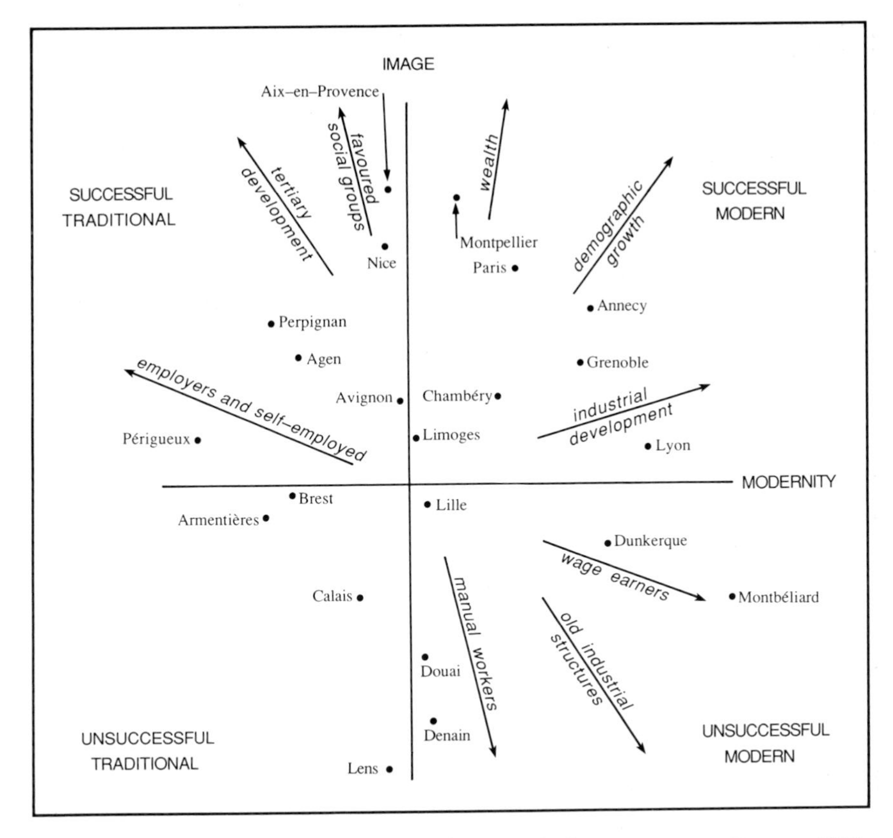

Fig. 5.8 Classification of major cities by principal components, France, 1982. (Source: Pumain and St Julien, 1984: 319)

The pattern of urban classification reflects the impact of history, of distance from Paris and of recent changes in the distribution of production and population. Contemporary processes of industrial restructuring and decentralization, the development of the service sector, and counterurbanization are serving to reduce the disparities between cities, whereas until the 1960s these were being accentuated (Pumain and St Julien, 1984). Most French cities, both individually and as parts of the urban system, bear the imprint of historic change. All French cities, whether originating in Roman times or after the Second World War, show the impact of twentieth-century urban processes and post-war urban planning.

CHAPTER 6

Paris

> The last thing a solid and virtuous citizen of central France desires to do in Paris is to Parisianize himself.
>
> Locke (1916: 220)

Paris dominates France. It is overwhelmingly influential in all aspects of the nation's affairs. The agglomeration houses 16 per cent of the nation's population, but contains 75 per cent of the headquarters of major companies and 66 per cent of postgraduate students. Many other indicators demonstrate the city's domination of political, economic and social activity. This stranglehold has resulted from a lengthy period of political centralization at the expense of other regions, which has established the capital city as the most desirable place in France in which to live and work.

Political control became centred in Paris throughout the nineteenth and twentieth centuries. In the years immediately following the French Revolution, the nation's administrative system of *départements* accountable to Paris was established. This inflated the significance of Paris at the expense of the other cities of France (see pages 150; Fig. 5.3). This administrative centralization still exists, but in a modified form since the 1980s (see pages 214–18). The construction of the national railway network between 1832 and 1855, a spider's web of lines strategically focused on Paris, reinforced the city's central role. The railways provided a mechanism for channelling migrants into the city from the provinces and allowed the widespread dispersal of manufactured goods, fashions, ideas and regulations from the capital. Political centralization brought in its wake a concentration of commerce and industry within Paris, fed by a huge in-migration of population from the provinces throughout the nineteenth century and the early part of the twentieth century.

The growth of Paris

The growth of the population of Paris occurred in response to waves of migration, especially in the 1860s, 1890s and 1930s, when people flooded in to the capital from the countryside (see pages 49–51). Permanent migration to the capital was stimulated by a number of factors in the nineteenth century; Weber (1977) cited the railways, the wider use of the French language and the impact of military service. However, Zeldin (1973) stressed that most migrations were 'migrations of poverty', away from the bleak conditions of the over-populated and under-productive rural areas rather than a result of a positive attraction to the growing capital.

Provincial migration to Paris resulted in continuous population growth in the city from the 1840s until the 1950s. In the second half of the nineteenth century, the built-up area began to spread beyond the constraining walls built in 1840–1. Paris at this time was the second largest city in Europe after London, a position it maintained until 1950 (De Vries, 1981). By the early 1950s, Paris contained 17 per cent of the population of France, a proportion which has subsequently remained stable or slightly diminished. In 1990, the City of Paris, that is, the densely built-up area within the ring road, the *boulevard périphérique*, contained 2 146 900 people. The agglomeration, which spreads well beyond the *boulevard périphérique*, covered an area 50 km by 70 km and contained a further 8 661 700 people. Since the 1950s, the urban region has undergone clear patterns of population change (Table 6.1).

Table 6.1 Population change in the City of Paris and the Parisian agglomeration, 1954–90

Year	City of Paris total population	Parisian agglomeration (excluding City of Paris) total population
1954	2 850 000	6 979 000
1962	2 790 000	7 370 000
1968	2 591 000	7 917 000
1975	2 299 800	8 424 092
1982	2 188 900	8 706 963
1990	2 146 900	8 661 676*
	City of Paris population change (% p.a.)	**Parisian agglomeration (excluding City of Paris) population change (% p.a.)**
1954–62	−0.3	+0.7
1962–68	−1.2	+1.2
1968–75	−1.6	+0.9
1975–82	−0.7	+0.5
1982–90	−0.2	−0.1

Source: INSEE, 1982; 1990.

*Data for the Parisian *Zone de Peuplement Industriel ou Urbain (ZPIU)*

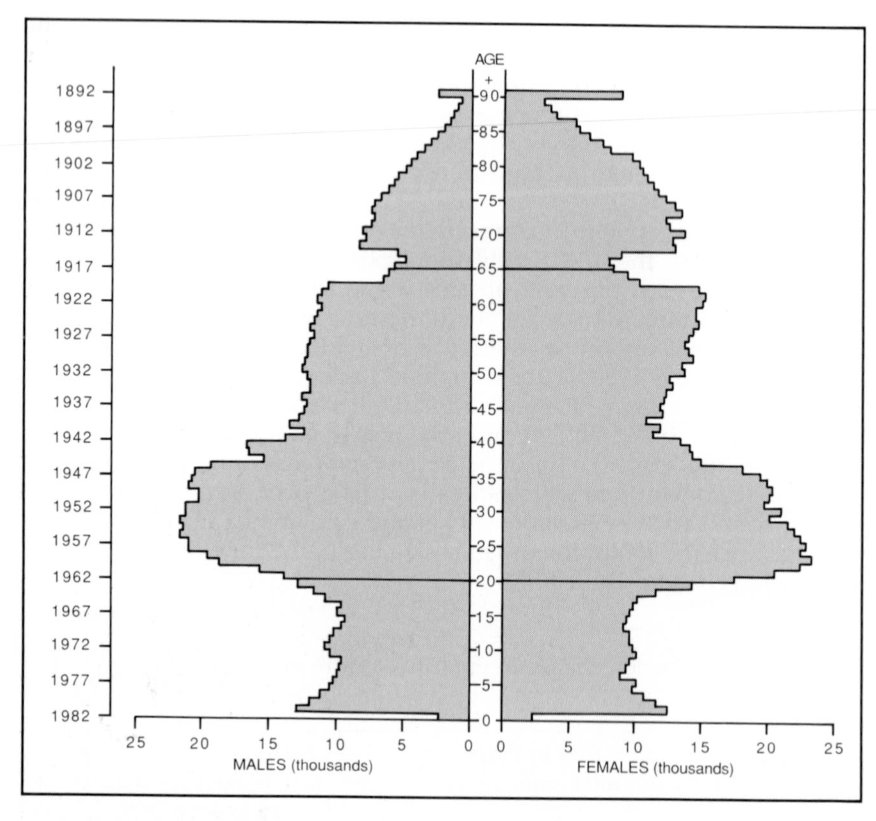

Fig. 6.1 Age-sex pyramid, City of Paris, 1982. (Source: INSEE, 1982)

The main characteristic of population movement has been decline in the city centre and growth in the rest of the agglomeration. This trend, evident throughout the period 1954–82, has been modified since 1982. The conurbation gained population continuously from 1954 to 1982, mainly from in-migration, but the rate of growth slowed after the late 1960s. Preliminary estimates from the 1990 census showed a minor population loss in the built-up area between 1982 and 1990 (INSEE, 1990). However, this apparent loss may conceal continued population deconcentration, as growth may still have taken place in the rural *communes* at the edge of the conurbation.

While the Parisian agglomeration has generally experienced population growth, the City of Paris has lost population in every intercensal period since 1954. The preliminary estimates of the 1990 census indicated only very minor loss of central city population in the period 1982–90 (INSEE, 1990). This relative stabilization may be related to the completion of a number of major redevelopment projects, and to the greater affordability of housing in a time of recession for those people in good jobs.

The population of Paris is distinctive in a number of ways from the population of the rest of France. First, the age-sex structure of the population is very imbalanced (compare Fig. 6.1 with Fig. 2.3). The population is mainly adult, with many more women than men in the older age groups. The imbalanced age groups result from differential migration for employment purposes. Secondly, the population of Paris is much more cosmopolitan than that of France as a whole (compare Fig. 6.2 with Fig. 2.6). The high proportion of foreign population results from the attraction of the city for labour migrants, diplomats and students. These groups live in very different parts of the city. The major student concentration occurs on the southern border of the city, the diplomats live in the expensive

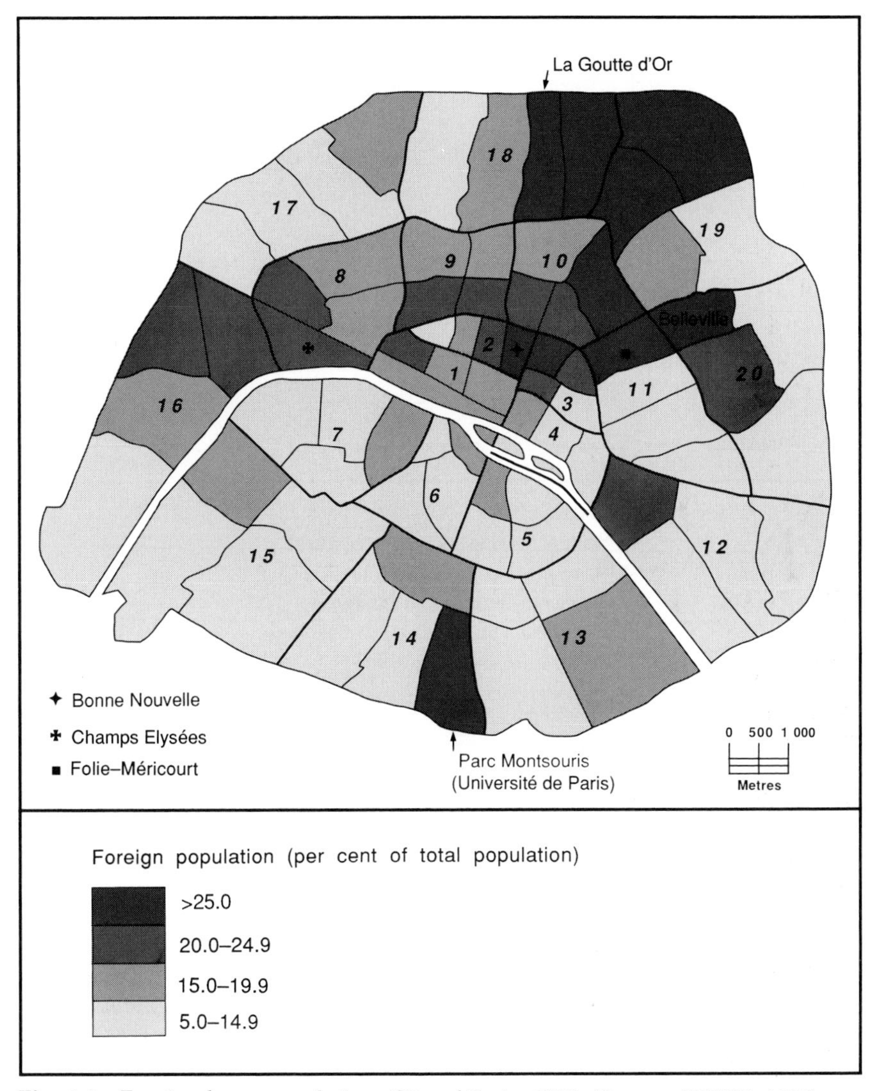

Fig. 6.2 Foreign-born population, City of Paris, 1982. (Source: INSEE, 1982)

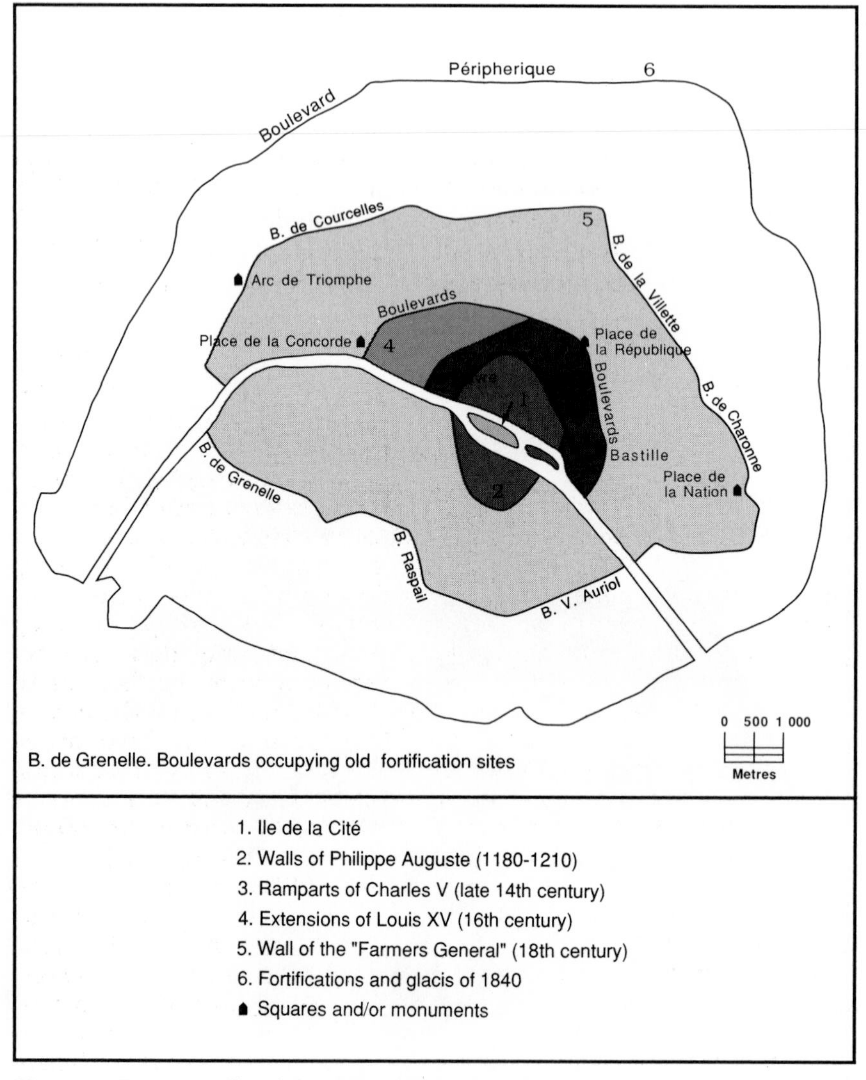

Fig. 6.3 Former walls of the City of Paris and their impact on urban structure. (After De Martonne, 1933: 7; White, unpublished)

west, while the quasi-ghetto of north Africans and other labour migrants is located in the Goutte d'Or in the *18^e^ arrondissement*. A recent concentration of migrants from Indo-China has developed in the *13^e^ arrondissement* (White *et al.*, 1987).

The building of Paris

> And when a girl walks around and reads all of the signs with all of the famous historical names it really makes you hold your breath . . . So I said to Dorothy, does it not really give you a thrill to realize that that is the historical spot where Mr Coty makes all the perfume? So then Dorothy said that she supposed Mr Coty came to Paris and he smelled Paris and he realized that something had to be done.
>
> Loos (1925: 78)

The site of the City of Paris, on twin islands in the Seine, the Ile de la Cité and the Ile St Louis, has been occupied since pre-Roman times. The topography of the site and even the number of islands has been substantially altered in the last two thousand years. Paris has grown outwards from the core; subsequent growth has occurred in a roughly concentric fashion. This growth was accommodated by a series of walls which were successively enlarged and replaced as the city grew (see also pages 148–9). Although the walls have long since been removed, they have had a significant impact on the structure of the city (Fig. 6.3). There are now only fragmentary traces left of the twelfth century walls of Philippe Auguste, but the lines of the fourteenth and sixteenth century ramparts are evident in the large boulevards which form a semi-circle from the Place de la Concorde to the Place de la Bastille; they also form the northern boundary of the *2ᵉ* and *3ᵉ arrondissements*, the internal administrative divisions of the City of Paris. Similarly, the lines of the walls of the late eighteenth century are now fossilized as the wide boulevards of the north of Paris, as *arrondissement* boundaries and as two of the *Métro* lines of the city. (*Arrondissement* boundaries are shown on the inset of the location map, Fig. 6.7.)

The most recent fortifications, constructed in 1840 and removed in 1919 when their obsolescence had been proved beyond doubt, have left the most profound impact on the physical structure of the city. This is a reflection of the width of the military zone and the recency of the demolition of the fortifications. After 1919, the *glacis* became available for development; this was the 400 metres of land left clear for a direct line of fire from the fortifications. This area of open space had no permanent buildings, but in 1919 was by no means unoccupied, being the location of squatter settlements, housing rag-pickers, *chiffoniers*, and homeless people. After clearance, the zone was used for a number of purposes which are visually distinctive. A major use is for public housing, typically older red-brick blocks of flats of five to eight storeys, solidly built but without lifts or bathrooms, and more recent tower blocks. There is also a mixture of land uses requiring the open space which was in short supply in the densely built-up area of the City of Paris itself; such uses include cemeteries, parks, football pitches and running tracks. The most significant land use, however, is the ring road, the *boulevard périphérique*, which occupies the outer edge of the zone and effectively as well as administratively demarcates the City of Paris from its suburbs.

Prestige space and the activities of Haussmann

The urban form of the City of Paris has been greatly influenced by a number of grandiose projects undertaken by its ruling authorities. A characteristic feature of the densely built-up area of central Paris is the presence of imposing boulevards, broad enough for multiple lanes of traffic, for pavements lined with stalls and for avenues of trees. The urban fabric is also punctuated by major squares, monuments and public buildings, often approached by wide avenues offering excellent views and providing a display case for prestigious architecture of national or international significance. The boulevards, squares and monuments together comprise areas of prestige space where the landscape has been designed to enhance the status of particular elements and to emphasize the power of the controlling authority.

The major boulevards are particularly associated with the street-widening activities of Baron Haussmann who was *préfet* of the *département* of Seine from 1859 to 1870. The effects of Haussmannization are particularly prominent in the western sector of the city, where boulevards radiate from the Arc de Triomphe-Etoile. These boulevards were designed as the basis for the building of new high-status residential areas. The eastern *quartiers* were also affected by Haussmannization but for different reasons. By the late nineteenth century, these districts, originally beyond the city walls, had become working class suburbs teeming with industrial workers housed in squalid and substandard accommodation. The construction of avenues was designed to serve multiple purposes; the worst of the insalubrious areas were to be demolished to allow air and light into those fetid buildings that were left; the work would provide employment for the labouring classes; and the radical insurgents of the *Commune* of the 1870s would be rehoused in areas safely removed from the city centre. If this strategy failed, the avenues, which were too wide to be successfully blockaded, provided a straight line of fire against rioters. The boulevards also linked the major barracks at the Place de la République and the Place de la Bastille for the rapid deployment of troops (Evenson, 1979). Undoubtedly, the wide boulevards improved the general appearance and status of the eastern *quartiers*, although in some places, such as the Porte St Denis, the demolition of existing housing has left very acute street angles and strange wedge-shaped corner buildings.

The major areas of prestige space, particularly monuments with vistas, are concentrated in the city centre. The Arc de Triomphe with its radiating boulevards was constructed on the orders of Napoleon I; the Opéra, the great parks and the Haussmann boulevards are the legacy of Napoleon III. Paris is structured around a triumphal axis which runs from southeast to northwest from the Royal Palace of the Louvre on the right bank of the Seine to the Arc de Triomphe-Etoile (Fig. 6.4). Along this axis, the Tuileries Gardens provide the required open space to form vistas both east to the Louvre and west to the Place de la Concorde. The Place de la Concorde is linked to the Arc de Triomphe-Etoile by the most famous of the Paris boulevards, the Champs Elysées. Outside the present city limits, the new

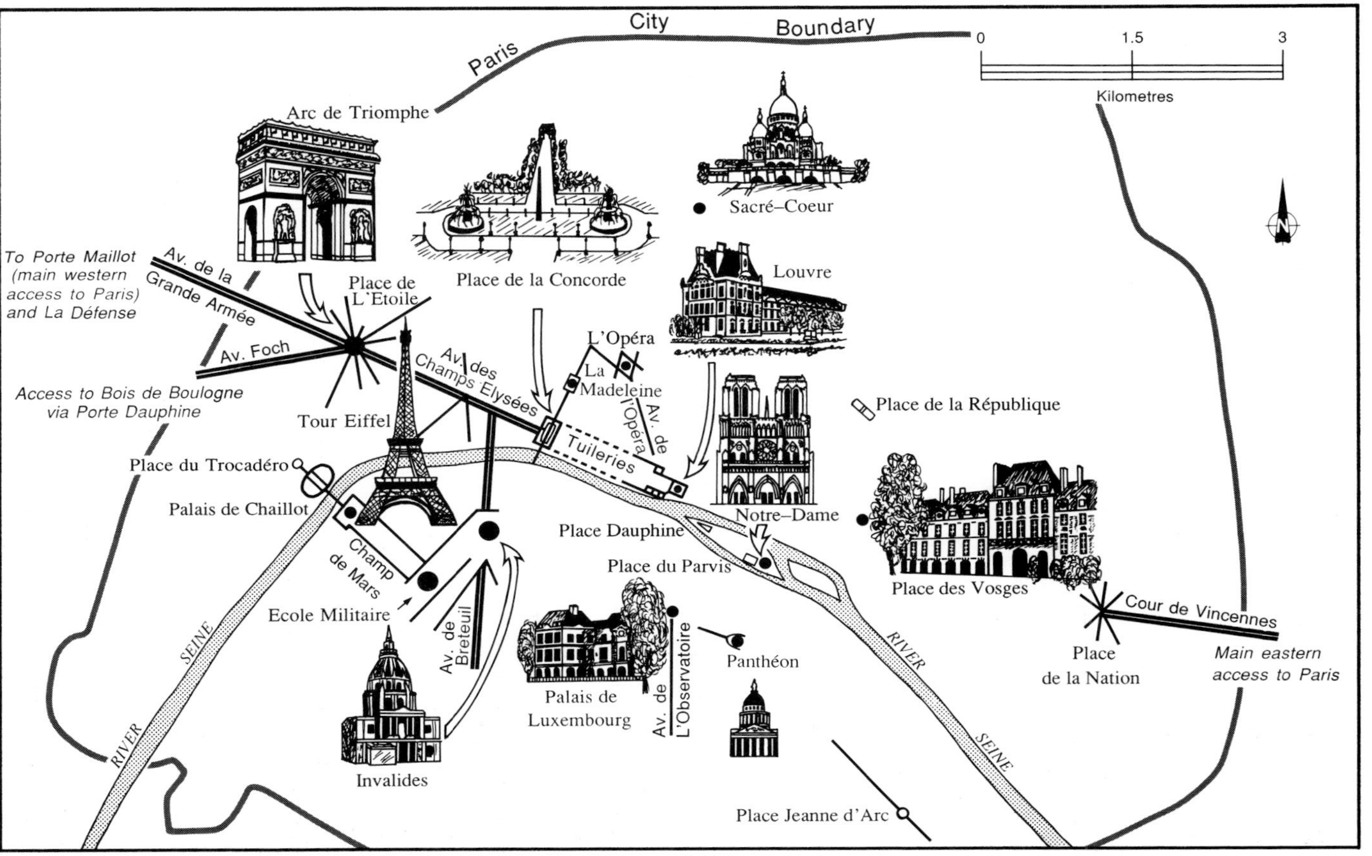

Fig. 6.4 Prestige space in the City of Paris, showing the triumphal axis and major vistas and monuments.

prestigious office development of La Défense (see pages 194–6) has been constructed directly in line with this axis. The axis, which is the backbone of the city's structure, is used mainly as public open space and is primarily symbolic in function; military parades up and down the axis by the occupying forces in 1940 epitomized the subjugation of France to German control.

Other areas of prestige space within the City of Paris are similar in execution but were built for rather different reasons. The Sacré-Coeur and the Tour Eiffel (Fig. 6.4) are two examples of prestige space constructed in the late nineteenth century, both of which have come to be associated with the outsider's image of Paris. The Sacré-Coeur on the hill of Montmartre was built after the rising of the Paris *Commune* in 1871 as a symbol of thanksgiving. However, its naming after one of the most reactionary objects of veneration (the Sacred Heart of Jesus) and its location on the site of the martyrdom of popular heroes of the *Commune* has also given it a reputation and an underlying meaning as a monument to monarchy, Catholicism and authoritarianism, and as a symbol of repression of the masses (Harvey, 1979). It was not finally dedicated and consecrated until 1919. Despite its importance in the tourist image of Paris, the basilica is disliked by many Parisians and is frequently referred to as a cream cheese, a *fromage blanc*, a pale edifice that flatters to deceive. The Tour Eiffel, the other internationally recognized tourist symbol of Paris, was designed as a temporary construction to celebrate the World's Fair held in the city in 1889. This fantastic futuristic construction, as it was at the time, demonstrated to the world the technological and scientific achievements of the French as a nation.

Building heights

The residential districts of the City of Paris typically consist of apartment blocks, some five or six storeys high, which line the boulevards and form densely packed areas along the side streets. The building heights within the city are very uniform, rising towards the city boundary. The pressure on space, caused by urban containment within the fortifications and by population growth, did not initially lead to the construction of taller dwellings. It did, however, lead to the greater utilization of ground space, often by subdivision of existing dwellings and the addition of sheds, lean-tos, workshops and other temporary structures. The height of buildings has been regulated since 1784; overall heights and the angles and heights of roof spaces were related to the width of the road (Evenson, 1979). The changes of regulations at various times allow reasonably accurate dating of many buildings within the city. By 1967, three general zones of height limits had been established, with a 25 metre limit in the city centre, and limits of 31 metres and 37 metres further out towards the city boundary (although with lower limits in some old village centres such as Montmartre and Charonne). From a vantage point such as the Arc de Triomphe, these zones can be clearly distinguished and the overall impression of a saucer-shaped city is reinforced.

The building height regulations have been breached on a number of

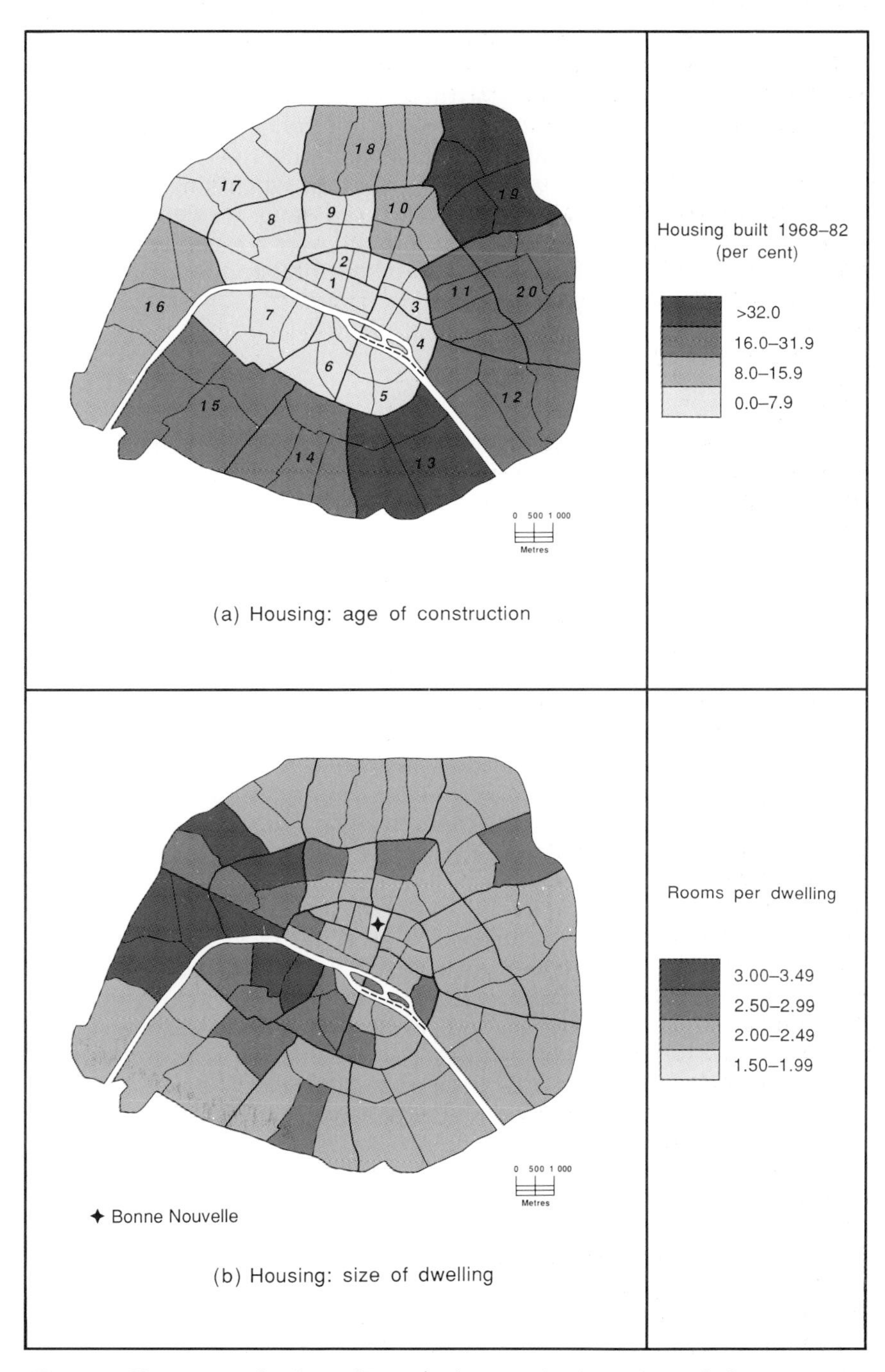

Fig. 6.5 Housing in the City of Paris (a) housing built, 1968–82; (b) housing size, 1982. (Source: INSEE, 1982)

occasions and early exceptions were made for the prestige buildings of the Sacré-Coeur and the Tour Eiffel. More recently, the prestige buildings of the post-war period have also been allowed to contravene the general regulations. Obvious examples of these are the towers of the Science building of the Université de Paris and of the office complex at Montparnasse, both on the Left Bank, and more centrally, the Centre National d'Art Moderne or Pompidou Centre, which forms a striking example of modern architecture amidst a sea of balconied stone residential blocks. Constructions of this type are visible from all over the city and may be viewed as symptoms of *gigantisme*, as architectural and planning disasters, or as symbols of the new France.

The housing stock of Paris

There are over a million dwellings in the City of Paris, most of which are flats. Only one per cent of dwellings consist of individual houses. Over two-thirds of the housing stock was built before the Second World War, but within the city there has been an acceleration of new building since 1968 (Fig. 6.5d). The size of housing units has increased only slightly due to constraints of space. Most Parisian dwellings consist of only two rooms although the average size in the degraded central *quartier* of Bonne-Nouvelle is even smaller (Fig. 6.5b). In 1982, almost 10 per cent of the housing stock was vacant, either awaiting demolition or in the process of renewal. As redevelopment and rehabilitation has occurred, housing standards in Paris have improved dramatically. This is particularly obvious in the general improvement in amenities, such as the addition of bathrooms and inside lavatories, central heating and telephones (Fig. 6.5c and 6.5d), but is also shown in the changing type and tenure of accommodation, notably the reduction in furnished lettings (Table 6.2).

In general terms, the housing stock of the City of Paris is poorer than might be expected of a world city. Nevertheless, there are enormous variations according to age, tenure and location, the most significant being the dichotomy between the wealthy west and the impoverished east (Fig. 6.5). The western sectors have the highest levels of amenity, the greatest space per occupant and the highest levels of owner-occupation, while parts of the centre, east and northeast are more likely to be overcrowded, cramped and rented, and lacking baths and telephones (Noin *et al.*, 1984).

The most favoured, desirable and expensive *quartiers* of the city are situated in the west, centre-west and on the Left Bank of the Seine (Fig. 6.5). The most extensive area of quality housing is in the *16^e arrondissement*, constructed in the late nineteenth century around Haussmann's boulevards. This housing was built around the pre-existing village centres of Chaillot, Passy and Auteuil, formerly important as summer residences for wealthy citizens and still perceived as villages by their residents. The other significant area of superb housing is in the Ile St Louis, at the very heart of the city, where upper-storey apartments built along the river banks in the seventeenth century give excellent views of the city as well as a prime location; this is a central city area which has never lost its attraction for

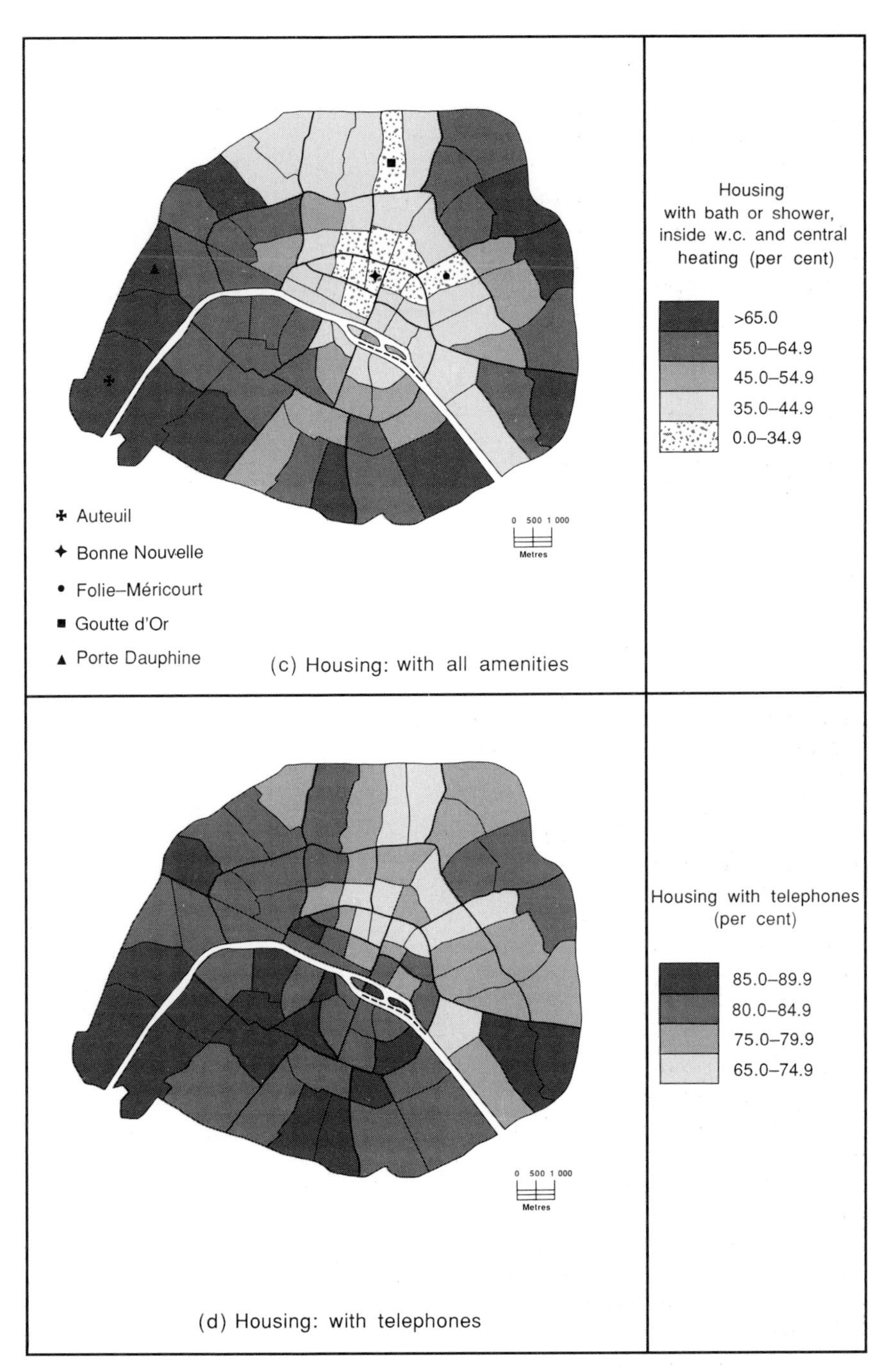

Fig. 6.5 Housing in the City of Paris (c) with amenities, 1982; (d) with telephones, 1982. (Source: INSEE, 1982)

Plate 6.1 Good quality housing in Paris: a traditional apartment block in the *17ᵉ arrondissement*. Note the shops beneath, the smaller windows in the top storey and the irregular building heights. (Photo: S J Gale)

Plate 6.2 Good quality housing in Paris: redevelopment in the *13ᵉ arrondissement*: this area now contains a new Parisian Chinatown. (Photo: P E White)

Parisians. The sought-after, *recherché*, quality of this housing may not always be evident from conventional indicators.

Good quality housing, within the price bracket of ordinary people rather than millionaires, is found in other districts of the west and centre, such as the *17e* and *8e arrondissements* (Fig 6.5; Plate 6.1). This housing consists of traditional balconied apartment blocks, similar in style and age to that of the *16^e^ arrondissement*, but it is more likely to be rented, to have fewer rooms per dwelling and to have lower levels of amenity. A rather different sort of good quality housing is found on the periphery of the south and east. This housing is relatively new, with very high amenity levels, but consists mainly of rented apartments of medium size. These are areas of post-war redevelopment, typified by the new blocks in the *13e arrondissement* around the Place d'Italie (Plate 6.2). Despite the quality of the amenities, this housing does not have the *recherché* status of the western and southwestern sectors.

The centre of the city contains the oldest housing, built before the 1870s, but its housing stock is quite heterogeneous. Old decaying working class housing co-exists with the former housing of the bourgeoisie and with areas of rehabilitation and gentrification (White and Winchester, 1991). A good example of this type of heterogeneous district is the Marais (see pages 210–11).

Table 6.2 Housing stock of the City of Paris, 1968–82

	1968 (%)	**1982 (%)**
Dwellings with amenities:		
Bath or shower	43.2	77.5
Inside w.c.	55.8	78.5
Central heating	47.9	64.0
Telephone	32.8	81.1
Housing tenure:		
Owner-occupied	21.8	25.4
Rented or sub-let	57.0	59.8
Rented furnished	11.2	5.4
By virtue of employment/free	10.0	9.4
Size of dwelling:		
1 room	31.6	26.4
2 rooms	33.7	33.0
3 rooms	19.8	22.6
>3 rooms	14.9	18.0
Mean number of rooms per dwelling	2.32	2.45
Mean number of persons per room	0.95	0.79

Source: INSEE, 1982.

The old working class districts of Paris lie mainly in the eastern and northern sectors (Fig. 6.5). In these areas, the dwellings are small, overcrowded and poorly provided with amenities. Some of the worst housing stock of the city is squeezed between the gentrifying areas of the inner city and the redevelopment of the suburban ring. Much of the housing of the *11^e*, *18^e*, *19^e* and *20^e arrondissements* was constructed in the nineteenth century; the tenement blocks, built without modern amenities, have been subsequently subdivided and infilled. Overcrowding is rife, the housing is broken down and the environment degraded by litter and graffiti. Whole streets lie empty and abandoned, or occupied by squatters, often as a result of years of planning blight (see also page 209). These districts, such as Folie-Méricourt (Fig. 6.5c), formerly deemed to be the most typically Parisian of *quartiers*, by the 1980s were occupied by a marginal population in which the elderly, the unemployed and the foreign-born were greatly over-represented (White and Winchester, 1991).

There is therefore a great variety of housing within Paris, ranging from the exquisite to the appalling. Despite the substantial improvement in housing since 1968 (Table 6.2), many Anglo-Americans would find the housing stock poor and cramped, yet relatively expensive, and would dislike the lack of individual houses and gardens, which are to be found only in the suburbs or in a few surviving village centres within the urban fabric. The quality of the housing has been constrained by lack of space, by height restrictions and by rent control fixed at levels which made adequate maintenance of the blocks uneconomic. Rent control, tied at 1948 levels for existing tenants, gave security of tenure to those residents, but also resulted in their immobility, with the result that the houses and their tenants aged together. The age of housing provides no real guide to its quality, although all recent housing is likely to contain basic amenities; nineteenth century housing, however, ranges from slum tenements to penthouses. Similarly, the tenure of housing is not a helpful indicator of quality; most apartments are rented and those that are owner-occupied are as often at the bottom of the pile as at the top.

Housing in the Parisian suburbs

Pre-war suburban housing

The housing of the suburbs is no less varied than that of the City of Paris, although the housing types are rather different. With the exception of former village and town centres, such as Choisy-le-Roi and Evry, the housing stock of the suburbs is rarely older than nineteenth century. The housing is generally at lower densities than in the cramped city, although there are exceptions. There are wide variations in style, tenure, age and environmental quality throughout the suburban ring (Fig. 6.6).

The housing constructed before the Second World War is located mainly in the inner suburbs and consists of both collective (apartment) housing and individual detached housing, *pavillons*. The collective housing blocks are mainly privately rented flats, which are stylistically similar to the earlier apartment blocks constructed in the city. The individual housing is

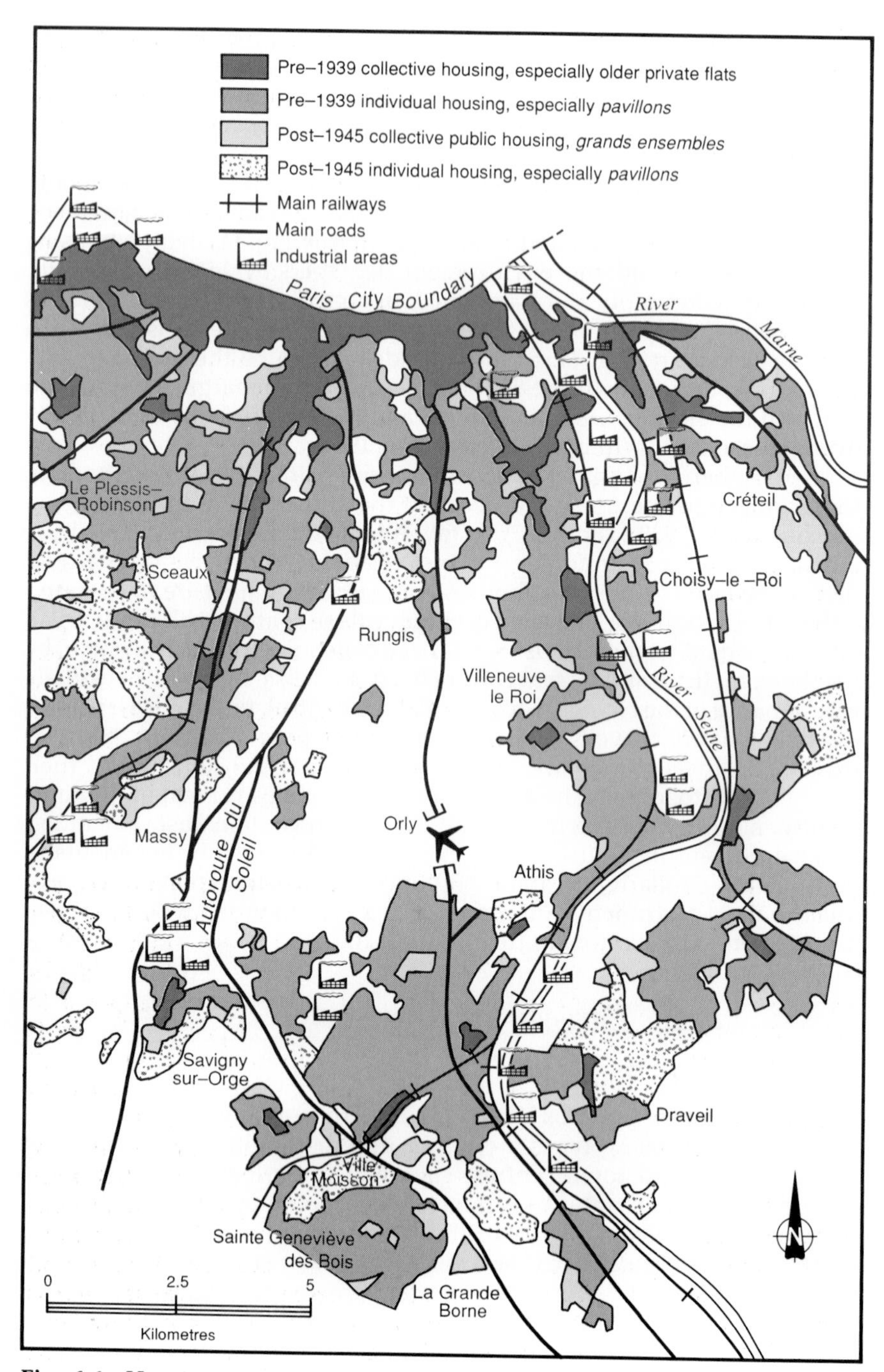

Fig. 6.6 Housing types in the southern suburbs of Paris, mid–1980s. (After White, unpublished)

of a completely different character and is located furthest from the city centre. At the time of construction, this housing was very poorly serviced by transport and facilities, and many of the dwellings themselves lacked basic amenities. The substandard construction of these dwellings gave them the label of *mal-lôtis*.

The explanation for the development of *mal-lôtis* lay in the unbearable pressure on the existing housing stock of the city caused by continued in-migration to an already overcrowded city centre. Furthermore, rent control legislation meant that there was little opportunity or incentive for developers to construct new housing within the City of Paris itself. Developers, therefore, turned their attention to the periphery and bought up large areas of farm land. This was advertized to Parisian residents desperate for housing, who were sold the dream of a house of their own in the country. In most cases, the plots of land were sold unseen, and the buyers were left to build their own dream homes without the benefit of mortgages, construction skills, or basic services such as paved roads, piped water, sewerage and electricity or gas. The reality of these homes turned out to be a series of tiny shacks and hovels, made mainly of wood, lacking the most basic comforts and remote from jobs and transport. Legislation was passed in 1928 to provide basic infrastructure or to improve that which existed. In subsequent years, many of the original houses have been improved, extended or replaced by much grander well-appointed suburban dwellings, in great contrast to their older neighbours. This mixture of solid *pavillons* with a few of the original garage-sized flimsy *mal-lôtis* can still be seen in areas such as Ste Geneviève des Bois in the south (Fig. 6.6), or at Bry-sur-Marne in the eastern suburbs.

Post-war suburban housing

Post-war housing in the Parisian suburbs consists of two main types, the collective housing of the *grands ensembles* and the individual housing of the *pavillons* (Fig. 6.6). The *grands ensembles*, which were also built in other large cities of France (see pages 160; 166–7), were huge apartment blocks constructed on the edge of the city in the 1950s and 1960s as a response to the exceptionally severe housing crisis of the post-war period. The existing housing stock was over-stretched and run-down by years of rent control and the growth of population nationally. This, combined with rapid in-migration to Paris, posed enormous housing problems.

The *grands ensembles* usually consist of serried ranks of tower blocks, thrown up in vacant areas of the periphery, often in interstitial areas lacking facilities and transport provision and not always structurally suitable for building. The large scale of these developments reflected a monumental approach to design, influenced by Le Corbusier's architecture and facilitated by industrialized construction processes. The Parisian *grands ensembles*, built from 1956 onwards, suffered from many initial problems and failings as a result of the scale of their construction. They epitomized the problem of suburban living; people isolated while living among great numbers of others, in cramped apartments, without facilities for shopping or child-care, with 'dead' indefensible space between the blocks. The *grand ensemble* at Sarcelles in the northern suburbs (Fig. 6.7) gave its name to an illness based on depression caused by these very difficult living conditions. More recently, structural problems, exacerbated by damage and neglect, have necessitated rehabilitation programmes.

However, of more concern to politicians and planners are the social problems of these huge estates. Some of the older estates such as at Massy-Antony have matured in population and service provision, whereas others have become 'sink' estates, populated almost entirely by foreign workers (Sporton and White, 1989). In 1991, a number of riots occurred in Parisian suburban estates (see also pages 166–7 for similar problems in other cities). In March 1991, for example, riots broke out at Sartrouville, following the shooting of an 18 year old north African by a security guard, himself a north African. In the disturbances, nine youths were arrested, five police were injured and shops and vehicles were damaged and burned. Two similar incidents occurred at the Val Fourée estate at Mantes-la-Jolie in the western suburbs in May-June 1991. The first of these was triggered by the death of a young Frenchman of north African descent while in police custody, the second by the killing of a youth by police in a violent incident in which a policewoman was also killed.

In June 1991, in response to the growing unrest in the suburbs, the then Prime Minister, Mme Cresson, announced an emergency package to help 400 disadvantaged estates. The package, costing around FF 140 million, aimed to provide 1000 more police, 12 advisory centres and 500 sports grounds within a period of 18 months; an unspecified number of youth training courses for the army, fire brigades and charities was also to be established. These measures were designed to remedy some of the service and infrastructural deficiencies of the *grands ensembles* and to offer a short-term solution to the problem of youth unemployment. The suburban unrest has stimulated growing concern over immigration and housing policy, both of which have become controversial political issues in the 1990s.

Most of the suburban housing constructed since 1965 has been on a smaller scale than the *grands ensembles* (whether public or private housing). The 1960s and 1970s were times of renewed economic prosperity in France, and those households which could afford to do so indulged in the dream of a house with a garden. The outer areas of the agglomeration have been infilled by new estates of privately-built *pavillons*, designed for the executive market and accessible to Paris by greatly improved rail connections (see pages 200–1). Over half of the individual housing of the Paris agglomeration has been built since 1948, mainly as infilling deep in the rural-urban fringe, but also as part of the most recent phase of new town construction (see pages 197–200).

Table 6.3 summarizes the housing conditions of the City of Paris and the Parisian agglomeration in 1982. The more recent suburban housing has higher standards of amenity, such as baths or showers and central heating. The suburbs also have higher levels of owner-occupation: 36 per cent compared with 25 per cent in the City of Paris. Similarly, a fifth of the suburban housing consists of individual housing, although apartment blocks still provide the majority of dwellings for Parisians.

Table 6.3 Housing stock in the City of Paris and the Parisian agglomeration, 1982

	City of Paris (%)	Parisian Agglomeration (excluding City of Paris) (%)
Dwellings with amenities:		
Bath or shower	77.5	85.1
Inside w.c.	78.5	85.7
Central heating	64.0	77.6
Telephone	81.1	79.9
Housing tenure:		
Owner-occupied	25.4	36.3
Rented or sub-let	59.8	53.2
Rented furnished	5.4	2.7
By virtue of employment/free	9.4	7.8
Housing age, type and use:		
Built after 1948	28.0	55.1
Apartment blocks	98.9	80.4
Second homes	4.4	2.6
Empty dwellings	9.8	7.7
Size of dwelling:		
Mean number of rooms per dwelling	2.45	2.99
Mean number of persons per room	0.79	0.82

Source: INSEE, 1982.

The planning of Paris

The growth of Paris and its suburbs has not been entirely haphazard, but city planning has been belated and controversial, despite the activities of Haussmann (see pages 176–8). By the mid-twentieth century, the problems of overcrowding, poor housing and traffic congestion caused by the growth of Paris and its agglomeration needed urgent solutions. Planning was made difficult by the scale and rapidity of growth, by constraints on development such as building regulations and rent control, and by the division of municipal responsibility between the City of Paris and the numerous *communes* of the suburbs. Furthermore, the Parisian and national authorities have been ambivalent in their attitude to planning; attempts to curtail growth were felt to be potentially harmful to the status of France's only world city, and the decentralization of people and jobs was perceived as a dilution of Parisian wealth, prestige and power. On the other hand, Parisians all recognized and deplored the tiresomeness of the daily grind of travel, work and sleep, *Métro, boulot, dodo,* inflicted by the unplanned development of housing, jobs and transport.

Early planning for Paris was patchy. The Haussmannization of the late nineteenth century was a form of urban planning which stemmed from a mixture of motives. One of the beneficial results of this process was the provision of the wide boulevards so essential to vehicular traffic (see pages 176–8). However, neither these nor the impact of building regulations (see pages 178–80) were felt beyond the city walls and few new boulevards were constructed after the turn of this century. A plan for the agglomeration drawn up in the inter-war years was shelved, unimplemented, at the start of the Second World War. Planning in the immediate post-war years was a lower priority than economic reconstruction and housing provision.

In 1960, the twin pressures of central congestion and suburban chaos gave rise to a variety of planning mechanisms and proposals for the region, the most significant of which was the PADOG, the *Plan d'Aménagement et d'Organisation Générale de la Région Parisienne*, a management plan for the Paris region. This plan was partly influenced by the Abercrombie Plan for London which had proposed a policy of urban containment. The underlying basis of PADOG was urban and industrial decentralization into the provinces, sufficiently distant from Paris to allow a rural zone to be maintained within which development would be restricted. At the same time, the services and the transport infrastructure of the suburbs were to be improved and integrated. These policies were overtaken by events; in particular, the population census of 1962 revealed rapid population growth in the agglomeration (Table 6.1). Expectations of continued growth led to the abandonment of the containment philosophy in favour of a pragmatic compromise solution which would tackle the continuing problems of the agglomeration while not inhibiting growth.

The Schéma Directeur

The *Schéma Directeur* (SD) or overall plan for Paris, published in 1965, is the foundation of post-war Parisian planning. Its implementation was assisted by the introduction of national spatial planning later in the decade, particularly the administrative reorganization of the Ile de France and the designation of 22 planning *régions* for the whole of France. The plan was based on an assumption of continued growth; it was predicted that by the mid-1980s France would have a population of 60 million and the Paris agglomeration a population of 11.6 million. (In fact, as a result of the slowing of population growth and the deconcentration of the urban area, by the mid-1980s the population of France was 54 million and of the Parisian agglomeration 10.8 million.) The other major assumption of the SD was that the requirements for space would be proportionally even greater than expected from population growth. Space would be needed for new offices, better housing, improved transport facilities and recreational amenities, requirements brought about by increasing living standards, changing occupational structures, higher expectations and greater potential for mobility. Accordingly, the SD contained strategies to cope with the spatial expansion of the agglomeration.

Several optional spatial strategies were considered for planned expansion, including the development of a single parallel city located some distance away from Paris and a London-style 1940s solution, with a ring of new towns located in the rural periphery. However, the chosen strategy was to reorganize the city region on two parallel axes running in a north-westerly-southeasterly direction on either side of the Seine Valley. These axes were designed to perpetuate the unity of the agglomeration but would break away from the traditional urban concentric structure. The proposals were more far-reaching and long-term than any earlier plans. Space and amenity requirements were to be met by the construction of new towns along the linear axes, which would absorb most of the expected population growth. The plan required improvements in mass transport systems to enable the agglomeration to function as a unit.

The strategy of corridor growth, which has been maintained since 1965, is a bold and imaginative policy which maintains flexibility. A corridor design allows the development of a polycentric urban structure with an interconnecting web of transport links, thereby removing some of the causes of congestion at the traditional centre. A second advantage in the Parisian case was that the orientation along the Seine followed the existing pattern of development and enabled areas of natural beauty along the river to be accessible for amenity and recreation. Furthermore, selected zones along the axes could be developed at different rates and at different scales without invalidating the overall scheme. In theory, therefore, the corridor growth principle allowed urban and industrial nodes to be defined with intervening green open space, producing a spatially discontinuous but functionally integrated urban network. In practice, the linear developments absorbed urban and industrial functions which were intended for decentralization beyond the Paris Basin, and urban sprawl looks much the same whether concentrically or linearly focused.

The Parisian SD has undergone two major revisions in 1969 and 1975. The 1969 revision, undertaken only four years after initial publication, was stimulated by a number of factors, particularly the events of 1968. 1968 was a year of political unrest which necessitated urgent government action on a number of fronts. It was also a census year, and preliminary results indicated changing conditions with an accelerated loss of population from the city centre and an increasing decentralization of population to the suburbs (Table 6.1). At the same time, an east-west socio-spatial polarization of Parisian society was becoming obvious. The east of the city and the inner suburbs were losing their industrial jobs but maintaining their working class status and at the same time attracting large numbers of foreign migrants. In the centre and west of the city, new office jobs were being created which were attractive to relatively wealthy professional people.

The 1969 revision, which did not include planning for the City of Paris itself, shifted the emphasis away from brand new developments on the periphery to the provision of improved facilities in the existing suburbs. Restructuring poles were designated at key points in the suburbs, at La Défense, Le Bourget, St Denis, Bobigny, Rosny, Créteil, Rungis and Vélizy to provide nodes for office space, new housing and improved services and shops (Fig. 6.7). The new towns proposed at the edge of the agglomeration were pruned from eight to five and their population targets

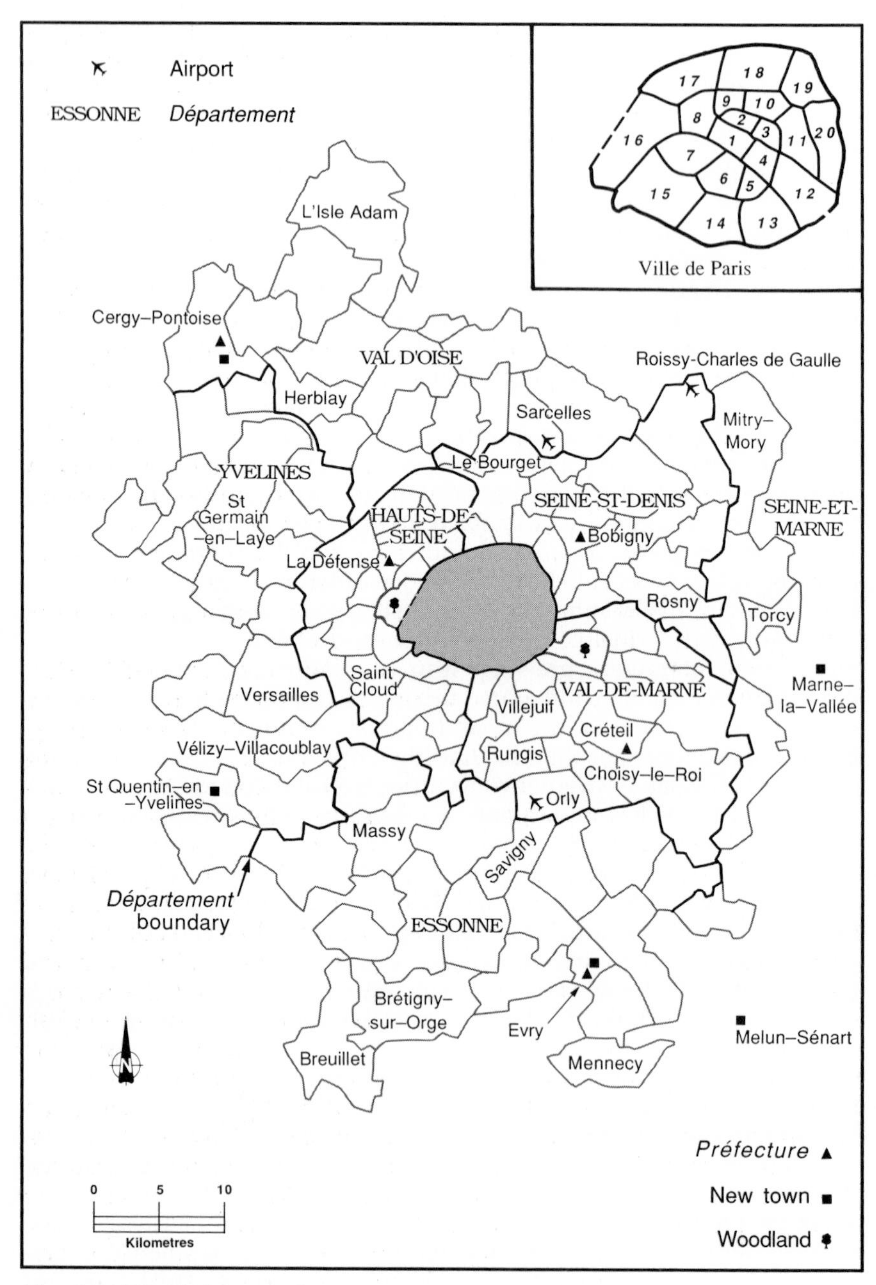

Fig. 6.7 The Parisian agglomeration, 1982. (After Noin *et al.*, 1984)

were revised downwards in accordance with changes in demographic projections. The five remaining new towns are Cergy-Pontoise and St Quentin-en-Yvelines in the west and Marne-la-Vallée, Melun-Sénart and Evry in the east (Fig. 6.7). The development of these towns did not effectively start until after 1969 (see pages 197–200). The overall aim of the new proposals was to provide a more balanced pattern of urban development, with the inner suburbs gaining as much benefit as the periphery and the city centre, while maintaining the basic unity and the axial strategy of the 1965 plan.

The 1975 revision, using the preliminary results of the 1975 census, further trimmed the population targets of the remaining new towns in line with slower population growth, while the suburban restructuring poles were confirmed and new services were allocated to them. The plan proposed a more careful protection of those outer rural areas still in agricultural use by their designation as environmental zones, *zones naturelles d'équilibre*, which would constitute a discontinuous green belt for amenity and recreation. The 1975 revision also recognized the desire for individual housing, which is reflected in the later housing developments in the new towns. A third feature of the 1975 revision was a greater concern for social welfare and for quality of life. The three versions of the SD have thus developed progressively towards a comprehensive overall structure plan, although always excluding the City of Paris itself.

A further revision was proposed in 1980. This was described by Dagnaud (1984) as a counter-plan. It was rejected by the Mitterrand government after its election in 1981. The 1980 scheme is nonetheless interesting, as although it proposed no new developments, it reflected the prevailing conditions of slower growth in population and industry. It was anticipated that the new towns would reach a maximum total population of less than a million by 1990 and that losses of industrial jobs would not be counterbalanced by gains in the tertiary sector. (Both these predictions had been proved accurate by 1990.) The revision suggested meeting local housing needs in areas where infrastructure was already available, and priority was to be given to energy saving in housing and to improved transport between the suburbs. The report also stressed the need to control social imbalances and to equalize the disparity in housing and jobs between east and west. The ideas put forward in the document were the antithesis of large-scale interventionist projects; instead they were concerned with local representation and decision making, in the designation of conservation areas and environmental zones, for example. These proposals were probably rejected because they did not sufficiently enhance the status of Paris as the national capital and a world city. In the absence of an accepted revision of the *Schéma Directeur*, the restructuring of the suburbs in the 1980s broadly followed the plans of 1969 and 1975, with major developments at existing suburban nodes and at the five designated new towns.

The SD was finally replaced in 1990 by a new plan contained in the *Livre Blanc*. The publication of this plan emphasizes the French commitment to urban planning as a positive mechanism for structuring city growth to the year 2015. The plan focuses on environmental protection, technological change and continued growth sectors. The agricultural and forest land on the periphery of the city is to be protected as a series of green wedges to provide recreational opportunities and access to rural areas for the Parisian population. The technological impetus for the city is represented in the huge investment in rapid transit systems, both inter- and intra-urban, and in the development of high-technology centres at La Défense and at Saclay-Massy in the southwest of the city (Burtenshaw *et al.*, 1991: 268). The main growth sectors designated are west, southeast and north from the edge of the City of Paris. The western sector extends westwards from La Défense, the southeast from major office redevelopments at Bercy in the *12ᵉ arrondissement* and the northern sector from La Villette in the *19ᵉ arrondissement* (Fig. 6.7); the major developments at these sites are considered on pages 206–8. Furthermore, two new urban centres within the suburbs will be the target for future development; one as a technological centre at Saclay-Palaiseau, the other around the international airport at Roissy-Charles de Gaulle.

Restructuring the suburbs

The haphazard growth of the Parisian agglomeration before 1965 left the majority of the suburbs with inadequate and ill-distributed facilities and services. Furthermore, the suburbs developed mainly as residential areas for a population working in Paris and so contained only limited employment opportunities. In 1969, eight suburban centres were designated as restructuring poles to remedy these deficiencies (see pages 190–1). Major public investment was channelled into facilities such as parks, health centres, sports and administrative facilities; private investment was attracted to new shopping centres and office employment; and in some areas, run-down residential districts were upgraded or replaced. The pace of development was slowed in the 1980s, both as a result of the high costs of reconstruction within the built-up area and by competition from development schemes in the new towns and elsewhere within the agglomeration.

La Défense

The earliest, most sensational, modernistic and controversial redevelopment occurred in the inner western suburbs at the site known as La Défense, where the French made their last stand against the Prussians in 1870. This area, in the *communes* of Puteaux and Courbevoie adjacent to Nanterre, was planned in the late 1950s and early 1960s as a major centre for businesses and offices. The redevelopment resulted in the demolition of over 9000 dwellings and the removal of 480 workshops and small factories (Clout, 1988). At that time the whole district in the vicinity of Nanterre was typical of the inner Parisian suburbs; a shambles of tenement blocks and small businesses, with the largest shanty town, *bidonville*, in Paris occupied by thousands of migrant workers. This was a prime site for redevelopment for a number of reasons. First, its location provided a westward extension of the triumphal axis of the city (see pages 176; Fig. 6.4) and was within easy reach of the high-status western suburbs such as

Versailles and St Germain-en-Laye. Once cleared, the site would therefore be excellently located, both for businesses seeking a prestigious address and for business employees who were likely to reside in the more exclusive western districts. Secondly, the demand for business space within the city could not be met because of historic building height regulations (see pages 178–80). This pressure had already caused business encroachment into the residential areas of the *16e arrondissement*. The development of a new business centre in the suburbs could solve a number of problems at a stroke. Skyscraper office blocks could be built at La Défense because the suburbs were not controlled by building height regulations, and these would be able to accommodate businesses searching for space in the *16e arrondissement*. The insalubrious areas of the *bidonvilles* would be removed and replaced with a project which would dazzle the world and which would, incidentally, provide a restructuring pole for the suburbs. The mixture of motives in the project is reminiscent of the Haussmannization of the city in the nineteenth century (see pages 176–8).

The jumble of housing and small industry was swept away to make way for a gigantic scheme, consisting of a projected 1.5 million square metres of office space in enormous multi-coloured skyscrapers; a massive covered underground shopping centre on four levels; jobs for 100 000 people and housing for 20 000; restaurants; cultural and entertainment facilities, including a national exhibition centre; a major interchange of the railway, the *Métro* and the improved suburban rail network of the RER; and motorway access segregated from pedestrians and linked to labyrinthine underground car parks. The initial development was slower than expected because of the procedural difficulties caused by operating such a scheme in two *communes* rather than one, but by the 1970s great progress had been made. The recession induced by the oil crisis led to the underutilization of office space at that time, but La Défense now forms a major employment centre in its own right with the offices of multinational firms such as Citibank, IBM, Esso and Fiat, and close by, the new administrative offices of the *préfecture* of Hauts-de-Seine (Fig. 6.7) and part of the Université de Paris at Nanterre.

La Défense is a concrete expression of enormous confidence in Paris and in France and it is designed to impress both by its scale and its location. The symbolic nature of the location is made clear by the decision to construct another arch at La Défense itself. This arch represents the western end of the axis which runs from the Louvre through the Arc de Triomphe (Fig. 6.4). The arch, which was the subject of considerable controversy and debate (Clout, 1988), is designed to provide a spectacular focus to the district and since its completion in 1989 has housed ministries and museums.

The whole redevelopment of La Défense is a spectacular project which is now complete but is still controversial. It has undoubtedly served its purpose in providing for Paris a prestigious alternative CBD, which has been successful in attracting multinational firms and commerce (Bastié, 1984). However, Clout (1988) has argued that La Défense has proved almost too successful at a regional scale, bringing considerable problems of traffic congestion at peak times. Furthermore, although La Défense contains residential accommodation, especially in the new park *quartier*, it still lacks the range of activities of a real city centre. The residents find the underground malls and car parks and the windswept pedestrian precincts rather oppressive and threatening when deserted at night. The location of the site has exacerbated the east-west imbalance of the city and promoted new office development in adjacent western suburbs. At the same time, the eastern suburbs have lost industrial employment. There are therefore many clerical workers and shop assistants who cannot afford the inflated rents around La Défense but cannot find work in the eastern suburbs and so are faced with long daily journeys across the city. The authorities of the City of Paris are also concerned by the success of La Défense, because of the loss of employment and potential tax revenue from the city itself. Nonetheless, the 1990 *Livre Blanc* for Paris envisaged La Défense as a focus for continued employment growth in the twenty-first century, and its extension further west to Gennevilliers and Montesson.

Restructuring poles

Of the other restructuring poles, Créteil in the eastern suburbs is the most developed, with a successful shopping centre, new industrial concerns, a relocated university site and a quarter of a million square metres of office space, including the administrative offices of the *préfecture* of Val-de-Marne (Fig. 6.7). Its early development in the 1950s took the form of a *grand ensemble*, but since then the energetic use of the development and planning mechanisms (see pages 160–1) has attracted other urban functions apart from the residential. Great efforts have also been made to produce a landscaped environment on a relatively human scale. Créteil lacks the visual impact and scale of La Défense, but it nonetheless provides a significant node within the eastern suburbs which is accessible to Paris by RER. It is less intimidating than La Défense and has a more familiar structure, with a much higher proportion of residential area and a shopping centre with a hypermarket and department store; it is a new town in all but name.

Rungis, in the Orly area of the southern suburbs, is a most distinctive zone. The wholesale market of Paris was relocated here from its central site at Les Halles, and construction work commenced in 1964. The development consists essentially of a very extensive entrepôt for the warehousing, marketing and transporting of food. Rungis is very well served by motorways; it has a regional commercial centre for retail shopping, and numerous hotels serving the market, the airport and the region as a whole. The relocation of the wholesale market has been a huge economic benefit to the southern suburbs by providing a multiplier effect into the local economy. It has also substantially reduced the traffic congestion at the site of Les Halles (see pages 204–6).

The other suburban restructuring centres are less distinctive, but have nonetheless attracted a variety of locally significant schemes, such as the shopping centres and industrial plants at Vélizy and Rosny, the new residential apartments at Bobigny and the redeveloped airport site at Le Bourget (for locations, see Fig. 6.7). These developments have helped upgrade the often dismal industrial suburbs of northeast Paris, although not to the same extent as the dramatic changes at Créteil or La Défense. Indeed, it is arguable that the northern suburbs of Seine-St-Denis still lack the type of major service centre which has been developed in other sectors of the Parisian conurbation. This will be corrected by the new sectoral development extending from La Villette through the old industrial suburbs of St Denis proposed in the *Livre Blanc* of 1990.

The Parisian new towns

The Parisian new towns form a major element in the planned growth of the Paris agglomeration (Fig. 6.7). They post-date the designation of the London new towns and differ from them in a number of significant respects. First, they were planned as centres of growth within the agglomeration, rather than being centres designed to take 'overspill' populations. Secondly, all the Parisian new towns were designated within 35 km of the centre of Paris and without an intervening green belt; there was therefore no real intention of creating self-contained communities (Merlin, 1969). Thirdly, their transport links with the rest of the conurbation and with Paris in particular were to be vastly improved by new road and rail links to facilitate commuting. In these ways, the designation of the Parisian new towns would help transform the agglomeration into a polycentric urban region, whereas the London model was designed to maintain a contained monocentric region with the new towns as satellites.

The administrative arrangements of the new towns did not always run smoothly. Evry, probably the best-known and most innovative of the new towns, survived an ignominious start, when nine of the original 14 *communes* seceded from the planning syndicate, accusing it of domination by external interests and authoritarianism. Melun-Sénart, located very close to Evry, experienced even greater initial difficulties. These were compounded by the choice of site, which included land in no less than 18 different *communes* and two different *départements*.

Transport connections to the earlier new towns were much inferior to those that have been developed subsequently. Cergy-Pontoise, under construction by 1969, was poorly connected to Paris for many years until the new rapid suburban rail link (RER) was built. Journeys to Cergy-Pontoise still require both RER and rail connections, but these have been greatly improved. The later new towns had RER connections earlier in their construction and in both St Quentin-en-Yvelines and Marne-la-Vallée, residential neighbourhoods are being developed around the RER stations. Marne-la-Vallée is only 19 minutes away from Paris by RER and is the most likely of the new towns to develop into a dormitory centre for commuters. Evry has good motorway connections to Paris by the A6 (Autoroute du Soleil) and RN7, but has no direct RER service; Melun-Sénart has the poorest public transport links with Paris.

The new towns have provided opportunities for innovative architecture and design. Evry has become famous for the adventurous architecture of both its public buildings and its varied types of housing. Although Evry contains a high proportion of apartments rather than individual dwellings, the designers have moved away from the slab architecture characteristic of the *grands ensembles*; the flats known as the *pyramides*, for example, show an imaginative use of colour, terraces and angular structures to create personal and individual spaces within a large block. Similarly, the major complex of public buildings in Evry centre constitutes a bold architectural statement of a type sorely lacking in the suburbs. At Marne-la-Vallée, collective housing in the western neighbourhoods was designed in monumental style by the architect Bofill, while nearby is a show-piece development designed by Nunez, known as the *Arènes de Picasso*. This consists of two enormous cartwheel blocks of flats arranged around a landscaped courtyard; they form a stark contrast with the small inter-war *pavillons* in the adjacent suburb of Bry-sur-Marne. The town centre, although less impressive than that of Evry, nonetheless contains a multi-tiered, multi-coloured commercial centre with 80 shops at Les Arcades.

The new towns have experienced varied success in attracting employment. Evry and Cergy-Pontoise have benefited from their designation as the *préfectures* of their respective *départements*, as this has brought significant administrative and clerical employment. St Quentin-en-Yvelines does not have a *préfecture*, but it has received some decentralized offices, including the census office and a number of major banks. Marne-la-Vallée had attracted over 16 000 new jobs by 1987, about a third of which were clerical and administrative. Industrial employment was located on zoned industrial estates, including a high technology park (see pages 138–41), and was concentrated in electronics, light industry, distributive trades and assembly work. Employers have been attracted here by the good communications to Paris as well as to Germany via the *Autoroute de l'Est*; accordingly, branch plants of multinationals such as Philips, IBM and Kodak have established here. The opening of Euro-Disneyland in April 1992 has provided a variety of jobs for the town as well as bringing visitors to the area. Clout (1988) estimated that up to 20 000 jobs would be created, but White (1991) considered that these would provide only limited employment for local people, as many of the jobs would be taken by foreigners. Disneyland was certainly actively recruiting labour in Britain in the early 1990s. The development of Euro-Disneyland may be seen as another example of a prestige project, drawn to Paris to enhance the status of the city and located in the eastern suburbs as a counterweight to the attractions of La Défense. By 1987, Marne-la-Vallée boasted 37 000 jobs in total, but had an economically active population twice that size. This pattern of imbalance between jobs and population is found with slight variations in all the Parisian new towns; commuting is clearly part of the way of life for many new town dwellers.

The housing in the new towns varies with the age of their construction. The earlier developments were mainly apartments in relatively tall and extensive blocks, although Cergy-Pontoise adopted a maximum height of eight storeys in the early 1970s. More recent housing consists of individual dwellings, some quite large and exclusive, as in the new neighbourhoods of St Quentin-en-Yvelines. St Quentin benefits from its location in the sought-after western suburbs, with direct access to the business centre of La Défense. Similarly, the newer suburbs of

Table 6.4 Growth of population in the Parisian new towns, 1968–82

	1968	1975	1982	Growth 1968–75 (% p.a.)	Growth 1975–82 (% p.a.)
Cergy-Pontoise	53 445	82 993	113 329	7.9	5.2
Evry	33 180	51 226	77 510	7.8	7.3
Marne-la-Vallée	58 085	66 879	115 751	2.2	10.4
Melun-Sénart	65 709	92 335	115 273	5.8	3.5
St Quentin-en-Yvelines	41 415	96 838	151 376	19.1	8.0

Source: INSEE, 1982.

Table 6.5 Parisian new towns: expenditure and construction from designation until 1982

	Budget	Housing	Industry	Offices
Cergy-Pontoise	1163	28 200	286	225 000
Evry	791	16 500	158	288 000
Marne-la-Vallée	1389	24 060	189	190 000
Melun-Sénart	412	15 820	162	25 000
St Quentin-en-Yvelines	1154	30 800	288	230 000
TOTAL	4909	115 380	1083	958 000

Source: INSEE, 1982
Budget: total expenditure in FF millions
Housing: number of housing units completed or under construction
Industry: area developed (ha)
Offices: area of offices constructed (m^2)

Marne-la-Vallée have excellent access to the city centre and are the site of new hamlets, *hameaux*, of low density 'executive' housing.

The Parisian new towns may be considered to be reasonably successful, although they have varied considerably in their rates of growth, and in general have provided homes for fewer people than was ever anticipated (Table 6.4). They have been criticized for the unbalanced age structure of their populations and for their lack of employment (Wilkes *et al.*, 1987). However, the population of young families is typical of many newer suburbs and will alter gradually. The provision of employment may be less than envisaged in the boom days of their designation, but their initial location close to the centre of Paris and the recent development of neighbourhoods around RER stations imply an acceptance of commuting by the planning authorities. The new towns have been successful in their use of modern technology, such as the adoption of solar energy at Marne-la-Vallée, the installation of teledistributors which make individual television aerials unnecessary and the segregation of wheeled from pedestrian transport (Ploegarts, 1986). The towns may also be considered a success from the point of view of providing a focus for growth in the suburbs and have contributed greatly to an improvement in living conditions and service provision in the Parisian agglomeration. The five new towns around Paris have provided over 100 000 new dwellings (Table 6.5), predominantly social housing for young working families, in living environments greatly superior to those of almost any other part of the suburbs.

Transport planning in the Paris region

> 'Yet, as it is necessary for him to learn something, suppose he studies geography?'
> 'O, noble marquis! what can a young man of quality have to do with geography? Will not his postilions know the road from Paris to Versailles . . .?'
>
> Voltaire (1805: 86)

One of the key elements in the planning of the Paris region since 1965 has been the provision of new and improved transport links. As the Parisian agglomeration grew and transport demands increased, mass transit provision lagged behind requirements. The earliest method of mass transport was horse-drawn omnibuses, which commenced operation in 1828; and by 1873 the General Bus Company, the *Compagnie Générale des Omnibus,* was operating 32 routes carrying 111 million passengers a year (Evenson, 1979). The buses suffered competition first from tram services in the late nineteenth century and then from motor buses in the 1930s.

An underground rail system, the Parisian *Métro*, the most efficient method of mass transit in a large city, was built long after those of London (1863) and New York (1868). Prolonged public discussion and argument surrounded its construction and there were widespread fears that digging into 'the bowels of the city' would release all sorts of foul pestilences, as well as being an inherently difficult and dangerous activity. The Parisian authorities did not have powers to plan the *Métro* until 1895, when the approaching exhibition of 1900 put pressure on the city to install a modern showpiece system of mass transit. The system was constructed very close to the ground surface along the main Haussmann boulevards (see pages 176–8). In 1900, 10 kilometres were opened and by 1914, 80 km of track were laid. The new system was an immediate popular success and by 1914, 400 million journeys were being made annually (Evenson, 1979).

Paradoxically, the establishment of the *Métro* did not relieve surface traffic congestion. Instead, it encouraged greater mobility and stimulated the need for connecting and feeder services. From 1914 onwards, as the working day contracted to a more manageable eight hours and as the suburbs spread, so the suburban rail lines constructed in the mid-nineteenth century came into more general use and more feeder services were needed. One major problem with the *Métro* and bus services was that their routes stopped at the physical and political boundary of the City of Paris, so any journey to or from the suburbs necessitated at least one change of transport service. A further problem was that the transport services were not co-ordinated by any one authority. As a result, there were severe problems of both competition and omissions in services, causing delays and inconvenience to passengers.

The pressures on the transport system continued to mount and became desperate during the Second World War. During the period of wartime occupation, the shortage of petrol forced 1500 million travellers a year on to the *Métro*, making it overcrowded, overheated and smelly. In the post-war period, the Parisian Transport Authority was formed, the *Régie Autonome des Transports Parisiens* (RATP), to co-ordinate a unified system of public transport and to improve and extend the system. The *Métro* of the 1990s consists of 13 lines criss-crossing the city. During the 1980s, these lines were extended to the inner suburbs and the *Métro* now connects Créteil in the east, St Denis in the north and Villejuif in the south (for locations see Fig. 6.7). By contrast, bus connections between the City of Paris and most suburbs are still poor and usually involve a change of buses at the *boulevard périphérique*, along which buses run every few minutes, providing a circular service around the whole of the ring road. In the 1980s, the *Métro* services were further improved by the refurbishment of many of the main stations, with brightly coloured seats and historical educational murals. The station at the Hôtel de Ville, for example, is patriotically kitted out in red, white and blue, and that at the Louvre has replicas of museum exhibits with interpretive signs.

Connections to the outer suburbs have improved enormously since the mid-1960s when the *Réseau Express Régional* (RER) was begun. Further extensions to the *Métro* were considered to be impractical, as the number of stops would have involved unacceptably lengthy journey times. The RER consists of automatically-operated high speed services on new or specially adapted tracks. This new network has transformed journey times to the suburbs. There are now three RER lines. The major interchange of the east-west line (line A) and the north-south line (line B) is at Châtelet-Les Halles; there are only a limited number of stops within the City itself. Line C serves the southern suburbs, including St Quentin-en-Yvelines in the west and Rungis and Orly in the south. Further major improvements are proposed. These include new RER links between the major railway stations, specifically Châtelet-Les Halles and the Gare de Lyon, and between the major main-line stations in northern Paris, that is the Gare St Lazare, the Gare du Nord and the Gare de l'Est. The construction of a new line D is also proposed by RATP. This line will run southeast to northwest, improving the service in the southeast from Melun-Sénart and duplicating line A in the city centre to relieve congestion (Bollotte, 1991). These schemes, funded by the state, the *région* and by the City of Paris, will not be completed until the mid-late 1990s. They will contribute significantly to more efficient transport within the agglomeration, although journeys across the suburbs are still difficult and time-consuming by public transport. Nonetheless, the continued improvement to the rail and RER network has enabled public transport to maintain and even improve its market share of all journeys made within the Paris region.

Roads

Road traffic in Paris has been notoriously congested for many years, despite the wide Haussmann boulevards, which, like many other transport improvements, may have stimulated further traffic. A major improvement in traffic flow was brought about by the construction of the *boulevard périphérique* (see page 175; Fig. 6.3), but by 1989 this ring road accounted for 55 per cent of the traffic jams in the Paris region (Bollotte, 1991: 165). Motor traffic is heavy and growing, and illegal car parking is an enormous problem in the City of Paris, with only 750 000 legal car parking spaces, but 1 350 000 cars entering the city each working day.

In the early 1990s, road building programmes will focus on completion of the A86 peripheral/radial roads and the Great Eastern Highway, the *Francilienne*. The A86 will form an outer ring road approximately 15 km outside the *boulevard périphérique*; the 30 km link between the A1 to Lille and the A6 to the south has recently been completed. This connecting motorway link will form a vital part of the national motorway network and will relieve the *boulevard périphérique* of through traffic. Its completion, particularly in the western suburbs, will be difficult and expensive, because of the elevated price of land and because of opposition from affluent residents. The completion of an improved motorway network is seen as part of a national strategy to retain Parisian centrality and competitiveness in Europe, in order to take full advantage of the benefits of the completion of the European Single Market by the end of 1992 (see pages 241–5).

Airports

Until 1975, Paris had two major airports at Orly and Le Bourget which were working at full capacity, dealing with 17 million passengers a year. In 1975, the new Paris-Nord airport at Roissy, named after Charles de Gaulle, increased the capacity by a further 10 million passengers annually. Roissy was built on an undeveloped rural site, with space for expansion. A second terminal was brought into service in 1981–2 and by 1983, the three airports were handling 30 million passengers a year, almost half of them at Roissy. Roissy-Charles de Gaulle is now the major international airport for Paris, while most domestic flights are scheduled from Orly. There are no plans for more airports, but there is a policy of continued upgrading, for example, a rail connection between Orly and RER line B is under construction, as is a one kilometre link from the RER line directly to the passenger terminal at Roissy-Charles de Gaulle.

The construction of the Roissy-Charles de Gaulle airport has had a major impact on the local area, although the freight zone is not yet working at full capacity. The airport directly employs 22 000 people and there is also a significant multiplier effect in the connecting transport networks, the new hotels and the spin-off industries. The airport is linked to the city centre by motorway, the *Autoroute du Nord* and by RER, enabling transfers to the city centre to be made quickly and easily. These improved road connections have incidentally facilitated access to and from some of the major *grands ensembles*, including Sarcelles (see page 188). Adjacent to the airport site is the national exhibition centre, the *Parc des Expositions*, located there because of the excellent access by all means of transport. The national exhibition centre provides a suitable showcase for French and Parisian achievements. This highly accessible area is scheduled for further expansion and development in the 1990 *Livre Blanc*.

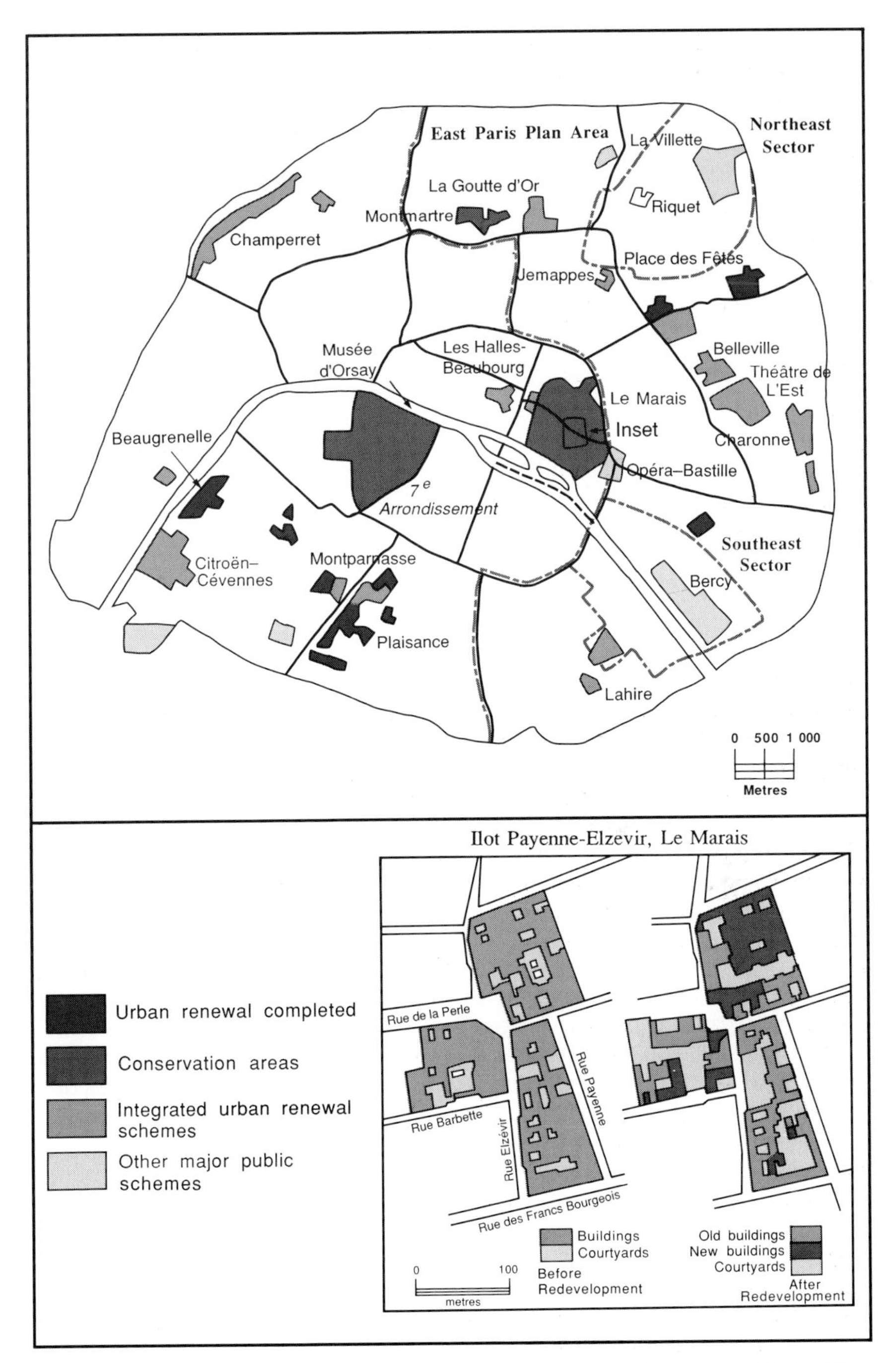

Fig. 6.8 Redevelopment and conservation zones, City of Paris, late 1980s. The unbroken lines represent *arrondissement* boundaries. Redevelopment zones in the East Paris Plan area are indicative of widespread renewal (after White, unpublished).

Conservation and redevelopment in the city centre

The regional SD did not include the planning of the City of Paris, but guidelines for the future development of the city were laid out in a separate structure and land-use plan in 1967, which was revised in 1976. Conservation programmes in the city were established under the *Loi Malraux* (see pages 162–3) and two major *secteurs sauvegardés* were designated (Fig. 6.8). Although these two conservation areas, the Marais and the *7e arrondissement*, are large in comparison to similar zones in provincial cities, they extend over only a small part of the historic centre of Paris. There are also many listed sites and buildings in the city with a lesser degree of conservation protection (Kain, 1980).

While conservation schemes have been established in some areas, the City of Paris has undergone major redevelopment, ranging from the replacement or renovation of individual buildings to the large-scale clearance and redevelopment of whole districts (Fig. 6.8). In 1964, Bastié estimated that even since the Second World War, a quarter of the city and a third of its population had been affected by urban renewal (Bastié, 1964: 45). The population of the City of Paris fell during this time (Table 6.1), but the completion of many renewal schemes has resulted in a stabilization of population in the 1980s.

Redevelopment and renewal

The extent of renewal and redevelopment in the City of Paris is indicated by Fig. 6.5a which shows the proportion of housing built in the period 1968–82. Large-scale residential redevelopment has already occurred in the outer *arrondissements* of the south and east, and is still in progress in much of the east and north. Total redevelopment has also occurred in the *1er arrondissement* around the old wholesale market of Les Halles: the limited amount of new housing in this area reflects the predominantly commercial nature of this development. Further major non-residential projects are in train in much of eastern Paris, which since the 1980s has been the subject of an urban regeneration programme. This consists of a number of prestige projects, for example, at La Villette, Bercy and Opéra-Bastille (Fig. 6.8), which are designed to rectify the status imbalance between the west and the east of Paris. These grandiose projects are being undertaken as a lasting memorial to the significance of the Mitterrand era in Paris (Chaslin, 1985) and to enhance the role of Paris as the major European cultural capital.

Les Halles

The redevelopment of Les Halles was one of the most controversial of all the urban renewal projects undertaken in Paris because of its prime location at the very centre of the city. Les Halles was the site of the national wholesale food market, built in the mid-nineteenth century to serve a city of approximately two million. The 32 hectare site gradually became extremely congested with traders, choked by large trucks and physically dilapidated. The trading of meat, fish and vegetables produced an accumulation of filth in the streets and a characteristic smell, worse in summer, which was increasingly felt to be unacceptable in the heart of a major world city. In 1971, the food halls, covered by the famous Baltard pavilions, were demolished, despite public opposition, and the food market was relocated at a site ten times the size at Rungis in the southern suburbs (see page 196). Before demolition, the district had been an extremely active and animated *quartier*, with a large number of traders and small businesses of all sorts, ranging from food industry suppliers and truck repairers to all-night cafés and brothels, competing for space in the densely-packed streets and seedy tenement blocks. The former character of the area can still be recognized in the streets immediately to the north of the redeveloped site, and prostitution still dominates the nearby Rue St Denis (Ashworth *et al.*, 1988).

In the early 1970s, Les Halles formed the biggest hole in the ground in Paris, while politicians, planners and architects debated the type and form of the redevelopment. The proposals included a scheme for an international trade centre, which was abandoned in 1974. The actual redevelopment used the hole for the largest underground station in Paris at Châtelet-Les Halles. This forms the major interchange of the *Métro*-RER rail system, the focus of two RER and four *Métro* lines. The remainder of the redevelopment is known as the Forum, completed in 1979. This forms a vast underground shopping centre on four levels. The Forum was intended as a high-class pedestrianized commercial centre, capitalizing on its improved accessibility. Other aspects of the redevelopment include the provision of much-needed green space, a concert hall and an international hotel adjacent to the Forum, all completed by 1987. Approximately half the site has been totally redeveloped for commerce and transport, and in its wake block-by-block upgrading is occurring in the vicinity.

The redevelopment has not been wholly successful. The *Métro* interchange is superbly efficient and is used by a million passengers each year. The rail access and the covered shopping centre have not only attracted high-status customers, but also have provided meeting places for other social groups, ranging from schoolchildren and punks, to migrant workers and the homeless (Winchester and White, 1988). The Parisian authorities have attempted to combat this use of the Forum, but a strong police presence is no more appealing to tourists and shoppers than the presence of potentially threatening groups lurking in the long concrete underground walkways. Some of the up-market shops have moved out of the Forum, but it is still a major centre for women's clothing boutiques, for exhibitions, for street artists and for tourists and travellers.

The remainder of the site has not been totally redeveloped but has undergone extensive renewal and upgrading. In the adjacent Horloge district, 750 homes have been renewed, but fewer than a third of the original inhabitants have been rehoused in the vicinity, the others having been forced to relocate in more distant *quartiers*. Because of their quality

Plate 6.3 Mime artists near Les Halles–Beaubourg follow an unsuspecting victim. (Photo: S J Gale)

and their location, the newly renovated apartments are extremely expensive and are beyond the means of the previously resident artisanal population.

Non-residential redevelopment schemes in the City of Paris

Slightly to the east of the Les Halles-Forum complex is the *Centre National d'Art Moderne*, usually known as the Beaubourg Arts Centre or Pompidou Centre. This was built on the site of one of the most degraded blocks in Paris, known as *îlot insalubre #1*, which decades earlier was designated as unfit for human habitation because of the high levels of disease found there. The buildings on the block were demolished in the 1930s, but it was not until 1977 that they were replaced by the cultural and popular arts complex of the Beaubourg Centre. This unique building, with its red, white and blue pipework and external infrastructure, embarrassingly prominent in its multicoloured glory, sits among the workaday drab of the surrounding tenement blocks, enlivened by the performances of novelty fountains and street artists (Plates 6.3 and 6.4). The Parisian population is not indifferent to this architecture; it is loved for its impact and originality as well as for the facilities and open space that it provides, but the centre is also considered to be an ugly building which is out of place and which is displacing people from the city centre.

Other cultural schemes occupying prime central sites include the Impressionist Museum at the Gare d'Orsay, opened in 1986, and the third

Plate 6.4 A cosmopolitan crowd appreciates the entertainment. (Photo: S J Gale)

opera house at Bastille, opened in 1990. The Musée d'Orsay utilizes the disused railway station on the Left Bank of the Seine and, as well as the largest collection of Impressionist paintings in the world, has exhibitions of photographs, architecture and furniture. It is easily reached by RER line C and has been enormously successful from the point of view of both tourists and Parisians alike. The Opéra-Bastille replaced an area of run-down housing adjacent to the Place de la Bastille with another prestigious cultural edifice. A third very large-scale scheme is under way in the northeast of Paris at La Villette (Fig. 6.8), where the former city abattoirs are being totally redeveloped to provide a leisure, entertainment and educational complex. The main features of the site are for music and science. The Music Centre incorporates music halls and rock and pop concert venues. The huge science park includes the National Museum of Science, Technology and Industry. There is much newly landscaped open space. This will form the largest park in the City of Paris and the first to be created for over a century (Clout, 1985). The residential area along the neighbouring canals is also undergoing upgrading. However, for many years the adjacent housing suffered from planning blight and was occupied by squatters (Winchester and White, 1988). La Villette still suffers from relatively poor public transport connections, but, as with the Opéra-Bastille and the new Théâtre de l'Est (Fig. 6.8), is designed to provide a major cultural facility in the eastern sector of Paris.

Large-scale development of offices and commerce has occurred in the outer *arrondissements* of the city since the mid-1960s. This has particularly been the case in the southern sector, from Gare de Lyon-Bercy in the southeast to Beaugrenelle, Montparnasse and the Fronts de Seine in the southwest (Fig. 6.8). In each case, the redevelopment has involved the demolition of old housing or derelict factories which had become increasingly degraded as a result of planning blight and construction delays. The older accommodation has generally been replaced by tower blocks; Montparnasse is dominated by a 200 m black rectangular tower which dwarfs the adjacent commercial centre and railway. The Fronts de Seine redevelopment, completed in the early 1970s, and the as yet incomplete Gare de Lyon-Bercy complex both consist of dozens of individual tower blocks owned or leased by major commercial companies. The Gare de Lyon-Bercy project also includes major public investment, notably in the new Ministry of Finance building, which it is hoped will stimulate further relocation of private industry and commerce. Both Bercy and La Villette have been designated in the 1990 *Livre Blanc* as the focal points of planned new growth poles extending from the city boundaries into the suburbs.

Residential redevelopment in the City of Paris

Approximately 20 to 25 per cent of the housing stock of the City of Paris has been renewed or rehabilitated since 1968. Large-scale residential redevelopment occurred in the 1970s and 1980s in the southern sector of the city, much of it in apartment blocks in association with commercial redevelopment, for example at Montparnasse and Place d'Italie. Redevelopment in the eastern *arrondissements* (the *11^{e}*, *19^{e}* and *20^{e}*) has consisted of large-scale slum clearance and its replacement by public housing, with much less emphasis on commercial and office development. One early scheme in the Place des Fêtes area replaced slum housing with a group of five enormous tower blocks, incongruously sited next to small terraced houses.

Recent redevelopment has taken place on a massive scale in Belleville. The original housing in this area was thrown up in the nineteenth century outside the existing city walls and consisted of piecemeal construction on individual plots, with extra storeys and additions jammed on higgledy-piggledy in the ensuing decades. By the late twentieth century, this housing formed some of the worst in Paris, providing furnished rooms, known as *meublés*, for immigrant workers, and slum conditions for the working classes. Large-scale clearance has been the only solution for the jerry-built and overcrowded tenements of Belleville. Between a third and a half of the reconstruction has consisted of public housing for lower-income groups. In the 1980s, the face of the area changed completely, with new medium-rise public housing in sparkling blocks of concrete and glass located around the grassed slopes and trellised walkways of the new Belleville Park. There have been significant improvements in landscaping and in the amenity levels of property, but the area has lost much of its character with the displacement of immigrant groups and ethnic shops. Similar massive redevelopment of the immigrant district of the Goutte d'Or commenced in the late 1980s, the likely consequence of which will be the social upgrading of the *quartier* and the break-up of the north African quasi-ghetto (Vuddamalay *et al.*, 1991).

Residential redevelopment is not always on this vast scale. Throughout the eastern *arrondissements*, piecemeal upgrading of individual blocks has taken place, with the shells being gutted and reconstructed, often for public housing (HLM) apartments. The Riquet redevelopment in the *19ᵉ arrondissement* and the district around Charonne in the *20ᵉ arrondissement* have both been more sensitively redeveloped. This is in part related to the more favourable locations of these two districts, the former along the canal and the latter on a hilly site around the former village centre of Charonne (Kain, 1980). The redevelopments affect relatively small areas and the buildings are lower in height and more architecturally varied than in Belleville. There is also a greater proportion of housing for a middle-class clientele, both private housing, and public housing for middle-income groups, *immeubles à loyer normal* (ILNs). In these cases, redevelopment by the city council using the ZAC procedure (see page 160) has sparked off private gentrification of relatively desirable areas, shown by the introduction of specialist shops and expensive restaurants to serve the new apartment dwellers.

Conservation in the City of Paris

The City of Paris contains two large conservation zones, the *7ᵉ arrondissement* and the Marais and a number of smaller ones, including Montmartre (Fig. 6.8). The *7ᵉ arrondissement* on the Left Bank consists mainly of ministries and public buildings on the one hand and a series of fashionable shopping streets such as the Boulevard St Germain on the other. Montmartre, the *quartier* which contains the Sacré-Coeur (see page 178), is an old village centre with a jumble of hilly streets and terraced houses, originally famous for its bohemian residents and village atmosphere, but now the haunt of coachloads of tourists. Neither of these areas contains a substantial residential population, so conservation policy has directly affected fewer people than in the other major conservation zone, the Marais.

The Marais

The Marais straddles the *3ᵉ* and *4ᵉ arrondissements* in the east central part of the city (Fig. 6.8). The area was originally a marshland on the floodplain of the Seine, which was drained and built on during the fifteenth and sixteenth centuries. Many imposing individual mansions, *hôtels particuliers*, were built here by bishops and nobles close to the Royal Palace at the Place des Vosges. By the eighteenth century, the relocation of the Palace to Versailles had caused an out-movement of the aristocracy. Progressively, and particularly after the Revolution, the grand houses became subdivided and rundown and their courtyards were infilled with workshops, lean-to sheds and other accretionary structures. The tall storeys and large rooms were divided both vertically and horizontally and gradually the population density increased as the housing quality declined. By the early twentieth century, the district was at its physical and social nadir, but was nonetheless an active working *quartier* with small-scale intensive industries such as jewellery, leather working, clothing and optical work, and contained a cosmopolitan population swollen by the immigration of refugee Jews from eastern Europe. At the census of 1968, the Marais had the highest population density in Paris (1483 people per hectare), the lowest amount of open space (less than two per cent) and the worst living conditions.

The designation of the Marais conservation zone in 1965 has stimulated transformation and improvement without wholesale demolition or redevelopment; this renewal *in situ* was made possible by the high quality of the original buildings (Kain, 1981). The large mansions are being renovated and restored to public use, for example, as libraries or museums. *Hôtels* and houses have been cleared and refurbished, notably around the Place des Vosges. Complete blocks have been stripped of their accretionary structures and workshops, and have been restored to residential use after renovation and the addition of modern facilities. Renewal of this type, exemplified by the *îlot* Payenne-Elzevir (Fig. 6.8), has improved access to the block, transformed amenity levels, extended the area of open space and reduced the population density of the area.

A further inevitable result of the conservation and renewal programme has been a gentrification of the population. The Marais is now one of the most sought-after locations in Paris, because of its centrality and because the conservation programme has enabled it to maintain much of its charm and character, with no major thoroughfares and no tower blocks. Prices for apartments rival those in the *16^{e} arrondissement* and the Ile St Louis (see pages 180–4), and although some of the original resident population is protected by rent control legislation, many of the new apartments have been taken by incoming professional people. The northern part of the Marais is not protected by conservation control and here the character of the area grades almost imperceptibly from specialist artisanal streets to the ordinary working class districts of inner eastern Paris.

Paris Euro-city

Redevelopment and renewal has affected a significant proportion of the City of Paris since the 1960s. The areas least affected have been the higher status districts of the west, which are of relatively recent and high standard construction. Redevelopment has changed part of the historic centre, particularly the area around Les Halles, but change has been most marked in the outer districts of the south and east. New residential and non-residential construction has replaced slum housing and has upgraded vacant or derelict land, such as that formerly occupied by the old abattoirs at La Villette or the site of the former Citroën factory in the southwest of the city. The vacant land has been used to provide both significant cultural facilities and vastly improved residential accommodation.

The purpose and impact of these major redevelopments may be viewed at a number of levels. One object of the redevelopment schemes has been to redress the imbalance between the east and the west of the city. However, the construction of extensive public (HLM) housing in the eastern sector will help perpetuate the existing working class concentrations in the east. A further consequence of these schemes has been the displacement of much of the original population to areas more distant from the centre and its replacement by professional groups with more spending power. Redevelopment has improved the housing conditions of all social groups but has also provided the opportunity for the elite to benefit from the development of prime sites. An underlying consideration in the grand schemes for the improvement of Paris has been the maintenance of its status and prestige not just as the national capital but as the leading cultural centre of Europe. When the European Single Market is finally achieved (see pages 241–5), the French expect that Paris will be the undisputed capital of Europe.

CHAPTER 7

Region, state and nation

On ne peut pas rassembler à froid un pays qui compte 265 spécialités de fromages.
Nobody can simply bring together a country that has 265 kinds of cheese.
Charles de Gaulle, speech after the elections, 1951

France as a nation

The formation of the French nation

The nation of France is a political unit which has evolved over a period of at least a thousand years. It is not a natural geographical unit and its boundaries have changed many times during its history (George, 1967; Pinchemel, 1969). Although more than half the national boundary follows the topographic features of the coast and the crest lines of the Alps and Pyrenees, the northern land border has no natural element. Even the coast has not necessarily proved to be a natural boundary, as several parts of France have been under English rule at one time or another.

France has grown from its medieval beginnings as a small royal domain around the City of Paris into a full nation state by a process of conquest and reconquest (Whittlesey, 1939) (Fig. 7.1). In medieval times, the territory which is now France consisted of numerous small provinces under the control of various bishops, counts and dukes, while much of eastern France formed part of the Holy Roman Empire. It was not until the early sixteenth century that the outline of the French nation became recognizable, although the provinces of Auvergne and Limousin and many of the peripheral parts of the present hexagon such as Bretagne, Lorraine and Corse, were still wholly or partly independent of the French crown (Fig. 7.1). Nevertheless, France achieved political hegemony over its territory significantly earlier than either Germany or Italy, incorporating most of the provinces of contemporary France by the end of the eighteenth century (Braudel, 1984: 315). During the nineteenth century, Savoie was finally acquired from Italy, although Alsace and part of Lorraine were temporarily lost to Germany between 1871 and 1919 (Pounds, 1954: 61).

The most recent boundary changes occurred in 1947 when the Franco-Italian border was relocated to the Alpine crest line around the *communes* of Tende and La Brigue (House, 1959).

Traditional regional differences predating the provincial boundaries exist between the north and south of France. Variations in climate between the warm Mediterranean and the cooler north (see pages 9–11) helped to initiate agricultural differentiation several thousand years ago, when wheat, olives and vines were first introduced to the Mediterranean (Grigg, 1974). At around the same time, the south was profoundly influenced by the customs, laws and architectural styles of the Romans. The

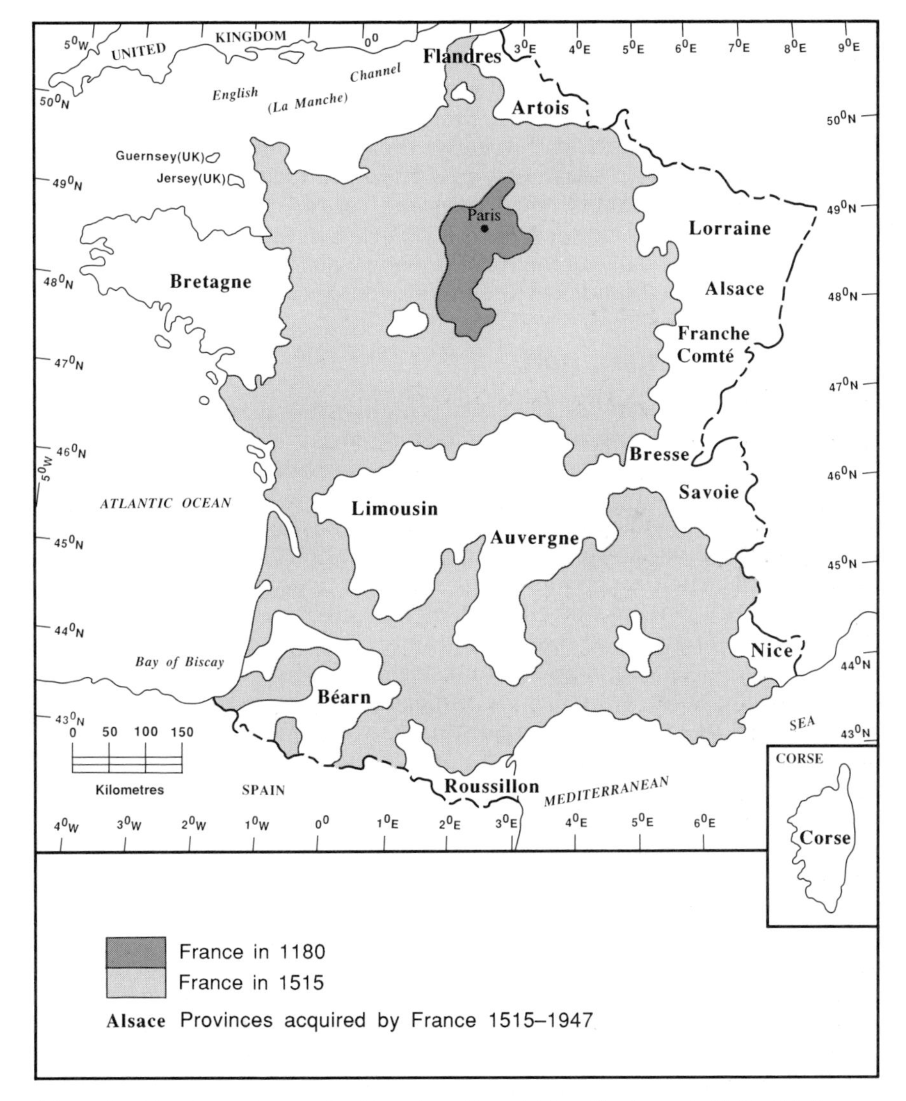

Fig. 7.1 The formation of the French nation. (After Pinchemel, 1969; 1986)

distinctive southern language, the *langue d'oc*, arose from the merging of Latin with the local dialect and was widespread south of a line from Bordeaux to Briançon. Much of the ancient civilization of the south was destroyed in the Middle Ages in the name of religious purity during the crusades against heretic groups such as the Cathars and Albigensians.

The process of the formation of the French nation has left its mark on the geography of the country in many ways. Most significantly, those provinces drawn most recently into the French ambit still maintain a high level of regional consciousness. Regional cultural differences, for example in dress and housing types, are much less marked than formerly, but they are still manifest in the continued use of minority languages and the growth of separatist and nationalist movements (see pages 230–3). Other effects of history are also clear. The location of towns is often due to factors which no longer operate; for example the towns in the Loire Valley at the borders of the provinces of Anjou, Bretagne and Poitou owe their location to historical strategic factors. Similarly, the fortification of towns such as Bastia and Bonifacio in Corse was a result of protracted warfare between Corsicans, Genoese and French over the control of the island. Some ancient provincial boundaries are still reflected in the location of forests, such as that of Arrouaise in the Nord–Pas-de-Calais, a remnant of formerly unoccupied marchlands between rival territories (Pinchemel, 1969).

National administration

The national administration of France is highly centralized in Paris, and is characterized by a powerful bureaucracy. France is a democratic republic led by a President directly elected by universal suffrage. The country is governed by a Council of Ministers headed by a Prime Minister. The Council is responsible to Parliament which has an upper house, the Senate, and a lower house, the National Assembly. The members of both houses are elected by universal suffrage for a five-year term. However, much power is wielded by the non-elected civil service, especially the ministerial cabinets. Most civil servants are graduates of the *école nationale d'administration* (ENA); these administrators form a technocratic elite, often moving sideways from the public service to head private-sector corporations or mass-media organizations (Birnbaum, 1980: 102). As a result, they are extremely influential in national affairs. Some devolution of power from the centre occurred under the Socialists in 1982, although it has been argued that this may have been more symbolic than real (Duboscq, 1989: 323).

There is a nested hierarchy of administration below the national level. This consists, in descending order, of *régions, départements, arrondissements, cantons* and *communes*, although the *arrondissements* and *cantons* have little administrative function. The separation of each level of government is by no means complete, as many local politicians take on national political roles, in a system known as the accumulation of offices, the *cumul des mandats*. For example, M. Gaston Defferre, the architect of regional decentralization policy in 1982, was also mayor of Marseille. The *cumul des mandats*, which has now been limited, results in a convergence of interests between the centre and the lower levels of administration (Mény, 1987: 63).

The basic administrative units of France are the *communes*, which were formed after the Revolution of 1789, mostly from pre-existing parishes. There are 36 494 *communes* in France, about 30 000 of which have populations of less than 2000 (Preteceille, 1988: 415). These very small *communes* have been described as 'out-of-touch with economic and social realities' (House, 1978: 44). The *commune*, led by an elected mayor, is primarily concerned with land-use, urban planning and with local services. Many small *communes* are unable to raise sufficient revenue from local taxes to pay for the services they require. The reform of *commune* boundaries has been on the political agenda for a long time, but local loyalties and prejudices are often hard to overcome. In practice, many *communes*, especially those in larger urban areas, have combined for the administration of joint projects (see, for example, pages 153; 161).

The main level of local administration is that of the *département*, a unit also formed after the Revolution of 1789. From that time to the 1980s, the chief officer of each *département*, the *préfet*, was the local representative of central government. The *départements* were designed to be of a similar size and of approximately equal population (see Frontispiece). When instituted in 1789, they were deliberately not coincident with former provinces, although many of their constituent *arrondissements* corresponded to distinctive cultural landscapes, known as *pays* (Boucher, 1973), and most were named after physical features. France now has 96 metropolitan *départements*; new ones were created in the Ile de France in the 1970s as a response to the growth of Paris (see pages 170–2). There is also a small number of overseas *départements* (see pages 237–9). Traditionally, the *départements* were responsible mainly for infrastructure provision such as roads and electricity supply, but since the Second World War they have been increasingly involved with the implementation of urban and regional planning programmes and with social services. The powers of the *département* were strengthened in 1972.

The *région* is the most recently introduced level of administration and has become increasingly powerful in recent years. The *régions de programme* were established in 1955 and were required to produce regional planning documents. In 1960, they were renamed as *circonscriptions d'action régionale* and during the 1960s their administrative machinery became fully established, complete with cabinets, civil servants and consultative councils. In 1972, they became simply *régions*, with no boundary changes but with a clearer consultative role on planning and development issues. There are now 22 *régions*, Corse having been designated as a *région* separate from Provence-Côte d'Azur in 1970 (Fig. 7.2). In the post-war period, the cause of regionalism moved to a central place on the political agenda. In particular, the Left adopted policies of regionalism and decentralization in the 1970s (Safran, 1984). The administrative reforms of 1972 changed little, as the *département* remained the principal level of administration, with the *région* merely adopting a co-ordinating role for local economic development. The *régions* were in any case just collections of the *départements* which already existed, and although some, such as Alsace and

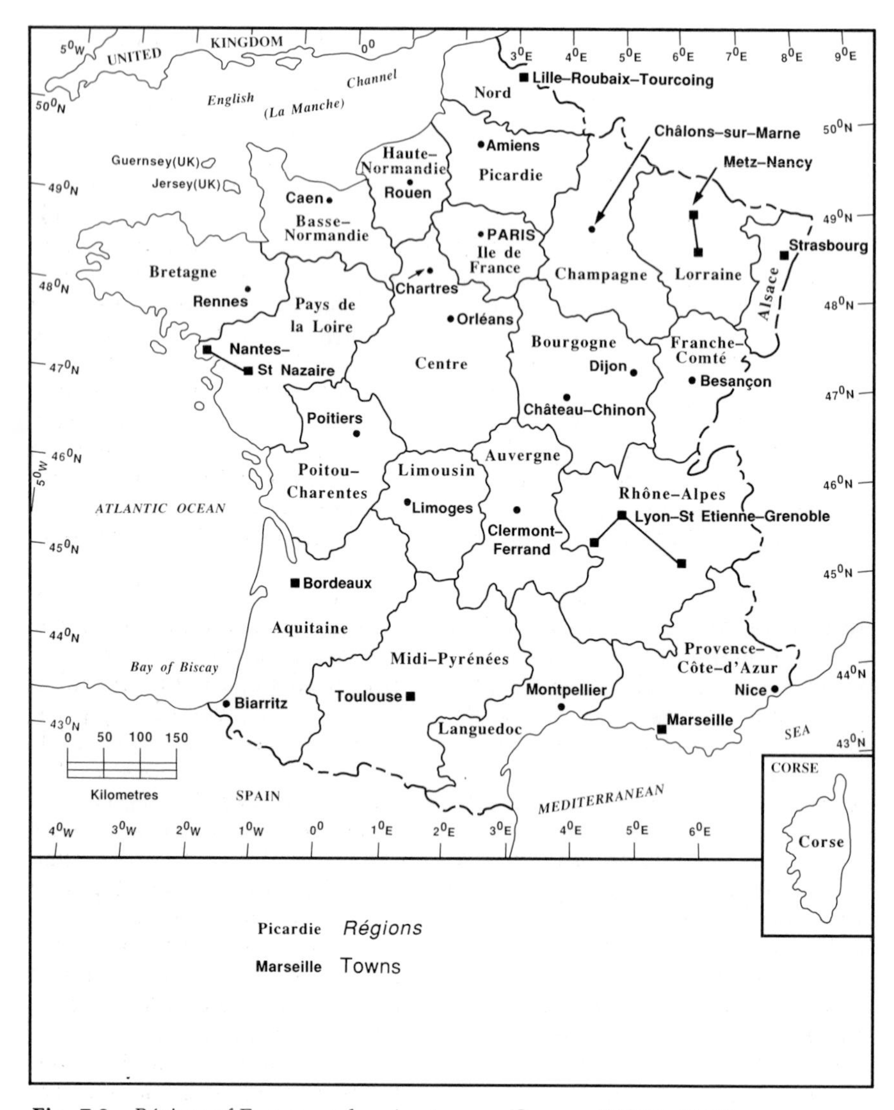

Fig. 7.2 *Régions* of France and major towns. (Source: INSEE, 1989)

Corse, had a real cultural identity, others, such as Centre, had none. Other regional boundaries failed to correspond to cultural regions. Thus, the *département* of Loire-Atlantique with its major centre of Nantes was not included in the *région* of Bretagne (Fig. 7.2). The formation of *régions* at this time was not so much a recognition of regional consciousness, but was rather a technocratic response to bring regional development into line with the rest of France (Coulon, 1978: 86–7). Although the legislation which established the *régions* was very restrictive, some of the regional councils nonetheless took on wider roles, such as the management of regional parks; the very creation of *régions* stimulated a regional awareness (Kofman, 1985; Mény, 1985).

When the Socialist government came into office under Mitterrand in 1981, one of its first commitments was to the decentralization of power. In 1982, the *Loi Defferre* was passed which gave the *régions* both the status and the funding to operate planning, economic and cultural activities. The *régions* were also to help co-ordinate local initiatives between lower levels of government. Examples of such co-ordination have occurred in urban planning, which is primarily the responsibility of the *commune*, and in social services, the responsibility of the *département*. There were two major differences between the *circonscriptions d'action régionale* of 1960 and the *régions* of 1982, even though the boundaries of the two units remained unchanged (Fig. 7.2). The first was that the *régions* were given sufficient finance to make a real impact on planning and economic development, even though the direction of plans had to be agreed by contract between *régions* and nation. The second and particularly important change was that the regional assemblies were to be elected bodies, which would in turn elect their own regional presidents. The first regional elections were held in 1983.

The planning contracts between national government and *régions* are an innovation of the 1980s, which stressed local initiative and needs. The contract was of a form which had been tested with the *villes moyennes* policy (see pages 155–6). It is generally conceded that the first round of regional planning contracts (1984–8) was relatively uninspired. The plans were rushed, and often consisted of little more than politically acceptable shopping lists (de Gaudemar, in Brunet *et al.*, 1989: 283). However, the second round (1989–93) was better thought out and broader in scope.

The devolution of power to the *régions* in 1982 has been accorded a very mixed reception. Some commentators have argued that the devolution has allowed new actors to take part in the planning and development process (Frémont, in Brunet *et al.*, 1989: 285). In many cases the local authorities have been able to acquire and use expertise formerly available only to agencies of central government. The *régions* may therefore be able to use technocratic arguments to further their own ends (Wachter, in Brunet *et al.*, 1989: 279). However, in a complex analysis of the politics of decentralization, Preteceille (1988: 413) has argued that decentralization is a way of consolidating centralized hegemonic control. The massive centralization of the post-war years, in everything from the welfare state to the location of the new towns, was increasingly subject to criticism in a period of economic austerity. The devolution offered an opportunity for local groups to participate in the political process, thereby providing a consensus for difficult economic decisions. Furthermore, the decentralization of the management of austerity interposed a protective layer of authority (the *région*) between the citizens and the state. Preteceille (1988) considered that the election of right wing officials in regional elections in 1983 and 1986 allowed neo-conservative policies to be reintroduced and legitimized, thereby reaffirming existing ruling class alliances in a slightly different form. He suggested that, in the long term, decentralization 'may create new conditions for local political debate and citizen's control' (Preteceille, 1988: 422), but that the Left's vision of a new citizenship formed by grass-roots participation was still an ideal rather than reality. Similarly, Flockton and Kofman (1989: 111) considered that the decentralization was a change at the institutional level which gave power to the bureaucracy rather than facilitating genuine public participation, radical local initiatives or new forms of relations between citizens and state.

National planning

In the post-war period, national planning by the national government has been primarily economic in nature. The idea of a sequence of National Plans was introduced in 1947. The First National Plan was prescriptive and was primarily concerned with the redevelopment of manufacturing industry for post-war reconstruction. Subsequent plans were more flexible and were used as indicators of future directions rather than as prescriptions. Unfortunately, the Ministry of Finance was never committed to the financial estimates of the Plans and so their targets were rarely met (Estrin and Holmes, 1983).

The National Plans have been increasingly overtaken by events (House, 1978: 22–7). The internal crisis of 1968, the oil crises of the 1970s and the enlargement of the EC in 1981 each caused enormous internal adjustments to be made to the French economy and consequently to the Plans (Duboscq, 1989). For example, when Spain and Portugal joined the EC, the Eighth National Plan was effectively abandoned and a new Plan prepared to protect the southwest from new economic competition (Duboscq, 1989: 323–4). The increasingly flexible nature of the National Plans reflects an increasing uncertainty about the role of centralized planning given the changing role of the region and the nation in Europe in the 1990s. The Tenth National Plan (1989–92) addressed some of the issues for the nation in the period until the single European market came into operation (Duboscq, 1989).

From the late 1970s, it became more difficult for the state to plan industrial development, because, with the exception of the nationalized sector, industry was increasingly under external control (see page 121). Planning was rarely used to interfere with market processes; instead the state itself became more entrepreneurial (Birnbaum, 1980: 110). In the 1980s, planning of industry necessarily changed from redistributing the fruits of economic growth to the management of economic crisis (de Gaudemar, in Brunet *et al.*, 1989: 282).

Increasingly centralized state control of regional planning and development was made possible by the creation of specialist technical commissions which circumvented the power of traditional local elites. The most significant of these bodies was the delegation for territorial planning and regional action, *Délégation à L'Aménagement du Territoire et à L'Action Régionale* (DATAR), established in 1963. This body controlled any investment of more than FF 10 million throughout France and therefore was instrumental in regional planning for tourism and industrial reconversion, and the urban planning of, for example, the *métropoles d'équilibre* (see pages 153–5). It is difficult to overstate the functions of DATAR, which oversaw enormous funds, established regional commissions and managed numerous special organizations. DATAR was the organization through which the state maintained its monopoly of planning and economic development (Lacour, in Brunet *et al.*, 1989: 278). This centralized planning was typical of the Gaullist regime and has been strongly criticized for its paternalism towards the regions and its preference for *gigantisme*, shown in monumental 'top-down' projects. The centralized nature of planning was modified in the 1970s to take more account of local initiatives. The most significant change was the introduction of the planning contract system which has changed the role of DATAR from an innovative planning body to an executive one (Wachter, in Brunet *et al.*, 1989: 283–4). From the 1970s, the national government came under increasing pressure from its own citizens to reduce the centralization of government, and increasing opposition was expressed to the imposition of planning from above (Flockton and Kofman, 1989: 60–3).

In the late 1980s, national planning appeared to be reaching a crisis point as France attempted to balance regional decentralization with the concerns of a united Europe (Lacour, in Brunet *et al.*, 1989). The dilemma for national economic planning stems from the apparent contradiction between needing to maintain a competitive national economy on the one hand and attempting to redress regional inequalities on the other (Brunet, in Brunet *et al.*, 1989: 273). French planning is being pulled two ways. At a national level, there is a continued desire to maintain the competitiveness of French industry to take full advantage of the single European market. Accordingly, substantial investment is being directed towards transport infrastructure, especially the extension of the TGV network, to ensure that France has the comparative advantage to attract footloose European capital (de Roo, in Brunet *et al.*, 1989: 280). Similarly, grandiose projects in Paris (see pages 206–8; 211) aim to maintain its primacy in the nation so that Paris will be the top ranking Euro-city of the 1990s and beyond (Riquet, in Brunet *et al.*, 1989: 286). However, planning at the regional level attempts to address local concerns which may not contribute to the overall efficiency of the French economy. Although many local projects will be too minor to attract European assistance, it appears to be in the regions' interests to stress structural regional disadvantages and spatial and economic problems in order to gain European regional development funding, especially in an EC enlarged by impoverished countries of the Mediterranean (Lacour, in Brunet *et al.*, 1989: 278).

National politics

The French political scene is complex, with a number of large political parties represented in power. Broadly, the parties can be divided into the left and right wings. The left wing, which consists mainly of the Socialist Party, the *Parti socialiste* (PS), and the Communist Party, *Parti communiste français* (PCF), came to national power under Mitterrand in 1981. Their major policies include social and regional equity and the nationalization of industry. Traditional left wing policies, for example, nationalization of industry, complete regional devolution and maintenance of employment in key industries such as steel, were partially abandoned in the mid-1980s because of the need for economic austerity (see pages 119–20; 135–6). The right wing consists of two major parties, the Union of French Democracy, *Union de la démocracie française* (UDF), a Liberal party led by Giscard d'Estaing, and the Rally for the Republic, *Rassemblement pour la république* (RPR), the Gaullist party led by Jacques Chirac. The right wing also contains numerous offshoots and affiliated groups of these two major parties. The right wing announced a new confederation between the UDF and RPR in 1990 to be known as the Union for France, *Union pour la France* (UPF), to run single candidates representing the centre-right in future elections. The right wing, which held national office from the end of the Second World War until 1981, tends to favour free-market policies and centralized government. Many of the post-war shifts in industrial and economic policy, for example, in economic planning or the nationalization of industry, can be related to the change in government which took place in the 1980s. Since 1983, the right wing has controlled many of the urban and regional authorities in France, including Paris, which has led to an uncomfortable balancing of power or *cohabitation* between national and local authorities.

In the 1980s, there has been a significant rise in support for the Far Right, notably the National Front, *Front National* (FN), which advocates extremist racist policies. In the 1984 and 1989 elections for the European Parliament, the FN captured between 11 and 12 per cent of the votes cast; in December 1989, the FN won a seat in a parliamentary by-election at Dreux, about 60 km west of Paris. This parliamentary success was achieved by a coalition with the Right which lent it respectability (Schain, 1987); a similar coalition in Le Luc, in the *département* of Var, early in 1990 caused the displacement of the incumbent socialist mayor. The subsequent formation of the UPF in 1990 is an attempt to exclude the FN from such coalitions and hence from mainstream right wing politics. However, in June 1990, in Villeurbanne, Lyon, although the left wing candidate won the seat, the FN candidate took 28 per cent of the vote, beating the UDF-RPR coalition into third place. The leader of the FN, Jean-Marie Le Pen, himself gained 15 per cent of the vote in the Presidential election in 1988, whereas in 1981 he had been struggling to find enough signatories to endorse his candidature (Singer, 1991).

The support for the FN is regionally based, being very strongly concentrated on the Mediterranean coast and in the Rhône Valley, Paris and major urban centres (Fig. 7.3). Analysis of the 1984 electoral support for the FN in Languedoc-Roussillon showed the highest level of support in the big cities of the region, Montpellier, Nîmes, Perpignan and Béziers, especially in areas with high unemployment and a high proportion of foreign workers; conversely it had little success in rural areas (Bernard and Carrière, 1986). Marseille is the particular heartland of the FN. Here immigrants form about 10 per cent of the population. More importantly, there are also very large numbers of returned migrants from north Africa, the *pieds-noirs*, many of whom adopt racist attitudes. Marseille also has high rates of unemployment, a declining population and massive social problems in the suburbs. Furthermore, the city has never fully recovered from the mistimed programme of economic expansion at Fos (see pages 130; 157) and consequently the vote for the FN is seen partly as a protest vote against the major parties.

Since the 1980s, there has been an upsurge of racist incidents in France. Such incidents include the desecration of a Jewish cemetery in Carpentras in 1990 and mass demonstrations organized by the FN in Marseille in 1991 against immigration. In 1990, Paul Bouchet, president of the national consultative commission on human rights, published a report on racism in

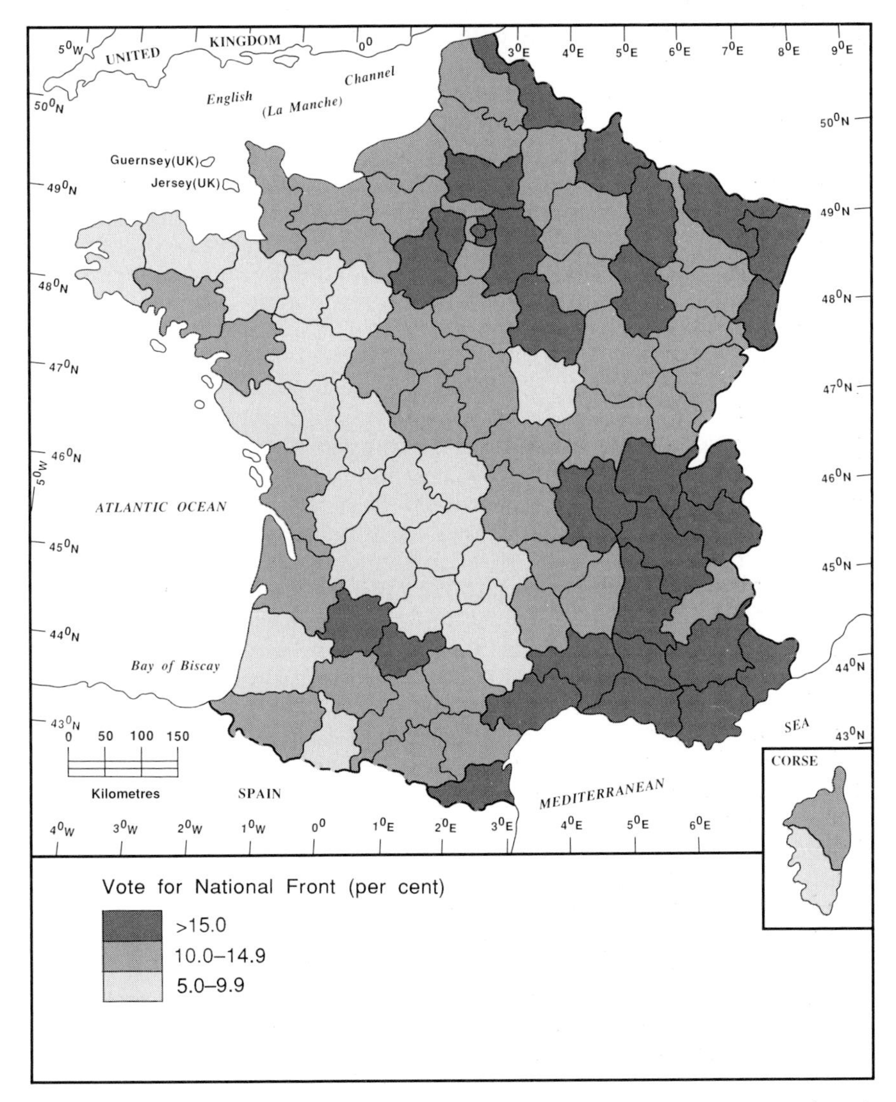

Fig. 7.3 Political support for the National Front, 1988, as a percentage of total votes cast. (Source: Flockton and Kofman, 1989: 36)

France. This report detected a sharp increase in verbal threats and in openly avowed and commonly accepted racism. As a result, new regulations imposing more stringent measures against racism and anti-semitism were passed in June 1990. This legislation followed the conviction of Le Pen on the grounds of incitement to racial hatred; in a well-publicized speech he described the Nazi gas chambers as a 'point of detail' in the Second World War.

The upsurge of support for the FN in the 1980s coincided with the election of the Left to national office. When the Right was in office, the problems of urban decay and immigration could be portrayed as part of an international trend. However, with the Left in office, their laxity could be blamed for these problems and people's fears about law and order, immigration and unemployment could be played upon (Oberhauser, 1991: 442). Indeed, Singer (1991) argued that the rise of the FN was only politically possible with the Left in power. The FN is a social movement of the extreme Right, but it also has elements of a single-issue movement (Mitra, 1988). It is thought that electoral support for the FN may diminish in the 1990s because of its limited political base on the issue of immigration, the removal of centre-right support from the FN and a change in the voting system away from proportional representation (Mitra, 1988).

National identity

The national identity of any country is a complex phenomenon based on a territorial and political reality. The creation of a national identity fulfils social, economic and political functions, the most obvious of which is the fostering of national unity. A national identity is unique; Smith (1991: 16) considered that certain events and places come to symbolize identity: '"sacred centres", objects of spiritual and historical pilgrimage, that reveal the uniqueness of their nation's "moral geography"'. For France, the obvious 'sacred centre' is Paris, which has been and is still deliberately constructed as a symbol of France. The centre of Paris reflects the deliberate centralization of the French nation.

Braudel (1986), in his major work entitled *L'identité de la France*, considered that the character of France was a product of environment, economy and society. These factors combined to produce the small distinct self-sufficient regions, the *pays*, which were so characteristic of pre-industrial France. A whole French school of human geography was based on the study of the regional differences between the *pays*, the most famous exponent of which was Paul Vidal de la Blache. Most geographers continue to stress this diversity within a powerfully centralized administrative framework, although Braudel (1986) admitted that this diversity may be no more than that which exists within the borders of Spain or Britain. Certainly, House (1978: 36–9) emphasized the dual themes of unity and diversity, of centripetal versus centrifugal forces, of a shared history, culture, government and destiny holding in balance the geographical and socio-economic differences evident at the regional and sub-regional level. The hexagon of France is seen as a marvellous symmetrical shape, a symbolically perfect territory which holds together this human and historical diversity.

Zeldin (1983) considered that essential Frenchness has undergone a significant change in the post-war period. This change corresponds primarily to the shift of emphasis from centralization to regionalism. Frenchness from the time of the Revolution to the Second World War used to be embodied in the nationalistic ideals (if not the reality) of liberty, equality and fraternity. During the post-war reconstruction and the European co-operation of the 1960s, nationalistic Frenchness was replaced by a sentiment of internationalism (Zeldin, 1983). Zeldin argued that the new Frenchness incorporated a commitment to humanistic and lofty ideals which transcended nationalism (although these ideals may have been far removed from France's international behaviour). This international idealism has subsequently been modified by an emphasis on the plurality which exists within the nation, a plurality which derives from traditional regional aspirations and from recent immigration. Zeldin (1983: 510) considered that the demise of nationalism would be followed by the demise of pluralism and that mature Frenchness would be less constrained by overarching ideologies but would allow individuals to be themselves.

The identity of the French nation is obviously greatly affected by changing international circumstances, in particular major post-war political events, such as the receipt of American aid after the Second World War, the formation of the European Community in 1960 and the dismantling of the Communist bloc in 1991. Great popular and official concern has been expressed about the extent of American influence on French culture, especially by the Americanization of television and language, and by the proliferation of consumer products such as Coca-Cola and McDonald's fast food. On the other hand, French distinctiveness has not been diminished by entry to the European Community; as a leading member, France has been very successful in maintaining its position. The preservation of national interests by the French has often been perceived as arrogance by other European countries: Ardagh (1968: 455) referred to this as 'old-style French insularity and disregard for other nations'. This reputation for arrogant individuality has been maintained by the French predilection for prestigious and profligate international projects, such as Concorde and the Channel Tunnel. French concern for its own interests has been consistently demonstrated by controversial participation in international affairs, ranging from nuclear testing in the Pacific to the lamb and cod wars with other EC countries.

The French character is thought to have a certain style, *élan*, linked with culture and intellectual life. Ardagh (1968) listed the 'real French virtues' but wondered whether these could last under the onslaught of Americanization and internationalization of culture: 'a flair for style and a care for quality; the honouring of individual prowess, the ethos of individual fulfilment; lucidity of thought, a passion for ideas, a certain concept of liberty, of human proportion, of harmony amid diversity; the enrichment of the present through the past' (Ardagh, 1968: 456–8). To foreign travellers, the essence of France is found in the timeless rural areas, the *pays*, differentiated in a myriad ways, notably by their food and wine. The trend

Plate 7.1 A major communication axis for France: barges on the River Seine at Rouen. (Photo: S J Gale)

towards regionalism has accentuated this impression. The outsiders' image of France remains a rural romantic one rather than that of a highly successful economic power. Certainly, problems such as those of the suburban housing estates or the national debate over immigration rarely impinge on visitors, because the suburbs and the quasi-ghettoes are not tourist destinations.

The regions of France

Regional inequalities

The ancient north-south divide between Roman Gaul and the feudal law (see page 213) has long since been replaced by other aspects of regional differentiation. By the seventeenth century, the south was relatively poor and backward compared with the north, but the dividing line was already shifting (Braudel, 1984: 336–7). From the eighteenth century, the impact of urbanization and industrialization affected the northeast much more than the southwest. Areas in the orbit of Paris and Lyon and the regions on the major coalfields attracted population from the declining west. In 1947, Gravier could characterize the west as the 'French desert'. Not only was the west of France depopulated and rural compared with the north and east; it also suffered from lower incomes, higher levels of illiteracy and poorer housing.

The division between northeast and southwest is still significant in contemporary France. Many of the main urban areas lie north and east of a line drawn between Rennes and Marseille. To the northeast there is a higher proportion of the labour force employed in industry (Fig. 4.1a) and levels of income and economic productivity are higher than in the southwest. By contrast, most parts of the south and west receive higher levels of assistance from France and from the EC for agricultural improvement (Fig. 3.7) and from the French state for industrial and service development (Fig. 4.2). This socio-economic division is accentuated by the location of the main axes of communication. Within France, these run from the Channel coast to Paris and from Paris to Marseille, and within Europe they lie between Paris, Strasbourg, Bruxelles and southern Germany (Plate 7.1).

Regional inequalities in France have become less clear and less extreme since the 1950s (de Gaudemar, in Brunet *et al.*, 1989: 277). A quality of life index calculated by Knox and Scarth (1977: 16) revealed clusters of *départements* which did not conform to any clear north/south, east/west or administrative classification. The ancient northeast/southwest dichotomy has been greatly affected by patterns of demographic and economic change. A major factor has been the reversal of historic migration trends in the 1980s. Many rural areas have gained population from counterurbanization. This population change has not only brought wealth and new development, but is also symptomatic of the functional integration of formerly isolated areas into the wider urban and economic system. At the same time, former growth areas such as the Nord and Lorraine have suffered from out-

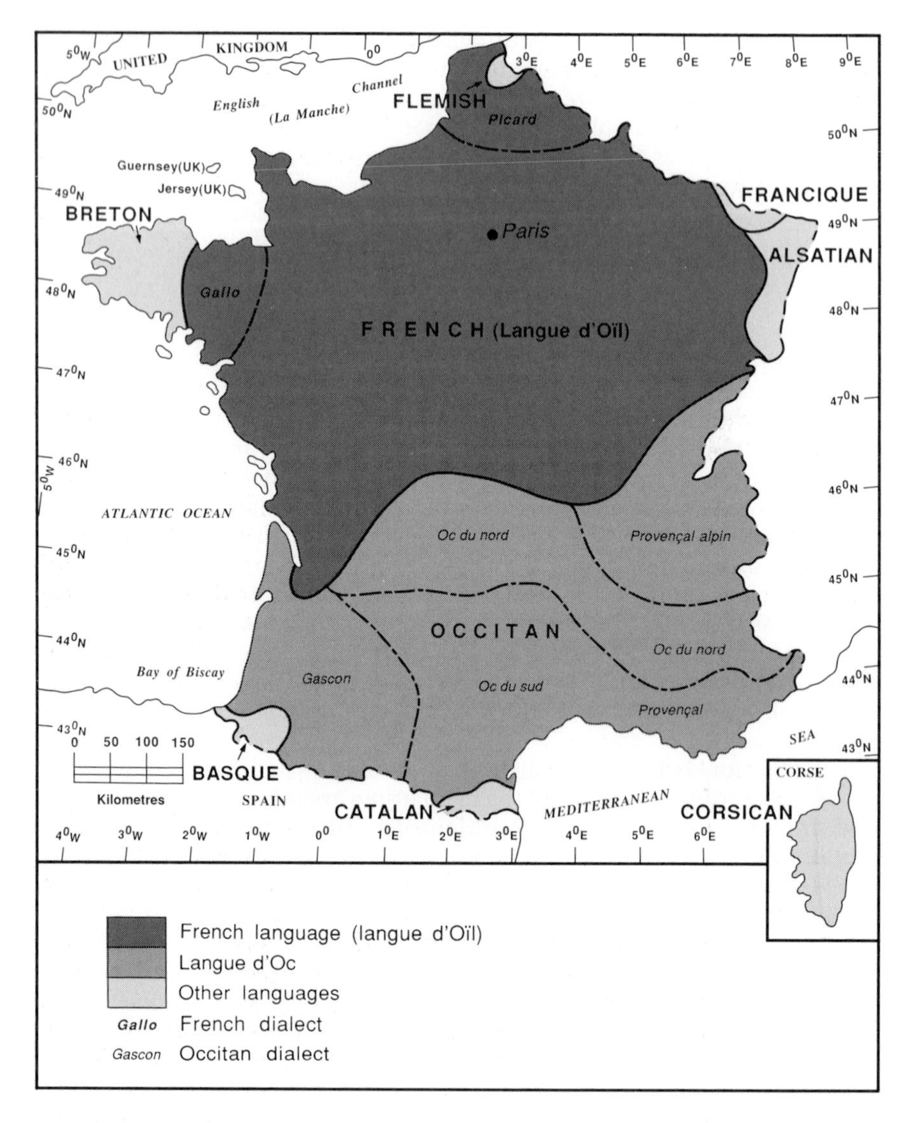

Fig. 7.4 Linguistic regions of France. (After Jacob and Gordon, 1985: 108; Offord, 1990: 144)

migration (Fig. 2.4c). Industrial restructuring has resulted in de-industrialization of the coalfields and decentralization from Paris, which in turn has brought about a more equitable distribution of manufacturing employment. New footloose high-technology industry is much more widely located throughout France than traditional heavy industry (Fig. 4.8). At a European scale, Keeble (1989: 8) argued that the impact of de-industrialization, restructuring and service development has been demonstrated in a reduction of core-periphery disparities. These patterns of industrial change have also reduced the economic disparities between the large cities of France (Pumain and St Julien, 1984), resulting in a more finely grained pattern of socio-economic inequality.

Problem regions have traditionally been characterized as underdeveloped rural areas, declining industrial areas and overdeveloped urban areas. In the 1990s, the formerly underdeveloped rural areas of the west are in many respects no longer problem regions, as their agriculture is efficient and competitive, tourism is highly developed and they have gained an influx of new people. Similarly, the old industrial regions have benefited from decentralization of growth industries and from programmes of environmental improvement. However, the employment deficit caused by the loss of the traditional industrial base has not been completely filled by new industry, resulting in the out-migration of skilled labour. Although regional disparities have been reduced, the inequality between Paris and the rest of France has been maintained or even exacerbated; during the 1980s, Paris gained half a million people and numerous prestigious developments and is still the control centre of the new decentralized administration. Despite the primacy of Paris, some of the most highly disadvantaged areas of France are contained within its suburbs. The new problem regions are to be found within the suburbs of Paris and large cities throughout France, which contain large concentrations of poor housing and disadvantaged people (see pages 166–7; 187–9).

Linguistic regions of France

Several minority languages are spoken in France, although French is the official language for the whole country. The main regional languages are Basque, Breton, Catalan, Corsican, Flemish, German (Françique and Alsatian) and Occitan (Provençal) (Fig. 7.4). France is therefore the most linguistically plural state in western Europe (Jacob and Gordon, 1985: 107). Each of these languages has affinities with the languages of neighbouring regions or nations. Thus, Breton is a Celtic language akin to Welsh, Occitan derives from Latin and Corsican from an admixture of Italian, while Basque is a distinctive language of non-European origin spoken mainly in the adjacent Basque territory in Spain. Regional dialects of French are spoken in much of France, usually in addition to standard French; the most important are Gallo, spoken in eastern Bretagne, and Picard, used in Picardie.

Regional languages were vigorously suppressed during the nineteenth and early twentieth centuries. The consolidation of the French language was part of the centralization of the French state; the existence of minority languages was felt to be inimical to the unity of the French nation. The main method of control was by the introduction of compulsory schooling through the medium of French. The use of regional languages was prohibited in schools and punished, while French was promoted as the language of progress, commerce, law and civilization and was compulsory in the armed services and the civil service (Offord, 1990). Other languages were equated with backwardness, stupidity and provincialness, in all its pejorative senses, so that even older people, the inheritors of the language, encouraged youngsters to learn and speak French, in order to 'get on'. It was expected therefore that after a few generations the regional languages would be used by only a few old people and would eventually die out altogether.

A resurgence of interest in regional culture and language first appeared in the late nineteenth century, as a reaction against the centralizing tendencies of the French state. In Provence, interest in Provençal as the language of the troubadours arose as an offshoot of the literary movement of Romanticism; the literary works of the region included the publication of a Provençal dictionary in the 1880s as well as numerous romantic and allegorical writings by Mistral (Hayes, 1930). The regional societies and journals that were founded about this time became more political than literary and, as early as 1892, a Provençal manifesto was published, demanding federalism rather than centralization and cultural and political autonomy for Provence.

Similarly, in Bretagne, where Breton had been widely spoken, a resurgence of linguistic interest occurred in the nineteenth century, when various epic poems in Breton and a Breton history were published and spellings and rules of grammar codified. At the turn of the century, the Breton Regionalist Union organized literary and theatrical prizes, and by the outbreak of the First World War, a number of rival societies had been founded, at least one of which demanded home rule and political independence from France and another which worked for a Pan-Celtic alliance. Similar linguistic and literary revivals which in turn have led to political action have been evident in the Basque country and in Corse.

The use of regional languages reached its nadir about the time of the Second World War. A prolonged period of suppression, the need for national unity in two world wars, and the opening of isolated regions to wider influences had reduced the speaking of minority languages to an increasingly residual and ageing population. After the Second World War, however, regional social movements were revived, spearheaded by a group of young left wing intellectuals, a group similar to those in the vanguard of student unrest and agricultural reform at that time. These groups campaigned vigorously for the right to use minority languages as the medium of instruction, and a number of educational reforms was introduced.

The *Loi Deixonne*, passed in 1951, permitted the optional teaching of Basque, Breton, Catalan and Occitan (and Corsican from 1974) in the regions in which they were spoken. However, the Ministry of Education often omitted to implement the necessary regulations and failed to permit the training of teachers in these languages (Safran, 1984: 451). The token nature of the legislation, which allowed only one hour per week of language instruction in secondary schools, is shown by the very small numbers of students who elected to study a minority language in the late 1970s and early 1980s. These averaged about 15 000 per year for all minority languages, of whom approximately 60 per cent learned Occitan and 20 per cent learned Breton (Jacob and Gordon, 1985: 123).

In 1982, the language teaching provisions were much extended and the Ministry of Education allowed bilingual education at primary level; in May 1983, six languages other than French were recognized for the *baccalauréat*: the five specified in the *Loi Deixonne*, with the strange addition of Gallo (Giordan, n.d.). Bilingual education is organized through joint co-operation between the Ministry of Education and regional linguistic bodies; in Bretagne this is the *Diwan*. Tertiary education in regional language and culture is available at regional universities, including the new Université de Corse at Corte. Additional diffusion of minority languages was made possible by the relaxation of broadcasting laws and the establishment of various regional cultural facilities, such as the regional museum at Rennes. In Corse, the new regional assembly formally adopted a policy of bilingualism in 1983, recognizing language as the 'cement of culture' (Fusina, n.d.).

Table 7.1 Minority languages used in France

Language	Number of speakers			
	1881[1]	**1953**[2]	**1968**[3]	**1989**[4]
Alsatian	n.d.	1 600 000	1 300 000	>1 000 000
Basque	140 000	100 000	85 000	60 000
Breton	1 340 000	1 000 000	800 000–900 000	700 000
Catalan	208 855	185 000	170 000	200 000
Corsican	272 639	300 000	200 000	<275 000
Flemish	176 860	200 000	200 000	80 000
Occitan	n.d.	9 515 000	7 000 000–8 000 000	several million

Source: [1]Groeber, 1902: 570 (quoted in Hayes, 1930).
[2]Dauzat, 1953: 122–38.
[3]Pottier, 1968: 1155–60.
[4]Rickard, 1989: 124.

n.d. no data.

No accurate census of non-French speakers in France is available. The data given in Table 7.1 are estimates and refer to passive rather than active users. Active users who use the language daily probably account for about one third of the totals given (Offord, 1990: 145). The data for 1968 and 1989 are roughly comparable, although there is an apparent dramatic decline in the number of Flemish speakers. Other languages are also spoken in France because of recent immigration, particularly Portuguese, Spanish, Italian and Arabic. The largest of these groups is the Arabic speakers from north Africa, who number well over a million and are chiefly located in Paris and other urban areas (see pages 62–3). Many of these people, especially women and the elderly, suffer from linguistic and cultural isolation, as it is estimated that approximately a quarter of France's four million recent immigrants from abroad speak little or no French.

Regional separatism

The objective of those regional groups which were revived in the 1950s and 1960s was not merely the promotion of minority languages. In the more extreme cases, they became liberation movements for complete independence. Whether independence or some lesser form of autonomy was the aim, in every case the regional groups questioned the way that economic and political practices affected their territory (Kofman, 1985). An example of such practices is the way that the new spatial division of labour has affected different parts of France (see pages 116; 121–3). Most regionalists would consider that the decentralization of production by the state and by large firms constituted a form of exploitation of the periphery, producing in the regions a branch plant economy, which had little control over its own destiny. Other controversial issues included environmental concerns, such as the location of nuclear power stations and the use of resources. For example, opposition to the Cattenom nuclear power station provided a focus of discontent for the Françique speakers of the *département* of Moselle (Laumesfeld *et al.*, n.d.). It was felt that the centralization of the French state had brought about a situation of internal colonialism, whereby the peripheral regions were economically exploited by the core of France in the same way that the overseas colonies had been ransacked for their raw materials (Hechter and Levi, 1979; Reece, 1979). Such separatist movements have been most active in Bretagne, Occitania (Provence-Languedoc), the Basque region and Corse. By the mid-1970s, regional ethnic violence within France reached unprecedented levels, culminating in 1978 with the bombing of the Palace de Versailles by Breton nationalists (Jacob and Gordon, 1985: 122).

The regional decentralization of 1982 did not wholly satisfy the groups wanting autonomy. Hopes of greatly extended regional powers, of flexible relations between state and region and of redrawn boundaries, such as the five-*département* Bretagne (see page 216) and new *régions* for Savoie and the Basque country, were dashed. Nonetheless, by 1984, all *régions* except Corse had signed planning contracts, which were renewed in 1988; these planning contracts were aligned with national development priorities.

Autonomist and separatist movements have continued their struggle, often violently, in the Pays Basque and in Corse. The island of Corse is a special case and has been given a special status, *statut particulier*, in the decentralization programme. The island has a bloody and tumultuous history and for long periods of time was under the control of, or at war with, the Genoese. For centuries it had been fought over by Italy, France, Spain and England, and has a history of resistance and independence (Turnbull, 1976). It was finally ceded to France by Genoa in 1768 and became a *département* of France at the Revolution, a process eased by the fact of Napoleon's Corsican heritage. Nonetheless, the historic struggle for independence never really died. Noin (1987) considered that Corse was not even on the periphery of French space, but was genuinely marginal, as it had no regional urban centre as a focus. Particular resentment is felt against the French for their economic colonialism of the island, in particular for 'parachute tourism' and for the lack of real economic development which has necessitated massive out-migration.

In 1983, the Corsican National Liberation Front, *Front de Libération Nationale de la Corse* (FLNC), was banned by the national government for its extremist activities. Despite the special statute granted to the island and its massive *per capita* funding, the nationalist movement has continued its violent attacks. During the 1970s, each wave of violent activity brought more resources to the island, but this no longer appears to be the case. The violence has included bombings of government buildings, such as the tax office at Bastia, which was totally destroyed by an explosion in 1987. Tourist developments have also been bomb targets; in December 1990, 40 holiday homes were blown up and a number of occupants temporarily taken hostage. These acts of violence have alienated the FLNC from a large proportion of the Corsican population. In 1987, during a spate of bombings, electoral support for autonomous movements reached a low point, recording less than nine per cent of the votes cast in a departmental election. The problem of regional autonomy in Corse is unlikely to be easily solved. It has been difficult to make the regional assembly work effectively, and proposals for greater devolution of power over education and transport scheduled for 1992 have been held up because of French opposition to what is seen as Corsican blackmail. Furthermore, the political system in Corse is dominated by traditional factions which are inflexible and self-serving. A fundamental problem appears to be that the Corsicans want autonomy but they want the French to provide the funds.

A rather different pattern of events has occurred in the Pays Basque and in Bretagne. The separatist movements in Bretagne were the fiercest and most vocal in France during the 1960s and 1970s, with widespread protests, daubing of slogans and publication of Breton nationalist literature. Breton autonomy movements before the Second World War emphasized the need to defend the Breton language and the historic independence of Bretagne, but in the post-war period, the emphasis of the argument shifted to economics. Liberationists felt strongly that Bretagne was disadvantaged because of its link with metropolitan France and that its relationship amounted to one of internal colonialism (Reece, 1979). The argument for decolonization of Bretagne was the more powerful because at that time France was withdrawing from its former overseas colonies; if Algeria, why not Bretagne?

The reduction in the intensity of the Breton nationalist movement during the 1980s may be attributed to a number of factors. First, the regional economic disparities between Bretagne and Paris which were so great in the 1940s (Gravier, 1947) had been massively altered by the 1970s as a result of agricultural modernization and counterurbanization (see pages 53–5; 99). Secondly, the teaching of Breton at all levels of the education system, the training of teachers of Breton since 1982, and the availability of Breton studies at tertiary level at universities in Brest and Rennes have significantly increased the status of the Breton language, although there is as yet no regional policy of bilingualism. Thirdly, the election of the Socialist government and the consequent regional reforms have provided an opportunity for greater political participation for Breton activists. The major difference between Bretagne and Corse in this respect is that Bretagne has a modernizing and mediating elite, a group of people who were in the forefront of agricultural change in the 1960s and who could see opportunities for themselves and for the region in the new political system. In Corse, on the other hand, no such group exists.

The separatist struggle of the Basque country has become more intense in the 1980s than in previous decades. In the 1970s, while the Spanish Basque liberation movement, ETA, was at its height, the French Basques were less politicized. However, Douglass and Zulaika (1990: 225) noted the extreme permeability of the border between Spain and France and the ability of ETA to use the French Basque country as a refuge. In a comparative study of Basque nationalism in France and Spain, Lancaster (1987: 568) found significant differences in self-identification; on the French side of the border, 85 per cent of respondents felt they were French or primarily French, whereas on the Spanish side, only 35 per cent considered themselves to be Spanish or primarily Spanish.

The Basque movement has focused less on language than the Breton, mainly because of the limited diffusion of the Basque language and its extreme fragmentation into dialects (Conversi, 1990: 63). In the Basque country, Lancaster (1987: 570) estimated that 73 per cent of the population spoke only French. The Basque movement, particularly in Spain, underwent its most violent period in the late 1970s when large-scale immigration threatened the dilution of Basque culture; accordingly, race rather than language became the basis of Basque separatism (Conversi, 1990: 63–4). However, neither the issue of language nor race is politically tenable; Basque is difficult to learn and French is widely spoken, while exclusion on the grounds of race or imputed origin is politically unacceptable to most Basque residents. Lancaster's (1987) survey found that only 25 per cent of French Basques desired regional autonomy.

Nonetheless, political violence in the Basque country has become deeply entrenched. This may be a function of the behaviour of a few individuals whose influence is out of all proportion to their numbers or status in the organization, as is the case in Spain (Douglass and Zulaika, 1990). Increasing violence in France and a hardening of government attitudes towards ethnic and cultural minorities has produced a number of extraditions of Spanish Basques to Spain since 1986. In 1987, the separatist group, *Iparretarrak*, was banned by the French government and its leader sentenced in his absence for the murder of two police officers.

The Basque movement differs from the Breton in three main ways, apart from the issue of language. The Basques have a greater sense of enduring regional socio-economic inequality; the movement is greatly influenced by the success of the Spanish Basques; and the leadership is less prepared to co-operate with the elected representatives in the national political process. Indeed, in these ways, the Basque movement bears more similarity to the Corsican than to the Breton; arguably however, the Basque movement appears to have even less public support.

The future of the regional separatist or nationalist movements is unclear. It is possible that their strength may diminish as a result of continued cultural integration and because of satisfaction with the limited level of autonomy already achieved. Their freedom from centralized suppression may also paradoxically induce decline. Mény (1987: 60) envisaged that the 1982 regional reforms had laid the groundwork for an integration based on consensus which would be much more effective than authoritarian measures. On the other hand, it seems highly likely that both internal and international affairs will combine to fuel the regionalist fires. Internally, national unemployment levels at around ten per cent mask regional unemployment rates which are twice as high. Continuing unemployment, if combined with further economic recession, will add to the sense of regional grievance. Internationally, reduced economic well-being may be brought about by the rationalization and reduction of CAP expenditure, which is a major financial support to the peripheral regions. Furthermore, the rise of ethnic movements in eastern Europe and the former USSR in the 1990s may provide the more militant regions with examples that they may wish to emulate. Although ethnic unrest may increase, the prospect of further devolution towards a completely federated structure within France nonetheless seems unlikely. France will resist regional autonomy because of fears of reducing the impact of the French presence in Europe and a reluctance to allow a concomitant increase in European control over internal French affairs.

The issues of immigration, of linguistic diversity and of regional separatism are common tensions generated within plural societies. These tensions have become significant political issues in late twentieth century France. Even before the Mitterrand era, France was reducing its centralizing tendencies and becoming more overtly pluralistic in its identity (Zeldin, 1983: 510). Often, however, the level of debate about these critical issues has been emotive, irrational and drastically over-simplified. For example, the simple equation of two million foreign workers and two million French unemployed put forward by the FN, fails to recognize the nature of immigrant employment (often in jobs the French will not do) and the nature of immigrant unemployment. Moreover, it is the problems rather than the benefits of plurality which have received most attention. The benefits are most clear in the retention of history and culture which could otherwise be subsumed, as well as in greater understanding and appreciation of the diversity of Frenchness. Clear recognition of the needs of ethnic minorities, whether territorially based or not, has been given by the Mitterrand regime. At the same time as the major minority languages were granted increased status in a plural society, France also recognized the special needs of non-territorial minorities, particularly the Armenians, the Gypsies and the Jews (Safran, 1985: 52). These needs may be practical, such as the designation of camping sites for Gypsies, linguistic, or mainly symbolic and protective, as in the case of French Jewry.

France in the world

Legacies of empire

France was the second largest colonial power in the late nineteenth century, with major colonial possessions in north, west and equatorial Africa, and lesser ones in Indo-China, the Pacific and scattered through every ocean. The French adopted politics of direct government and assimilation, such as education in French. In return, the recipients could claim French citizenship although they had little opportunity to take advantage of this gift (Wallerstein, 1961: 66–75). By 1945, this colonial edifice was crumbling and by 1960, decolonization was virtually complete. Nonetheless, France is still a significant international force and has been classed by Aldrich and Connell (1989a) as a mid-range power with a global presence:

> As a permanent member of the United Nations Security Council, the fifth largest economy in the world, the proprietor of the world's third largest maritime zone, the former administrator of the world's second largest colonial empire, the mother country of a language spoken by over 1 hundred million people, one of the five nuclear powers, the third largest arms vendor in the world, and a founding member of the European Community, France can justifiably claim to have an international presence, even if its might does not equal that of the superpowers. (Aldrich and Connell, 1989a: 1)

France's effectiveness as an international power derives from three sources. First, it retains considerable influence and support from continuing relationships with its former colonies and from the strategic retention of a number of overseas *départements* and territories. Secondly, it is a major economic power with a pivotal role in the European Community. The relative international status of France has been increased by the dismantling of the USSR in 1991. Thirdly, and less tangibly, its status derives from its cultural significance in the realms of cuisine, art, fashion and literature (Wackermann, 1992).

France, unlike Britain, opposed independence for most of its colonies and decolonization was often only achieved after protracted negotiation and overt hostility. A proposal for the tripartite division of Libya between Italy, Britain and France was rejected by the United Nations (UN), Libya receiving independence in 1951. The independence of Libya stimulated similar movements in other French north African possessions, especially in Tunisia, Morocco and Algeria. A plan for Tunisian autonomy was sabotaged by the Tunisian French community, many of whom had been established there for generations. After some years of guerrilla warfare and terrorism, Tunisia was granted independence in 1954, followed by Morocco in 1956. In the same year, the impotence of France and Britain as powers in the region was demonstrated by the failure of their intervention in Suez where their armies were withdrawn at the insistence of the UN (Tipton and Aldrich, 1987). During the same period, decolonization occurred in Indo-China, with independence being granted to Laos and Cambodia in 1954, while Vietnam was divided pending elections.

The most painful and protracted experience of decolonization occurred in Algeria, which was the major French settler colony. Algeria, colonized in 1830, was legally part of metropolitan France. Effectively, however, all the Algerians were disenfranchized. As part of France's assimilationist policy, Algerians and other non-French could only apply for citizenship if they renounced Moslem law. The independence struggle which started in 1954 was fuelled by the experience of Tunisia and Morocco and by demographic and urban growth in Algeria. Between 1956 and 1958, there was outright warfare, involving torture, executions and intimidation. The Algerian crisis brought General de Gaulle to power in 1958 and initiated the formation of the Fifth Republic. Algerian independence was finally guaranteed by the 1962 Evian agreement; in that year many of the *pieds-noirs* returned to France and the great migration of Algerian workers to France commenced (Adler, 1977). The key political issues of immigration, citizenship and racism (see pages 220–2) can be traced to this period.

The decolonization of west Africa and equatorial Africa was stimulated by the granting of independence to British colonies, in particular the Gold Coast (Ghana), about this time. In 1958, the French west and equatorial African colonies were offered the choice of independence, or autonomy within a French African community. With the exception of Guinea, all initially voted for a French community, but Guinea's refusal was enough to undermine the proposal. The idea of the community collapsed and, by 1960, the huge area stretching from the Sahara to the Congo had formed a galaxy of separate independent states. Some of the last colonies to achieve independence were the Comoro islands (except Mayotte) in 1974 and French Somaliland in 1977.

Since the time of independence, the relationships of France with its former colonies have become neo-colonial. Colonization had produced very little industrial investment in the colonies. These states were used to provide raw materials and to consume manufactured goods from France. To a great extent, these trading relationships continued. For example, the Evian agreement kept Algerian oil and natural gas under French control, where it remained until the industry was nationalized in 1971 (Bennamane, 1980). Current patterns of trade and aid are notably asymmetrical; about 40 per cent of imports and exports of the former colonies are traded with France, while about eight per cent of French imports and exports are traded with Africa. Meanwhile, the level of French aid to the former colonies has stagnated or has even fallen and now consists of less than the official target of 0.7 per cent of GDP (Martin, 1989: 107–8). French influence is maintained by a variety of mechanisms. These include the trade links within the franc zone and formal institutional groupings of the former colonies. There are also significant linguistic ties which unite the French-speaking world, *Francophonie*. Furthermore, France maintains a continued substantial military presence in its former colonies, particularly in Africa (Aldrich and Connell, 1989b; Martin, 1989).

The continued international influence of France is likely to be maintained, especially in Africa. The loosening of ties with Francophone countries (a policy of self-determination favoured by many of the Left) has not occurred under Mitterrand because of the weight of historical tradition

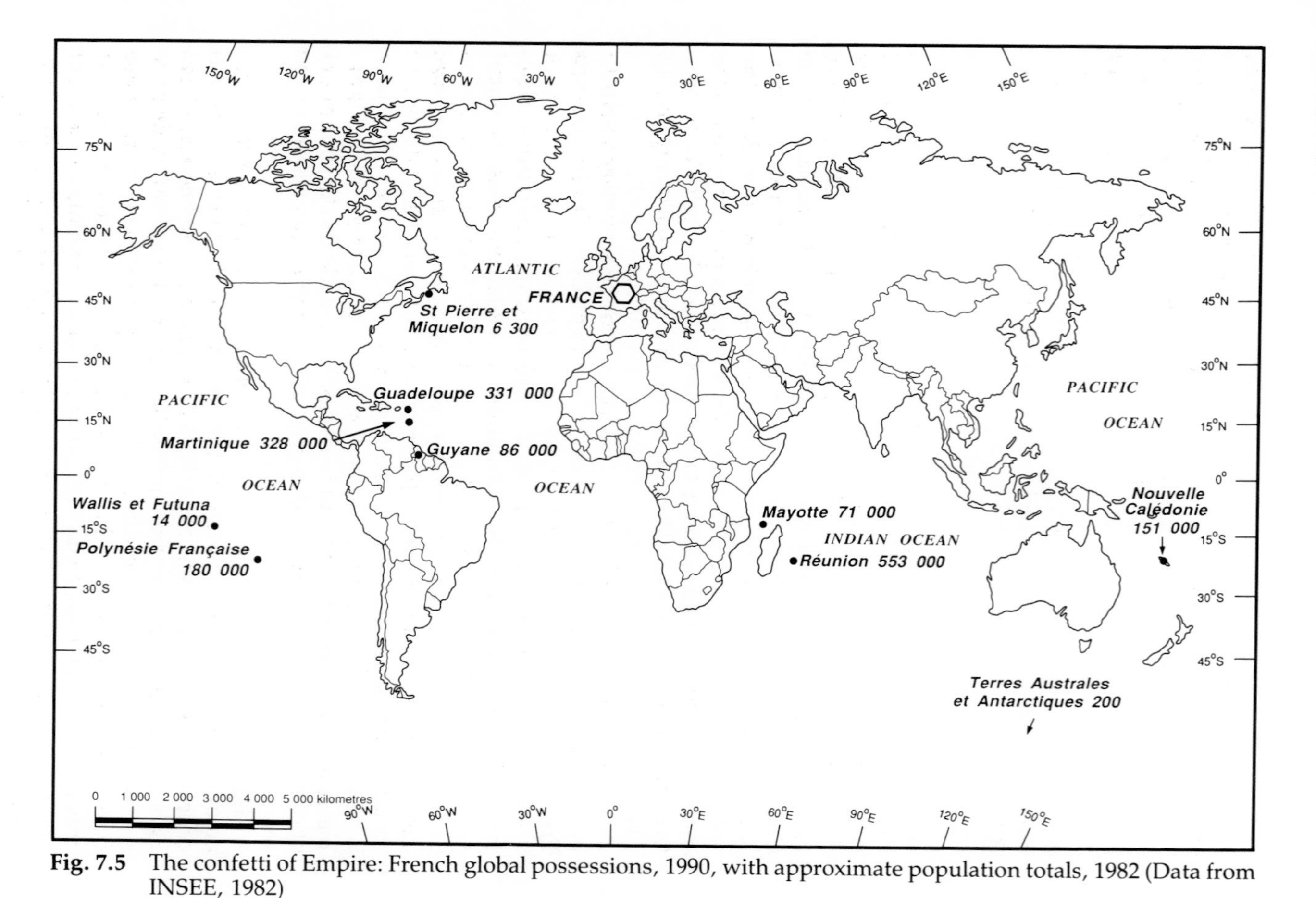

Fig. 7.5 The confetti of Empire: French global possessions, 1990, with approximate population totals, 1982 (Data from INSEE, 1982)

and the importance of nationalistic and strategic needs. Strategically, French influence in Africa and in the Pacific is seen as a powerful agent of western influence to counteract any potential threat either from the former USSR or from other imperialist countries. The integration of Francophone states with western economy and ideology is reinforced by links of language and culture (Wallerstein, 1961). However, the westernization of these countries is being reversed by the spread of Moslem fundamentalism in north Africa. French influence has also been used to protect the sea lanes for the transhipment of oil, although France itself has reduced its dependence on imported petroleum.

The DOM-TOM

France has four overseas *départements*, Guyane, Guadeloupe, Martinique and Réunion; two overseas *collectivités territoriales*, Mayotte and St Pierre et Miquelon; and four overseas territories, Nouvelle Calédonie, Polynésie Française, Terres Australes et Antarctiques, and Wallis et Futuna. These collectively make up the DOM-TOM, the *départements d'outre-mer* and the *territoires d'outre-mer* (Aldrich and Connell, 1992). The overseas territories have the greatest autonomy, while the overseas *départements* have the same status as metropolitan *départements*, sending representatives to the National Assembly and the Senate. The DOM-TOM, strategically scattered around the globe, are the remains of the French colonial empire (Fig. 7.5).

In 1946, four former colonies of France became overseas *départements*: the Caribbean islands of Martinique and Guadeloupe, Guyane (French Guiana) in South America and the Indian Ocean island of Réunion. In 1974, each of these also became *régions* in the new administrative structure. By contrast, the TOM have more independent general councils and Nouvelle Calédonie also has regional councils. Most of the former colonies still rely heavily on primary production: sugar in Réunion; sugar, bananas and rum in Martinique and Guadeloupe; copra in Polynésie; and strategic minerals, especially nickel and cobalt, in Nouvelle Calédonie. In other DOM-TOM, however, the primary base has shrunk to almost nothing. The fisheries production of Guyane and St Pierre et Miquelon, for example, and the subsistence economy of Mayotte have essentially disappeared.

The standards of living in the overseas *départements* and territories are much lower and more unequal than in metropolitan France, although generally higher than their independent neighbours. In the Caribbean colonies, slavery was not abolished until 1848 and living standards have consistently lagged behind those of France, a process accentuated since 1946 by high rates of population growth. The change from colonial status in 1946 brought with it a dismantling of the productive plantation sector of the economy and a growth in service occupations (Rollat, 1991). All of the DOM-TOM have enlarged their bureaucracies, which are often staffed by French expatriates, whose salaries are much higher than those of the indigenous populations and much higher than similar officials would receive in France. Although health services, education and public infrastructure such as electricity are now widely available, large shanty towns still exist in Pointe-à-Pitre (Guadeloupe), Fort-de-France (Martinique) and Cayenne (Guyane). The unemployment rate is two to three times higher than in metropolitan France and the proportion of people living below the poverty line is 15–20 times higher (Rollat, 1991). There is therefore a great socio-economic and ethnic inequality within the small populations of the DOM-TOM.

The economy of the DOM-TOM is essentially supported by France, either by direct subsidies, or by immigration to metropolitan France and the sending of migrant remittances. The destruction of the productive or subsistence economies of the islands has led Connell and Aldrich (1989: 157) to argue that the DOM-TOM have now turned into consumer colonies, in a novel twist to the classic dependency relationship. It is not surprising, therefore, that large numbers of French from the overseas *départements* have decided to improve their economic opportunities by migration to metropolitan France.

Immigration from the DOM-TOM peaked in the intercensal period 1975–82. By 1982, 282 300 migrants from the DOM-TOM lived in France, a 64 per cent increase since 1975. The majority of these were from Martinique (95 700), Guadeloupe (87 000) and Réunion (75 700) (Butcher and Ogden, 1984). This migration is particularly difficult to trace because it is recorded as internal migration, although some migrants have been recruited directly by state agencies. The migrants are characterized by their extreme youth, by their concentration in the service sector, especially the public service, and by their location in and around Paris (Condon and Ogden, 1991). The Pacific Islanders' migration destinations remain mainly within the Pacific, but the sending of remittances forms an important source of finance for the overseas territories. Many hundreds of islanders from St Pierre et Miquelon have emigrated to Canada; the islands of Wallis et Futuna are almost totally dependent on subsidy and remittances; and Mayotte has gone from subsistence to subsidy (Connell and Aldrich, 1989). The most economically viable of the DOM-TOM is Nouvelle Calédonie, which has a growing tourist industry in places such as Tahiti and Nouméa, drawing particularly on an Australian clientele. It also has potentially valuable mineral resources, including those of the vast ocean territory of the South Pacific, and has gained economic development as well as nuclear fallout from the nuclear testing on Mururoa Atoll (Aldrich, 1990).

The overseas *départements* form a particular case of regional inequality emphasized by a colonial history and linguistic and ethnic differences. Some of the DOM-TOM, however, have never expressed interest in autonomous movements. Even in Nouvelle Calédonie, where nationalist movements were extremely violent in the 1980s, a referendum in 1987 recorded a majority vote for staying with France. The arguments of the DOM-TOM for autonomy may be more pressing than those of many of the regions of metropolitan France, such as Bretagne, yet the overseas possessions are generally less in the public and political eye. However, riots in Réunion in 1991 brought the needs of the overseas *départements* into the political limelight. Many overseas residents do not wish to lose the protection of France, but wish to attain the practical as well as the theoretical advantages of being part of France. There seems very little likelihood of independence for any of the remnants of French empire; trade ties, legal barriers to secession and French strategic interests in maintaining a presence in every world ocean make the possibility remote (Connell and Aldrich, 1989).

France in Europe

> France does not ask what it can do for Europe, only what Europe can do for France . . .
>
> Simon Jenkins, *The Times* (1992)

The formation of the European Communities (EC)

The history of the post-war unification of European states is well-known and well-documented (see, for example, Kerr, 1977; Clout, 1987). In brief, the reconstruction of individual European economies was aided by the Marshall Plan, but a broader vision of a united Europe prompted the formation of major international organizations, particularly the European Coal and Steel Community (ECSC) in 1951, and the European Atomic Energy Community (Euratom) and the European Economic Community (EEC) in 1958. These organizations are now generally referred to as the European Communities (EC). Much of the credit for the organizations of European unity must go to France; not only were individuals, such as Jean Monnet, influential in setting up the early European structures, but France as a whole adopted an expansive view of European unity, in which a strong Europe was viewed as an extension of French national interests (Wise, 1989: 37–40).

Throughout the period of EC formation and expansion, France has protected its own national interests, as indeed have other countries. In the formation of the EEC, the unstated trade-off was that France would benefit from an enlarged market for farming products, while Germany would benefit from the expansion of industrial markets. It was therefore in French interests to have the Common Agricultural Policy (CAP) fully established in the formative years of the EEC and to block Britain's entry until after the procedures were set in place. Britain, the Republic of Ireland and Denmark finally joined the EC in the early 1970s (Fig. 7.6). Since that time, the French have continued to defend the CAP as the only common policy of the EC. During the expansion of the 1980s, when Greece was admitted to membership (1981) and later, Spain and Portugal (1985), there was a number of concerns for France, not least the effect on the southern agricultural regions of the country from competition with other Mediterranean producers (Story, 1990: 41). The concerns over French agriculture were allayed by the establishment of the Integrated Mediterranean Programmes (IMP), which gave massively increased restructuring subsidies to the Mediterranean agricultural regions (Winchester and Ilbery, 1988). Other concerns arising from the enlargement of the EC included fears of an influx of labour migrants from the Mediterranean to the core countries; this process, which has in any case been slowed by economic recession,

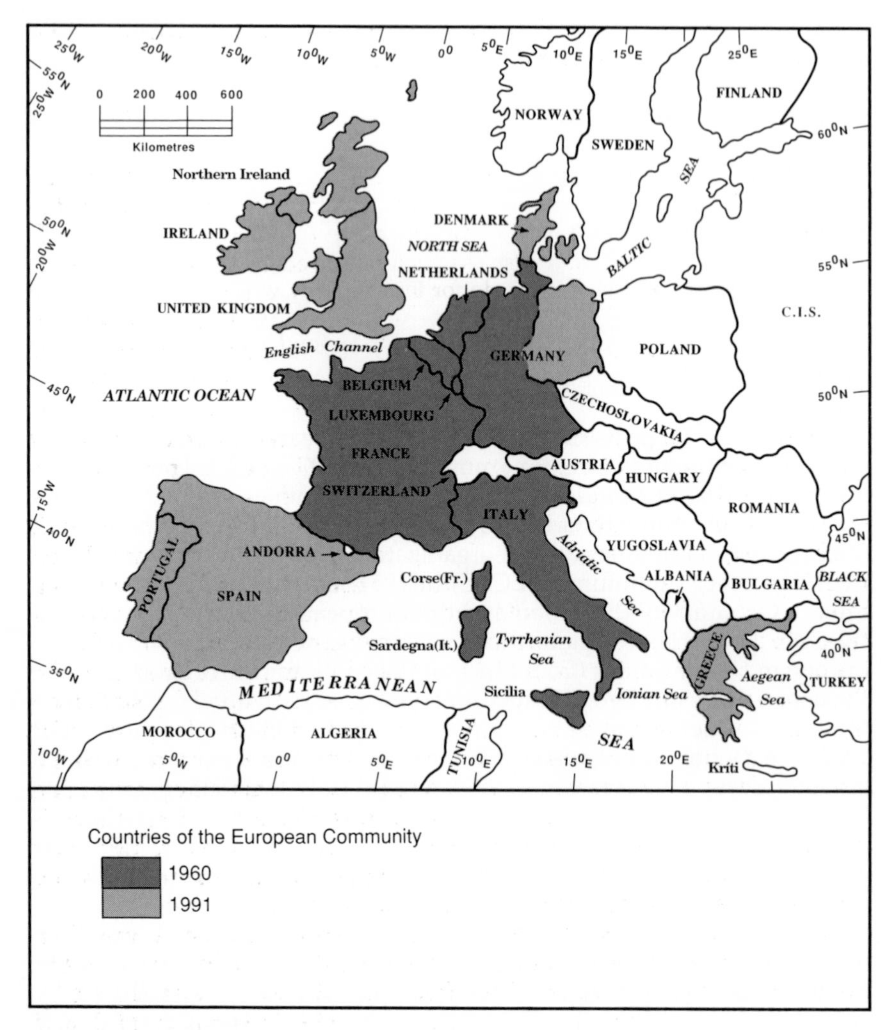

Fig. 7.6 The formation of the European Communities

was deliberately stalled by establishing an elaborate and protracted phasing-in period, to the extent that Europe was discussed in terms of a two-tier or two-speed community.

France has also acted independently of its other European partners on occasions and has developed alliances with other countries. Independent action is best exemplified by the French withdrawal from the North Atlantic Treaty Organization (NATO) to pursue its own nuclear strategic defence policy. France still refuses to consider nuclear disarmament in Europe and has continued nuclear testing in the Pacific (Howorth, 1989). Furthermore, France has sought non-EC alliances for some of its major industrial, technical and cultural projects, notably the Ariane rocket which has been developed under the aegis of the European Space Agency (including non-EC partners) and the Eureka projects established in 1985 for joint ventures in advanced technology.

Maastricht, 1992 and all that

The single European market which was scheduled to be achieved by the end of 1992 was envisaged in the Treaty of Rome in 1958. By 1968, the European Communities had already become a customs union, and tariffs and quotas on internal EC trade had been removed. However, numerous barriers to free trade still existed, ranging from the physical barriers of customs posts to the technical barriers of common standards for goods. In 1985, the European Commission produced a White Paper outlining 300 (later 279) legislative proposals to remove all barriers to the free flow of goods, services, capital and labour within the EC; these were not just to be incremental changes but the whole gamut of physical, technical and fiscal barriers (Cockfield, 1990). In December 1985, the Single European Act was passed; the deadline of 1992 for the achievement of the single market was chosen because this covered the lifetime of two Commissions (Cockfield, 1990). A major achievement was that the Single European Act introduced majority voting at the European Council of Ministers, rather than unanimous voting. This has prevented one country using its veto to delay or frustrate the passage of legislation. Two other major features of the reform are significant. First, it introduced the idea of 'essential requirements' for the devising of standards for manufactured goods. National legislation on a particular item would be mutually recognized within the EC provided it conformed to certain essential requirements. This has already reduced the amount of paperwork and bureaucracy required in Brussels by delegating detailed specifications to the individual nations (Cockfield, 1990). Secondly, services are specifically treated in the same way as goods; their increased mobility is likely to make its biggest impact on fast growing sectors of the economy such as finance and producer services.

Progress towards the single European market has been slower than anticipated, and has been hindered by resurgent nationalism from within many of the member states. The Treaty of Maastricht was signed by representatives of all 12 member nations of the EC in February 1992, and was intended to mark a new stage in European union, effective from January 1993. The union would create an area without internal frontiers, and ultimately a common currency and common defence policy. Citizenship of the European union would be open to anyone holding the nationality of a member state. The Treaty of Maastricht was intended to finalize the progress towards European union envisaged in the Single European Act.

Three countries, Denmark, France and Ireland, held referenda on the treaty during 1992. The French referendum, held in September 1992, resulted in a very narrow vote in favour of the treaty (51.1 per cent). The highest proportions of votes in favour were recorded in Alsace (65.6 per cent), Paris (62.5 per cent), Lyon (60.3 per cent) and Bretagne (59.9 per cent). The lowest proportions of votes in favour of closer European union were cast in Picardie (43.0 per cent), Corse (43.3 per cent), Nord–Pas-de-

Calais (44.3 per cent) and Provence (44.7 per cent). Although the opinion polls had predicted 'yes' votes from the urban middle classes and 'no' votes from the poorer suburbs and rural areas, it is difficult to detect any clear regional or class-based pattern from the aggregate data. What is apparent from the results of the ballot, is that almost half of French voters were concerned about the possible negative impact of closer European union. Their concerns were mainly of perceived threats to national sovereignty, both from the faceless bureaucracy at Brussels and from the increased citizenship rights that would be granted to many foreign workers.

The single European market has European and global significance. The removal of all internal barriers to trade and factor mobility necessitates major changes and co-ordination in areas as diverse as customs, excise and VAT rates; health, safety and transport legislation; government standards for goods; and the recognition of tertiary qualifications. The result will be the effective creation of a single market of 336 million consumers, with a combined annual GDP of 3.7 billion ECU. In comparison, the USA has a population of about 226 million with a GDP of 3.9 million ECU. The creation of such a market is globally significant. The single market is designed to revitalize the EC economy which has lagged in economic and employment growth behind both Japan and the USA in the 1980s. European industry has been diagnosed as suffering from Eurosclerosis (see pages 108; 126). Often its restructuring has taken unproductive forms, which have reduced labour but have not promoted new investment (Brandao, 1991). The opportunity of a single market will provide industry with greater economies of scale; it has been estimated that the immediate effects will be to increase GDP by up to seven per cent, consumer prices by up to six per cent and employment by up to five million (Cecchini *et al.*, 1988).

Views on the economic impact of the single market are polarized. In a report on the 'cost of non-Europe', it was estimated that the current wasteful bureaucracy and lack of co-ordination of legislation costs about five per cent of GDP (Cecchini *et al.*, 1988; Price Waterhouse, 1988). The report predicted dramatic economic benefits from the single market, especially for the weaker southern economies, through greater comparative advantage in labour intensive production. An alternative pessimistic view by Padoa-Schioppa *et al.* (1987) stressed the economic vulnerability of the southern economies. In particular, the relatively overpopulated agricultural areas of Greece, Spain and Portugal would be vulnerable to exploitation. Within Europe, a division of labour and occupational structures could occur which would replicate the present global international division of labour (see pages 121; 138–9). The periphery would contain the poorly-paid production processes, while the core would retain control functions and highly paid service occupations. Furthermore, small and medium-sized firms will be vulnerable to competition, extinction and takeover. The situation of the single market is likely to create circumstances similar to those already experienced by the industrial sector in Spain and Portugal on accession to the EC, in which small firms were vulnerable to economic competition (Brandao, 1991). A third assessment of the implications of the single market was much more low-key: Neven (1990) considered that the whole affair will be a damp squib. It is considered that the 'big six', which have the most industrial muscle and most potential to benefit, have already been economically integrated for so long that the further removal of barriers will make little difference.

The geographical impact of the single market is therefore unclear. However, it is likely that the increased scale of operation will allow large multinational firms to benefit most. Almost certainly another round of industrial restructuring and mergers will occur. The benefits of industrial restructuring will accrue to the core rather than the periphery and existing regional inequalities will be accentuated (Fig. 7.7). In recognition of the potential inequity, spending by the European Regional Development

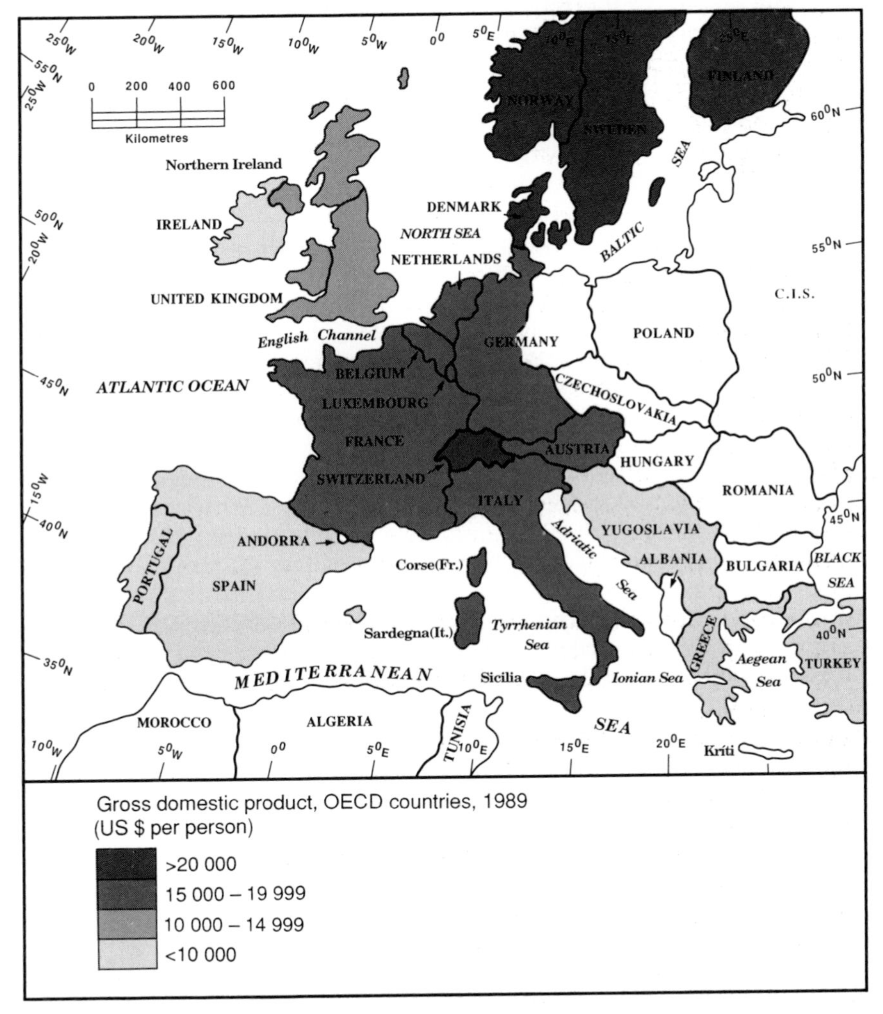

Fig. 7.7 Core-periphery inequalities in Gross Domestic Product, OECD countries of western Europe, 1989. (Data from OECD, 1991)

Fund (ERDF) and by the European Social Fund will be greatly increased. The budgets of the ERDF and Social Fund, most of which are spent in the economically backward regions, have been doubled between 1987 and 1992 (Keeble, 1989); together they now constitute 18 per cent of the European budget. The potential economic advantages of the single market can turn into diseconomies of scale if they are geographically over-concentrated; potential problems of inflation, housing quality, urban sprawl and environmental destruction could counterbalance purely financial advantages of income and employment (Liberatore, 1991).

The geographical impact of the single market is a critical one for France. Until the issue of the Treaty of Maastricht, it was generally considered that the French had enthusiastically adopted the idea of the single market as a means to galvanize the economy (Story, 1990: 52). The prosperity and status of France is seen to be intimately linked to the prosperity and status of Europe, which after union, should be a match for both Japan and the USA. The French have geared their planning towards maintaining their position at the centre of Europe, with Paris as the major Euro-city, served by new TGV and motorway links and the Channel Tunnel link with Britain. However, it is not completely clear whether France is part of the core or part of the periphery. In Ilbery's (1984: 294) analysis of core-periphery differentials within Europe, France, although scoring relatively highly on most indicators, was not included in the core of Europe. Certainly some of the peripheral areas of France still suffer from some locational and structural socio-economic disadvantages. This is the present dilemma of French regional and national planning policy. It is torn between the desire for economic advantages of the single market on the one hand and the desire to gain its full share of European Regional and Social Funds on the other (see pages 218–9).

Closer monetary union within Europe has traditionally been seen as the anchor of the single market. The Treaty of Maastricht is designed to achieve this, with a single currency as a long term aim. However, the precariousness of the existing European monetary system was demonstrated in September 1992 when the Irish, Italian and the British currencies were forced out of the fixed exchange rate bands. The French franc also suffered from severe speculation, but remained within the system. The single market negotiations included provisions for interest rates to be fixed by a European bank and for a limited rationalization of fiscal policy, especially VAT, within two major percentage bands. However, the overwhelming influence of the German currency and financial institutions has cast doubt on the likelihood of a successful monetary union, certainly in the short term. The financial crisis of September 1992 has resurrected fears of a two-speed or two-tier Europe (see page 240).

The formation of the single market will have a significant impact on patterns of global trade. Two issues are of particular importance to France. One is trade with the French former colonies and overseas territories, which has been protected by trade preference agreements under the Lomé Convention. The goods from these countries, which are traded mainly with France, have been imported without customs duties. However, these special relationships are under threat and are likely to be substantially changed when the Lomé Convention undergoes its quinquennial review. The second issue is a more general one which affects the whole of Europe's trading relations with the world. Non-European countries, such as the USA and Australia, are particularly concerned that the strong single market will develop into a self-sufficient trading block, with high levels of tariffs and protection which will exclude non-EC members. The fears that a 'Fortress Europe' would exclude non-EC products have encouraged US and Japanese industrialists to invest in Europe so they have a foothold in the European market before barriers close. A 'Fortress Europe' could lead to a segregation of the world into ever clearer trading blocs with greater protectionism and reduced trade.

At the same time as the single market is being developed, there is also increasing co-ordination on aspects of social policy. This, however, has proved one of the major stumbling blocks for the acceptance of the Treaty of Maastricht, particularly in the United Kingdom. The range of issues affected includes working hours and paid leave entitlements, co-ordination of educational standards based on the *baccaleauréat*, a European crime force and a European army rapid response unit.

The role of European single market will be rather different from that envisaged by the European policy makers in 1985. The unification of Germany produced short-term stresses in coping with massive migration flows and antiquated industrial structures, but will provide Germany and the rest of Europe with greater market opportunities. The collapse of the USSR, the dismantling of the Berlin wall and the moves within eastern Europe towards a democratic capitalism provide even greater economic opportunities for an efficient and streamlined European industry. The strengthening of Europe as a trading bloc will not only form a counterweight to the strength of the USA and Japan, but will also balance the rise of the Pacific nations as the new global focus of trade. France, as a major power within the Pacific as well as within Europe, cannot lose.

The moves towards a single market, a monetary and trade policy, and a social policy are highly significant steps in the progress of European union. Although political union will be far from complete, from 1993 there will be a new form of condominium of powers (Smith, 1991). Many nations, particularly Britain and Denmark, are concerned about potential loss of sovereignty, but Europe is moving rapidly along the path towards federation and towards a 'United States of Europe' (Wistrich, 1989). To a great extent, this is a Europe cast in a French mould, through which France can exercise national ambitions as a global power.

Glossary

agrément Government approval for industrial development
Ancien Régime The monarchy before the French Revolution of 1789
arrondissement An administrative division; there are 20 in Paris

baccalauréat School leaving certificate taken at age 18
bidonville Shanty town
bocage Land use pattern particularly found in Bretagne, characterized by small fields bounded by hedges and ditches
boulevard périphérique Ring road (especially that around Paris, following the line of the former city walls)

canton Administrative division usually consisting of several *communes*
Charbonnages de France (CDF) National coal board
chiffoniers Rag pickers, that is, people who live by salvaging from rubbish
CNRS Centre National de la Recherche Scientifique (Paris)
Code de la Famille Family code, a pronatalist system of family allowances introduced in the 1930s
cohabitation Cohabitation, that is, an uneasy alliance between the political left and right wings
commune The basic administrative division, roughly corresponding to a parish
contrat ville moyenne A contract between government and medium-sized towns for urban development
Crédit Agricole Large credit union/bank
cumul des mandats The accumulation of offices by politicians, often at both local and national levels

Délégation à l'Aménagement du Territoire et à l'Action Régionale (DATAR) Office for planning and regional development
département Administrative division, corresponding to a county or state
département d'outre-mer (DOM) Administrative division outside metropolitan France
Dotation aux Jeunes Agriculteurs (DJA) Subsidy to young farmers

Ecole Nationale d'Administration (ENA) National Business School (see also *énarchie*)
école supérieure Higher education institution
écomusée Ethnic, local or folkloric museum
Electricité de France (EDF) National electricity board
énarchie Elite group of graduates from *Ecole Nationale d'Administration*
Établissement Public d'Aménagement (EPA) Public development corporation (for new towns)
étang Lake, pond, lagoon
exode rural Rural exodus, rural depopulation

faiseuses d'anges Angel-makers (abortionists)
Francophonie French-speaking parts of the world
Front de Libération Nationale de la Corse (FLNC) Corsican National Liberation Front, an activist political group
Front National (FN) National Front, an extreme right wing political party

garrigue Mediterranean vegetation (low-growing)
gastarbeiter Guest workers, from foreign poorer countries
Gaz de France (GDF) National gas board
gigantisme A predeliction for large-scale high-status projects
glacis A clear zone around a city's walls for military purposes (historical)
grande culture Large-scale commercial agriculture, particularly around the Paris Basin
grande école National tertiary institutions in specialized disciplines
grand ensemble Huge apartment blocks built at the edge of large cities after the Second World War
Groupement Agricole en Commun (GAEC) A farming co-operative

habitat, vie sociale Urban policy to improve housing and social interaction
habitation à loyer modéré (HLM) Government-subsidized housing (at moderate rent)
hameau Village or hamlet, a term often used for new prestigious 'village-style' housing development
hôtel particulier Mansion owned by elite individual (historical)

îlot Block or census tract in the city
image de marque Image, brand name
immeubles à loyer normal (ILN) Government-provided housing (at market prices)

Indemnité Spéciale Montagne (ISM) Subsidy available for agriculture in designated mountain zones
Indemnité Viagère de Départ (IVD) Retirement or pre-retirement pension for farmers
INED Institut National d'Etudes Démographiques (Paris)
INSEE Institut National de la Statistique et des Etudes Economiques (Paris)

Jeunesse Agricole Chrétienne (JAC) Christian Agricultural Youth movement

langue d'oc Provençal language, spoken in south of France
limon Fertile soil formed on loess
Livre Blanc White paper (for urban planning)

mal-lôtis Badly-housed; self-built inter-war suburban developments around Paris
malthusien Malthusian; person in favour of restricting fertility
maquis Mediterranean vegetation (shrubby)
métayage Share-cropping
Métro Parisian underground system
métropoles d'équilibre Counter-magnets, large urban centres designated for growth as counter-magnets to Paris
meublés Furnished rooms
microbidonville Small shanty town
minette Phosphoric iron ore
Mistral Seasonal Alpine/Mediterranean wind

OECD Organization for Economic Co-operation and Development (Washington, D.C.)
Opérations Programmées d'Amélioration de l'Habitat (OPAHs) Programmes for housing improvements, often in central cities

Parti communiste français (PCF) French Communist Party
Parti socialiste (PS) French Socialist Party
pavillon Villa or suburban house
pays Countryside, distinctive rural landscape
pépinière d'enterprise Industrial incubator unit
petite culture Small-scale agriculture, often with some subsistence element
phylloxera Fungal disease affecting vines
pieds-noirs Ex-colonials returned to France especially after colonial independence (literally black feet)
Plan d'Aménagement et d'Organisation Générale de la Région (PADOG) Regional planning scheme for Paris region, 1960
plan d'occupation des sols (POS) Land-use plan
pôle de conversion Restructuring pole, area eligible for subsidy because of the need for economic restructuring
préfecture Site of administration for a *département* (see Frontispiece)
préfet Chief administrator of a *département*

Prime d'Aménagement du Territoire (PAT) Regional development grant
Prime Régionale à l'Emploi Small business grant payable by local authorities for regional development

quartier District; term used both in a general sense and specifically as a sub-division of the Parisian *arrondissements*

Rassemblement pour la république (RPR) Rally for the Republic, right wing political party
recherché Sought-after or desirable (area)
redévance Development tax on new industry to restrict Parisian development
Régie Autonome des Transports Parisiens (RATP) Parisian transport authority
région (de programme) Region of France (see Fig. 7.2)
remembrement Consolidation or enlargement of farm units from small fields
rurbanisation Counterurbanization

Schéma directeur (d'aménagement et d'urbanisme) (SD or SDAU) Master plan for urban development
secteur sauvegardé Conservation area
Société d'Amélioration Foncier et d'Etablissement Rurale (SAFER) Land management company
Société d'Initiative et de Coopération Agricole (SICA) Agricultural co-operative
statut particulier Special status
statut social de la mère de famille Subsidy for mothers at home with children

territoires d'outre-mer (TOM) Overseas territories
trains à grande vitesse (TGV) High-speed trains

Union de la démocracie française (UDF) Union for French Democracy, right wing political party
Union pour la France (UPF) Union for France, right wing political party
urbanisme Urban planning

Verts The Greens (Green Party)
villes moyennes Medium-sized towns (see *contrat ville moyenne*)

zone industrialo-portuaire Industrial port zone
zone à urbaniser à priorité (ZUP) Priority urban development zone
zone d'aménagement concertée (ZAC) Concerted development zone
zone d'aménagement deferrée (ZAD) Zone of deferred development with right of public pre-emption of land sales in zone
zone de rénovation rurale Rural renovation zone
zones naturelles d'équilibre (ZNE) Designated green spaces or wedges to prevent continuous urban sprawl

Bibliography

Adler S 1977 *International migration and dependence*. Saxon House, Farnborough
Agence Française pour la Maîtrise de l'Energie 1985 *La Maîtrise de l'énergie: une politique pour l'avenir*. Agence Française pour la Maîtrise de l'Energie, Paris
Aldrich R 1987 Late-comer or early starter? New views on French economic history. *The Journal of European Economic History* **16**: 89–100
Aldrich R 1990 *The French presence in the south Pacific, 1842–1940*. Macmillan, London
Aldrich R, Connell J 1989a Beyond the hexagon: France in world politics. In Aldrich R, Connell J (eds) *France in world politics*. Routledge, London: 1–15
Aldrich R, Connell J 1989b Francophonie: language, culture or politics. In Aldrich R, Connell J (eds) *France in world politics*. Routledge, London: 170–93
Aldrich R, Connell J 1992 *France's overseas frontier: départements et territoires d'outre-mer*. Cambridge University Press, Cambridge
Allen L 1984 Doing it better the French way. *Town and Country Planning* **53**: 303–5.
Anon 1976 Drought, sort of. *The Economist* **261** (6 Nov): 91
Anon 1986 *La SICA de Saint Pol de Léon*. Kerisnel, Saint Pol de Léon
Ardagh J 1968 *The new French revolution: a social and economic survey of France 1945–1967*. Secker and Warburg, London
Ardagh J 1970 *The new France*. Penguin, Harmondsworth
Ardagh J 1982 *France in the 1980's*. Penguin, Harmondsworth
Arléry R 1970 The climate of France, Belgium, The Netherlands and Luxembourg. In Wallén C C (ed) Climates of northern and western

Europe. *World Survey of Climatology* (15 vols). Elsevier, Amsterdam: vol 5: 148–59

Ashworth G J, White P E, Winchester H P M 1988 The red-light district in the west European city: a neglected aspect of the urban landscape. *Geoforum* **2**: 201–12

Baker A R H 1973 Adjustments to distance between farmstead and field: some findings from the south western Paris Basin in the early nineteenth century. *Canadian Geographer* **17**: 259–75

Baker A R H 1980 Ideological change and settlement continuity in the French countryside: the development of agricultural syndicalism in Loir-et-Cher during the late nineteenth century. *Journal of Historical Geography* **6**: 163–77

Barral P 1968 *Les agrariens français de Méline à Pisani*. Colin, Paris

Barry R G, Chorley R J 1992 *Atmosphere, weather and climate* 6th edn. Routledge, London

Bastié J 1964 *Paris en l'an 2000*. Sédimo, Paris (Collection 'mise au point')

Bastié J 1984 *Géographie du grand Paris*. Masson, Paris

Bateman M 1988 *Shops and offices: locational change in Britain and the EEC*. John Murray, London (Case studies in the developed world)

Bauer G, Roux J M 1976 *La rurbanisation ou la ville éparpillée*. Seuil, Paris

Bazin M 1990 Reims, de la croissance industrielle des années 60 aux perspectives européennes. *Travaux – Institut de Géographie de Reims* **77–78**: 77–95

Beaujeu-Garnier J 1975 *France*. Longman, London (The world's landscapes)

Bell D 1973 *The coming of post-industrial society: a venture in social forecasting*. Basic Books, New York

Belliard J C, Boyer J C 1983 Les 'nouveaux ruraux' en Ile-de-France. *Annales de Géographie* **92**: 433–51

Bellon B, Chevalier J-M 1983 *L'industrie en France*. Flammarion, Paris

Benko G B 1989 Géographie des mutations industrielles: le phénomène des pépinières d'entreprises. *Annales de Géographie* **550**: 628–45

Benko G B 1991 *Les technopoles*. Masson, Paris

Bennamane A 1980 *The Algerian development strategy and employment policy*. Discussion Paper **9**, Centre for Development Studies, University College, Swansea

Bennoune M 1975 The Maghribian migrant workers in France. *Race and Class* **17**: 39–56

Bernard M-C, Carrière P 1986 Mobilité démographique et comportement électoral: le front national en Languedoc-Roussillon aux Européennes de 1984. *Bulletin de la Société Languedocienne de Géographie* **20**: 81–99

Bernhard M 1981 Heavy metals and chlorinated hydrocarbons in the Mediterranean. In Geyer R A (ed) *Marine Environmental Pollution 2*. Elsevier, Amsterdam: 143–92

Besson L 1985 Les risques naturels. *Revue de Géographie Alpine* **73**: 321–33

Beyer J L 1974 Global summary of human response to natural hazards: floods. In White G F (ed) *Natural hazards: local, national, global*. Oxford University Press, Oxford

Billard A, Cosandey C, Muxart T 1985 L'érosion physique sur la montagne du Lingas en relation avec les pratiques agropastorales. In Spencer T (ed) *International geomorphology 1985: abstracts of papers for the first international conference on geomorphology*. School of Geography, University of Manchester, Manchester: 38

Birnbaum P 1980 The state in contemporary France. In Scase R (ed) *The state in western Europe*. St Martin's Press, New York: 94–114

Blaikie P, Brookfield H 1987 Questions from history in the Mediterranean and western Europe. In Blaikie P, Brookfield H (eds) *Land degradation and society*. Methuen, London: 122–42

Blayo C 1989 L'avortement en Europe. *Espace, Populations, Sociétés* **1989/2:** 225–38

Bloch M 1966 *French rural history: an essay on its basic characteristics* trans J. Sondheimer. Routledge and Kegan Paul, London

Boigeol A, Commaille J, Muñoz-Perez B 1984 Le divorce. *Données Sociales*: 428–46

Bollotte L 1991 Transport in Paris and the Ile de France. *Built Environment* **17:** 160–71

Boucher M 1973 *La région*. Cahiers Français, Paris.

Boudoul J, Faur J P 1982 Renaissance des communes rurales ou nouvelle forme d'urbanisation? *Economie et Statistique* **149:** 1–16

Bourgeois-Pichat J 1965 The general development of the population of France since the eighteenth century. In Glass D V, Eversley D E C (eds) *Population in history: essays in historical demography*. Edward Arnold, London: 474–506

Bourke A 1984 Impact of climatic fluctuations on agriculture. In Flohn H, Fantechi R (eds) *The climate of Europe: past present and future*. Reidel, Dordrecht: 269–313

Boussard I 1990 French political science and rural problems. In Lowe P, Bodiguel M (eds) *Rural studies in Britain and France*. Belhaven, London: 269–85

Bowler I R 1985 *Agriculture under the Common Agricultural Policy: a geography*. Manchester University Press, Manchester

Boyer R, Mistral J 1978 *Accumulation, inflation, crises*. Presses Universitaires de France, Paris

Braid Wilson I 1983 The preparation of local plans in France. *Town Planning Review* **54:** 155–73

Brandao G 1991 *The west European internal market analysis of selected impacts*. Discussion Paper **37**, Department of Geography, University of Toronto, Toronto

Braudel F 1981 The structures of everyday life: the limits of the possible. *Civilization and capitalism 15th–18th century* (3 vols) trans S. Reynolds. Collins, London: vol 1

Braudel F 1982 The wheels of commerce. *Civilization and capitalism 15th–18th century* (3 vols) trans S. Reynolds. Collins, London: vol 2

Braudel F 1984 The perspective of the world. *Civilization and capitalism 15th–18th century* (3 vols) trans S. Reynolds. Collins, London: vol 3

Braudel F 1986 *L'identité de la France: espace et histoire*. Arthaud-Flammarion, Paris

Braudel F 1988 History and environment. *The identity of France* (2 vols) trans S. Reynolds. Collins, London: vol 1

Bravard J P 1987 Rhône power: mighty rivers of the world. *Geographical Magazine* **59:** 537–42

Brocard M 1981 Aménagement du territoire et développement régional: le cas de la recherche scientifique. *L'Espace Géographique* **10:** 61–73

Brunat E, Reverdy B 1989 Linking university and industrial research in France. *Science and Public Policy* **16:** 283–93

Brunet R and 22 others 1989 L'aménagement du territoire: nouvelles donnes? *L'Espace Géographique* **18:** 273–90

Bruyelle P 1987 Un cas particulier de désindustrialisation: les villes à centrales électriques. *Hommes et Terres du Nord* **1989/1:** 51–9

Buller H, Lowe P 1990 Rural development in post-war Britain and France. In Lowe P, Bodiguel M (eds) *Rural studies in Britain and France*. Belhaven, London: 21–36

Burtenshaw D, Bateman M, Ashworth G 1991 *The European city: a western perspective*. David Fulton, London

Butcher I, Ogden P E 1984 West Indians in France: migration and demographic change. In Ogden P E (ed) *Migrants in modern France: four studies*. Occasional Paper **23**, Department of Geography, Queen Mary College, University of London, London: 43–66

Carnet C 1979 L'effet anti-érosif du bocage en région granitique de Bretagne. In *Colloque sur l'érosion agricole des sols en milieu tempéré non méditerranéen*. Université Louis Pasteur, Strasbourg: 153–6

Castells M 1983 *The city and the grassroots: a cross-cultural theory of social movements*. Edward Arnold, London

Castles S 1984 *Here for good: western Europe's new ethnic minorities*. Pluto Press, London

Castles S 1985 Immigrant workers and class struggles in advanced capitalism: the western European experience. *Politics and Society* **5:** 33–66

Cecchini P, Catinat M, Jacquemin A 1988 *The European challenge 1992: the benefits of a single market* trans J. Robinson. Gower, Aldershot

Chaline C H 1984 Contemporary trends and policies in French city planning: a chronicle of successive urban policies, 1950–1983. *Urban Geography* **5**: 326–36

Champion A G (ed) 1989 *Counterurbanization: the changing pace and nature of population deconcentration*. Edward Arnold, London

Chappal V, Martinez J C 1986 Notes sur l'amphithéâtre antique de Béziers. *Société Archéologique* **34**: 26–33

Charbonnages de France 1988 *Le charbon: matière énergétique*. Charbonnages de France, Paris

Charbonnages de France 1991 *Statistique annuelle édition 1991*. Charbonnages de France, Paris

Chardon M 1987 L'éboulement de Séchilienne (Isère): un exemple de catastrophe naturelle géomorphique en milieu alpin. *Revue de Géomorphologie Dynamique* **26**(3): 107–9

Chardon M, Castiglioni G-B 1984 Géomorphologie et risques naturels dans les Alpes. In Comité International d'Organisation de l'UGI (ed) *Les*

Alpes: 25 congrès international de géographie. Comité International d'Organisation de l'UGI, Caen: 13–41
Chaslin F 1985 *Les Paris de François Mitterrand: histoire des grands projets architecturaux*. Gallimard, Paris
Chauviré Y 1986 Nombre et taille de ménages en France: disparités géographiques et évolution de 1962 à 1982. *Espace, Populations, Sociétés* **1986/2**: 99–106
Chevreuil M, Chesterikoff A, Létolle R 1987 PCB pollution behaviour in the river Seine. *Water Resources Research* **21**: 427–34
Cleary M 1986 Patterns of transhumance in Languedoc. *Geography* **71**: 25–33
Cleary M 1989 *Peasants, politicians and producers: the organisation of agriculture in France since 1918*. Cambridge University Press, Cambridge
Clout H D 1983 *The land of France 1815–1914*. George Allen and Unwin, London
Clout H D 1984 Bordeaux: urban renovation, conservation and rehabilitation. *Planning Outlook* **27**: 84–92
Clout H D 1985 From marketplace to megastructure. *Town and Country Planning* **54**: 197
Clout H D (ed) 1987 *Regional development in western Europe* 3rd edn. David Fulton, London
Clout H D 1988 The chronicle of La Défense. *Erdkunde* **42**: 273–84
Cockfield Lord 1990 The real significance of 1992. In Crouch C, Marquand D (eds) *The politics of 1992: beyond the European single market*. Blackwell, Oxford: 1–8
Coleman D C 1983 Proto-industrialization: a concept too many. *Economic History Review* **36**: 435–48
Coles P 1989 Worldwide atmosphere authority revisited. *Nature, London* **338**: 31
Commune de Béziers 1987 *Commune de Béziers: plan d'occupation des sols révision*. Mimeo, Commune de Béziers, Hérault
Condon S A, Ogden P E 1991 Afro-Caribbean migrants in France: employment, state policy and the migration process. *Transactions of the Institute of British Geographers* **16**: 440–57
Connell J, Aldrich R 1989 Remnants of empire: France's overseas departments and territories. In Aldrich R, Connell J (eds) *France in world politics*. Routledge, London: 148–69
Conversi D 1990 Language or race?: the choice of core values in the development of Catalan and Basque nationalisms. *Ethnic and Racial Studies* **13**: 50–70
Coulon C 1978 French political science and regional diversity: a strategy of silence. *Ethnic and Racial Studies* **1**: 80–99
Crabbe D, McBride R (eds) 1978 *The world energy book*. Kogan Page, London
Dagnaud M 1984 A history of planning in the Paris region: from growth to crisis. *International Journal of Urban and Regional Research* **7**: 219–36
Danielsson B 1984 Under a cloud of secrecy: the French nuclear tests in the southeastern Pacific. *Ambio* **13**: 336–41
Dauzat A 1953 *L'Europe linguistique*. Payot, Paris

David J, Herbin J, Meriaudeau R 1986 La dynamique démographique de la zone de montagne française: le tournant historique des années 1970. *Espace, Populations, Sociétés* **1986/2**: 365–76
Dean K G 1986 Counterurbanisation continues in Brittany. *Geography* **71**: 151–4
Dean K G 1987 The disaggregation of migrant flows: the case of Brittany 1975–1982. *Regional Studies* **21**: 313–25
Debatisse M 1963 *La révolution silencieuse, le combat des paysans*. Calmann-Lévy, Paris
Delavalle M 1983 Prévention de l'enrichessement des eaux souterraines en nitrates d'origine agricole. In *Ground water in water resources planning. Proceedings of Koblenz symposium 1983*. International Association of Hydrological Sciences, Wallingford: vol 2: 1005–14
Demangeon A 1920 *Le déclin de l'Europe*. Payot, Paris
De Martonne E 1933 *Geographical regions of France* trans H. C. Benthall. Heinemann, London
De Saboulin M 1986 Aspects géographiques de la solitude en France. *Espace, Populations, Sociétés* **1986/2:** 117–24
Desbordes J M, Valadas B 1979 L'érosion agricole en milieu cristallin: processus, conséquences et dimension historique du phénomène à partir d'exemples pris dans le massif d'Ambazac, Limousin (France). In *Colloque sur l'érosion agricole des sols en milieu tempéré non méditerranéen*. Université Louis Pasteur, Strasbourg: 19–25
Desplanques G, De Saboulin M 1986 Première naissance et mariage de 1950 à nos jours. *Espace, Populations, Sociétés* **1986/2:** 47–56
Dettwiller J 1970 Deep soil temperature trends and urban effects at Paris. *Journal of Applied Meteorology* **9**(1): 178–80
Dettwiller J, Changnon S A 1976 Possible urban effects on maximum daily rainfall at Paris, St Louis and Chicago. *Journal of Applied Meteorology* **15:** 517–19
Deville J C, Naulleau E 1982 Les nouveaux enfants naturels et leurs parents. *Economie et Statistique* **145:** 61–81
De Vries J 1981 Patterns of urbanization in pre-industrial Europe 1500–1800. In Schmal H (ed) *Patterns of European urbanization since 1500*. Croom Helm, London: 77–109
De Vries J 1984 *European urbanisation 1500–1800*. Harvard University Press, Cambridge
Dézert B 1981 La lutte contre les déchets et les nuisances industriels: effets géographiques. *Bulletin de l'Association de Géographes Français* **481:** 269–72
Dézert B 1989 Parcs technologiques, nouveaux espaces industriels et péri-urbanisation. *Acta Géographica* **80:** 26–32
Dicken P 1986 *Global shift: industrial change in a turbulent world*. Paul Chapman, London
Di Méo G 1984 La crise du système industriel, en France, au début des années 1980. *Annales de Géographie* **517:** 326–49
Douglass W A, Zulaika J 1990 On the interpretation of terrorist violence: ETA and the Basque political process. *Comparative Studies in Society and History* **32:** 238–57

Dovland H 1987 Monitoring European transboundary air pollution. *Environment* **29:** 10–15, 27–8

Duboscq P 1989 Aménagement territorial, dialectique identitaire: la France d'après 1974. *L'Espace Géographique* **18:** 321–9

Dutailly J-C, Hannoun M 1980 Les secteurs sensibles de l'industrie. *Economie et Statistique* **120:** 3–23

Dyer C 1978 *Population and society in twentieth century France.* Hodder and Stoughton, London

Edouard J L, Vivian H 1984 Une hydrologie naturelle dans les Alpes du Nord. *Revue de Géographie Alpine* **72:** 165–88

Edworthy K J, Downing R A 1979 Artificial groundwater recharge and its relevance in Britain. *Journal of the Institution of Water Engineers and Scientists* **33:** 151–72

Electricité de France 1983 *Orédon-Aragnouet: le dernier maillon de l'aménagement hydraulique du massif de Néouvielle.* EDF, Paris

Electricité de France 1984 *Electricité de France et l'environnement.* EDF, Paris

Electricité de France 1991 *Résultats techniques d'exploitation 1990.* EDF, Paris

Electricité de France n.d. (a) *Le programme electro-nucléaire français.* EDF, Paris

Electricité de France n.d. (b) *Les barrages et les centrales hydrauliques.* EDF, Paris

Embleton C (ed) 1984 *Geomorphology of Europe.* Macmillan, London (Macmillan reference books)

Ernecq J M 1988 Quelques aspects de la politique de l'environnement dans une région de tradition industrielle: le cas de la région Nord–Pas-de-Calais en France. In Joyce F E, Schneider G (eds) *Environment and economic development in the regions of the European community.* Avebury, Aldershot: 39–62

Escourrou G 1984 La climatologie appliquée en France. *Annales de Géographie* **93:** 249–53

Estrin S, Holmes P 1983 *French planning in theory and practice.* George Allen and Unwin, London

Evenson N 1979 *Paris: a century of change, 1878–1978.* Yale University Press, New Haven

Fagnani J 1987 Organisation de l'espace et activité professionnelle des mères: le cas des nouvelles couches moyennes en région Île-de-France. *Cahiers de Géographie du Québec* **31:** 225–36

Fagnani J 1988 Les Parisiennes seraient-elles malthusiennes? *Regards sur l'Île-de-France* **2:** 4–8

Faidutti A M 1986 La réorganisation de la place des étrangers dans la population active française. *Espace, Populations, Sociétés* **1986/2**: 191–6

Fel A 1984 L'agriculture française en mouvement. *Annales de Géographie* **517**: 303–25

Fielding A J 1982 Counterurbanisation in western Europe. *Progress in Planning* **17**: 1–52

Fiessinger F, Mallevialle J 1978 A critical evaluation of the Seine river quality standards from the view point of potable water production. *Progress in Water Technology* **10**: 243–50

Flageollet J C, Helluin E 1985 Morphological investigations in the sliding area along the coast of Pays-d'Auge near Villerville, Normandy, France. In Spencer T (ed) *International geomorphology 1985: abstracts of papers for the first international conference on geomorphology*. School of Geography, University of Manchester, Manchester: 184

Flandrin C 1987 The land policy for the new towns in the Paris region 1965–1985. In Chaline C (ed) *Major urban landowners*. Association pour le Développement et la Diffusion des Études Foncières, Paris: 13–18

Flockton C, Kofman E 1989 *France*. Paul Chapman, London

Fouquet A, Morin A C 1984 Mariages, naissances, familles. *Données Sociales*: 408–27

Frank W 1983 Part-time farming, underemployment and double activity of farmers in the EEC. *Sociologia Ruralis* **23**: 20–7

Franklin S H 1969 *The European peasantry: the final phase*. Methuen, London

French Ministry of the Environment 1982 French coastal policy. *Ekistics* **293**: 128–30

Fruit J P 1985 Migrations résidentielles en milieu rural péri-urbain: le pays de Caux central. *Espace, Populations, Sociétés* **1985/1**: 150–9

Fusina J n.d. Le corse. In Giordan H (ed) *Par les langues de France*. CNRS, Paris: 83–91

Gabert P, Nicod J 1982 Inondations et urbanisation en milieu méditerranéen: l'exemple des crues récentes de l'Arc et de l'Huveaune. *Méditerranée* **3–4**: 11–24

Gachelin C 1987 Le Nord–Pas-de-Calais face à la révolution technique. *Hommes et Terres du Nord* **1987/1**: 29–38

Ganiage J 1980 La population de Beauvaisis: transformations économiques et mutations démographiques, 1790–1975. *Annales de Géographie* **89**: 1–36

Gaz de France 1985 *Le gaz naturel*. GDF, Paris

Gaz de France 1987 *Le stockage souterrain de gaz naturel de Chémery*. GDF, Paris

George P 1967 *La France*. Presses Universitaires de France, Paris

Gervais M, Jollivet M, Tavernier Y 1976 La fin de la France paysanne, de 1914 à nos jours. In Duby G, Wallon A (eds) *Histoire de la France rurale*. Seuil, Paris: vol 4

Giordan H n.d. Du folklore au quotidien. In Giordan H (ed) *Par les langues de France*. CNRS, Paris: 5–13

Grahl J, Teague P 1989 Labour market flexibility in West Germany, Britain and France. *West European Politics* **12**(2): 91–111

Gras F 1979 L'érosion des sols 'lessives' de Lorraine et son incidence sur les projets de remembrement rural. In *Colloque sur l'érosion agricole des sols en milieu tempéré non méditerranéen*. Université Louis Pasteur, Strasbourg: 89–94

Gravier J 1947 *Paris et le désert français*. Le Portulan, Paris

Grigg D B 1974 *The agricultural systems of the world: an evolutionary approach*. Cambridge University Press, Cambridge

Guillauchon B 1986 *La France contemporaine: une approche d'économie descriptive*. Economica, Paris

Guillon M 1986 Les étrangers dans les grandes agglomérations françaises. *Espace, Populations, Sociétés* **1986/2**: 179–90

Guillon M 1988 Les ménages étrangers en France: évolution et disparités spatiales. *Espace, Populations, Sociétés* **1988/1**: 53–68

Hajnal J 1965 European marriage patterns in perspective. In Glass D V, Eversley D E C (eds) *Population in history: essays in historical demography*. Edward Arnold, London: 101–43

Hall R 1986 Household trends within western Europe 1970–1980. In Findlay A M, White P E (eds) *Western European population change*. Croom Helm, London: 18–34

Hall R, Ogden P E 1983 *Europe's population in the 1970s and 1980s*. Occasional Paper **4**, Department of Geography, Queen Mary College, University of London, London

Hanagan M 1989 Nascent proletarians: migration patterns and class formation in the Stéphanois region, 1840–1880. In Ogden P E, White P E (eds) *Migrants in modern France: population mobility in the late 19th and 20th centuries*. Unwin Hyman, London: 74–96

Harvey D 1979 Monument and myth. *Annals of the Association of American Geographers* **69**: 362–81

Hayes C J H 1930 *France: a nation of patriots*. Columbia University Press, New York

Hechter M, Levi M 1979 The comparative analysis of ethnoregional movements. *Ethnic and Racial Studies* **2**: 260–74

Hémery S 1986 Etrangers et nouveaux immigrés par catégorie de communes. *Espace, Populations, Sociétés* **1986/2**: 171–8

Henin S 1979 L'érosion liée à l'activité agricole en France. In *Colloque sur l'érosion agricole des sols en milieu tempéré non méditerranéen*. Université Louis Pasteur, Strasbourg: 9–12

Heugas-Darraspen H 1985 *Le logement français et son financement*. La Documentation Française, Paris.

Higgitt S R, Oldfield F, Appleby P G 1991 The record of land use change and soil erosion in the late Holocene sediments of the Petit Lac d'Annecy, eastern France. *The Holocene* **1**: 14–28

Hill B E 1984 *The Common Agricultural Policy: past, present and future*. Methuen, London

Hohenberg P M, Lees L H 1985 *The making of urban Europe*. Harvard University Press, Cambridge

Holton R 1986 Industrial politics in France: nationalisation under Mitterrand. *West European Politics* **9**(1): 67–80

House J W 1959 The Franco-Italian boundary in the Alpes-Maritimes. *Transactions and Papers Institute of British Geographers* **26**: 107–31

House J W 1978 *France: an applied geography*. Methuen, London

Houston J M 1953 *A social geography of Europe*. Duckworth, London

Howorth J 1989 Consensus and mythology: security alternatives in post-Gaullist France. In Aldrich R, Connell J (eds) *France in world politics*. Routledge, London: 16–34

Huber M 1931 *La population de la France pendant la guerre*. Presses Universitaires de France, Paris

Hudson R, Sadler D 1989 *The international steel industry: restructuring, state politics and localities*. Routledge, London

Humbert M 1987 Rapport final sur la géomorphologie et les risques naturels. *Revue de Géomorphologie* **36**: 128–30
Hurel P 1992 Mais qui a escamoté les deux rapports officiels prevenant du danger? *Paris Match* **2263/8**: 86–7
Huss M M 1980 *Demography, public opinion and politics in France, 1974–80.* Occasional Paper **16**, Department of Geography, Queen Mary College, University of London, London
Ilbery B W 1984 Core-periphery contrasts in European social well-being. *Geography* **69**: 289–302
Ilbery B W 1985 *Agricultural geography: a social and economic analysis*. Oxford University Press, Oxford
INED 1975 *Rapport sur la situation démographique de la France en 1973*. INED, Paris
INED 1979 *Huitième rapport sur la situation démographique de la France*. INED, Paris
INSEE 1982 *Recensement général de la population de 1982*. INSEE, Paris
INSEE 1987 *Données sociales*. INSEE, Paris
INSEE 1989 *Annuaire statistique de la France 1989: résultats de 1988 vol 94.* Ministère de l'Economie des Finances et du Budget, Paris
INSEE 1990 *Annuaire statistique de la France 1990: résultats de 1989 vol 95.* Ministère de L'Economie des Finances et du Budget, Paris
Jackson M (ed) 1977 *The world guide to beer* 1st edn. Michell Beazley, London
Jacob J E, Gordon D C 1985 Language policy in France. In Beer W R, Jacob J E (eds) *Language policy and national unity*. Rowman and Allanheld, New Jersey: 106–33
Jeanneau J 1989 La naissance d'Angers-technopole. *Norois* **36**: 89–93
Joly F 1984 Aquitaine Basin trans C Embleton. In Embleton C (ed) *Geomorphology of Europe*. Macmillan, London: 161–4 (Macmillan reference books)
Jones H R 1981 *A population geography*. Harper and Row, London
Jones P C 1989 Aspects of the migrant housing experience: a study of workers' hostels in Lyon. In Ogden P E, White P E (eds) *Migrants in modern France: population mobility in the later 19th and 20th centuries.* Unwin Hyman, London: 177–94
Juste C, Menet M, Wilbert J 1979 Expérimentation contre l'érosion éolienne en forêt des Landes. In *Colloque sur l'érosion agricole des sols en milieu tempéré non méditerranéen*. Université Louis Pasteur, Strasbourg: 49–52
Kain R 1980 A Parisian village. *Town and Country Planning* **49**: 51–3
Kain R 1981 Conservation planning in France: policy and practice in the Marais, Paris. In Kain R (ed) *Planning for conservation*. Mansell, London: 199–233
Kain R 1982 Europe's model and exemplar still? The French approach to urban conservation, 1962–1981. *Town Planning Review* **53**: 403–22
Keeble D 1989 Core-periphery disparities, recession and new regional dynamism in the European community. *Geography* **74**: 1–11
Kerr A J C 1977 *The Common Market and how it works*. Pergamon, Oxford
Knox P, Scarth A 1977 The quality of life in France. *Geography* **62**: 9–16

Kofman E 1985 Regional autonomy and the one and indivisible French republic. *Environment and Planning C: Government and Policy* **3**: 11–25
Kriedte P 1981 *Industrialization before industrialization: rural industry in the genesis of capitalism*. Cambridge University Press, Cambridge
Labasse J 1985 Décentralisation financière et métropoles régionales. *L'Information Géographique* **49**: 177–9
Laborie J P, Langumier J F, De Roo P 1985 *La politique française d'aménagement du territoire de 1950 à 1985*. La Documentation Française, Paris
Lancaster T D 1987 Comparative nationalism: the Basques in Spain and France. *European Journal of Political Research* **15**: 561–90
Landes D S 1966 The structure of enterprise in the nineteenth century. In Landes D S (ed) *The rise of capitalism*. Macmillan, New York: 99–111
Landes D S 1969 *The unbound Prometheus: technological change 1750 to the present*. Cambridge University Press, Cambridge
Landsberg H E 1981 *The urban climate*. Academic Press, New York (International geophysics series)
Laumesfeld D, Rispail M, Atamaniuk H n.d. Le françique. In Giordan H (ed) *Par les langues de France*. CNRS, Paris: 40–6
LeFrou C, Bremond R 1975 Inventory of the degree of surface water pollution in France. *Progress in Water Technology* **7**(2): 93–7
Liberatore A 1991 Problems of transnational policymaking: environmental policy in the European Community. *European Journal of Political Research* **19**: 281–305
Limouzin P 1980 Les facteurs de dynamisme des communes rurales françaises: méthodes d'analyse et résultats. *Annales de Géographie* **89**: 549–87
Locke W J 1916 *The wonderful year*. John Lane, London
Loos A 1925 *Gentlemen prefer blondes: the illuminating diary of a professional lady*. Harper's Bazaar, New York
Lowe P, Buller H 1990 The historical and cultural contexts. In Lowe P, Bodiguel M (eds) *Rural studies in Britain and France*. Belhaven, London: 3–20
Mahaney W C 1987 Lichen trimlines and weathering features as indicators of mass/balance changes and successive retreat stages of the Mer de Glace in the western Alps. *Zeitschrift für Geomorphologie* **31**: 411–18
Malanson G P, Trabaud L 1988 Computer simulations of fire behaviour in garrigue in southern France. *Applied Geography* **8**: 53–64
Manickam S, Barbaroux L, Ottmann F 1985 Composition and mineralogy of suspended sediment in the fluvio-estuarine zone of the Loire river, France. *Sedimentology* **32**: 721–41
Margat J 1982 Les ressources en eau du bassin méditerranéen. *Méditerranée* **45**: 15–28
Marre A 1987 Processus d'évolution de l'escarpement de tête du glissement de Rilly la Montagne (Marne). *Revue de Géomorphologie* **4**: 113–15
Martin G 1989 France and Africa. In Aldrich R, Connell J (eds) *France in world politics*. Routledge, London: 101–25
Martin J E 1984 Réalisation d'une carte des mouvements de terrain dans les Alpes Maritimes. *Méditerranée* **1–2**: 93–7
Massabuau J C, Fritz B, Burtin B 1987 Mise en évidence de ruisseaux

acides (pH≤5) dans les Vosges. *Comptes Rendus de l'Academie des Sciences* **305**(5): 121–4

Mendels F 1972 Proto-industrialization: the first phase of the industrialization process. *Journal of Economic History* **32**: 241–61

Mendras H 1964 *The vanishing peasant: innovation and change in French agriculture* trans J. Lerner. Massachusetts Institute of Technology Press, Cambridge

Mény Y 1985 French regions in the European Community. In Keating M, Jones B (eds) *Regions in the European Community*. Clarendon Press, Oxford: 191–203

Mény Y 1987 France: the construction and reconstruction of the centre, 1945–86. *West European Politics* **10**(4): 52–69

Merlin J 1986 La réhabilitation des quartiers anciens de Béziers. *Bulletin de la Société Languedocienne de Géographie* **20**: 453–67

Merlin P 1969 *Les villes nouvelles*. Presses Universitaires de France, Paris

Merlin P 1971 *L'exode rural*. Presses Universitaires de France, Paris

Merriman J M 1982 Introduction: images of the nineteenth-century French city. In Merriman J M (ed) *French cities in the nineteenth century*. Hutchinson, London: 11–41

Merriman J M 1991 *On the margins of city life: exploration of the French urban frontier 1815–1851*. Oxford University Press, Oxford

Meybeck M 1982 Carbon, nitrogen, and phosphorus transport by world rivers. *American Journal of Science* **282**: 401–50

Ministère de l'Agriculture 1984 *Statistique agricole*. Ministère de l'Agriculture, Paris

Mitra S 1988 The National Front in France – a single issue movement? *West European Politics* **11**(2): 47–64

Mori A, Begon J C, Duclos G, Studer R 1983 First approximation of a national land evaluation system (France). In Haans J C F M, Steur G G L, Heide G (eds) *Progress in Land Evaluation*. A A Balkema, Rotterdam: 43–55

Muxart T, Billard A, Cohen J, Cosandey M, Denefle M *et al*. 1985 Occupation agropastorale et érosion des versants depuis le moyen-age, d'après les analyses polliniques et sédimentologiques de la pseudo-tourbière de l'Airette (Lingas-Cévennes France). In Spencer T (ed) *International geomorphology 1985: abstracts of papers for the first international conference on geomorphology*. School of Geography, University of Manchester, Manchester: 432

Nadot R 1970 Evolution de la mortalité infantile endogène en France dans la deuxième moitié du XIX siècle. *Population* **1**: 49–58

Naylor E L 1985 *Socio-structural policy in French agriculture*. O'Dell Memorial Monograph **18**, Department of Geography, University of Aberdeen, Aberdeen

Nefussi J 1990 The French food industry since the 1950s. *Food Policy* **15**: 145–51

Neven D J 1990 EEC integration towards 1992: some distributional aspects. *Economic Policy* **10**: 13–62

Nicod J 1980 Les ressources en eau de la région Provence-Alpes-Côte d'Azur: importance et rôle des réserves souterraines. *Méditerranée* **2–3**: 23–34

Nilsson S, Duinker P 1987 The extent of forest decline in Europe: a synthesis of survey results. *Environment* **29**(3): 4–9, 30–1

Noin D 1987 *L'espace français*. Armand Colin, Paris

Noin D, Chauviré Y 1987 *La population de la France*. Masson, Paris

Noin D, Chauviré Y, Gardien C M, Globet F, Guillon M *et al.* 1984 *Atlas des Parisiens*. Masson, Paris

Nonn H 1989 Chronique alsacienne, 1988. *Revue Géographique de l'Est* **29**: 139–52

Oberhauser A M 1987 Labour, production and the state: decentralization of the French automobile industry. *Regional Studies* **21**: 445–58

Oberhauser A M 1990 State policy, employment and the spatial organization of production: the establishment of an automobile plant in the north of France. *Tijdschrift voor Economische en Sociale Geografie* **81**: 58–68

Oberhauser A M 1991 The international mobility of labor: north African migrant workers in France. *The Professional Geographer* **43**: 431–45

O'Brien P K, Keyder C 1978 *Economic growth in Britain and France 1780–1914: two paths to the twentieth century*. George Allen and Unwin, London

OECD (Organization for Economic Co-operation and Development) 1972 *Structural reform measures in agriculture*. OECD, Paris

OECD (Organization for Economic Co-operation and Development) 1991 *France*. OECD, Paris (OECD economic surveys)

Offord M 1990 *Varieties of contemporary French*. Macmillan, London

Ogden P E 1982 France adapts to immigration with difficulty. *Geographical Magazine* **52**: 318–23

Ogden P E, Huss M M 1982 Demography and pronatalism in France in the nineteenth and twentieth centuries. *Journal of Historical Geography* **8**: 283–98

Ogden P E, Winchester H P M 1986 France. In Findlay A M, White P E (eds) *West European population change*. Croom Helm, London: 119–41

Ogden P E, Winchester S W C 1975 The residential segregation of provincial migrants in Paris in 1911. *Transactions of the Institute of British Geographers* **65**: 29–44

Oke T R 1979 *Review of urban climatology 1973–1976*. World Meteorological Organization, Geneva

Padoa-Schioppa T, Emerson M, King M, Milleron J-C, Paclunch J *et al.* 1987 *Efficiency stability and equity: a strategy for the evolution of the economic system of the European Community*. Oxford University Press, Oxford

Pailhé J 1986 Naissances hors mariage. *Espace, Populations, Sociétés* **1986/2**: 57–64

Park C C 1987 *Acid rain: rhetoric and reality*. Methuen, London

Pautard J 1965 *Les disparités régionales dans la croissance de l'agriculture française*. Gauthier-Villars, Paris

Perrouxe 1970 Note on the concept of 'growth poles' trans L. Gates, A. M. McDermott. In McKee D L, Dean R D, Leahy W H (eds) *Regional economics: theory and practice*. Collier Macmillan, London: 93–103

Piganiol C 1989 Industrial relations and enterprise restructuring in France. *International Labour Review* **128**: 621–38

Pihan J 1979 Risques climatiques d'érosion hydrique des sols en France. In *Colloque sur l'érosion agricole des sols en milieu tempéré non méditerranéen.* Université Louis Pasteur, Strasbourg: 13–18
Pinchemel P 1969 *France: a geographical survey.* Bell and Sons, London
Pinchemel P 1980 *La France* (2 vols). Armand Colin, Paris
Pinchemel P 1986 *France – a geographical, social and economic survey.* Cambridge University Press, Cambridge
Ploegarts L 1986 Les villes nouvelles françaises et l'innovation urbaine, vingt ans après. *Canadian Geographer* **30**: 324–36
Pomerol C 1980 *Geology of France with twelve itineraries* trans A. Scarth. Masson, Paris (Guides géologiques régionaux)
Pottier B 1968 La situation linguistique en France. In Martinet A (ed) *Le langage.* La Pléiade, Bruges: 1144–61
Pottier C 1987 The location of high technology industries in France. In Breheny M J, McQuaid R W (eds) *The development of high technology industries: an international survey.* Croom Helm, London: 199–222
Pounds N J G 1954 France and 'les limites naturelles' from the seventeenth to the twentieth centuries. *Annals of the Association of American Geographers* **44**: 51–62
Pounds N J G 1979 *An historical geography of Europe 1500–1840.* Cambridge University Press, Cambridge
Pourcher G 1964 *Le peuplement de Paris.* Presses Universitaires de France, Paris
Pourcher G 1970 The growing population of Paris. In Jansen C J (ed) *Readings in the sociology of migration.* Pergamon, Oxford: 171–202
Preteceille E 1988 Decentralisation in France: new citizenship or restructuring hegemony? *European Journal of Political Research* **16**: 409–24
Price Waterhouse 1988 *The 'cost of non-Europe' in financial services.* Office for Official Publications of the European Communities, Luxembourg
Prinz B 1987 Causes of forest damage in Europe: major hypotheses and factors. *Environment* **29**(9): 10–15, 32–7
Probst J L 1985 Nitrogen and phosphorus exportation in the Garonne basin (France). *Journal of Hydrology* **76**: 281–305
Pumain D, St Julien T 1978 *Les dimensions du changement urbain.* CNRS, Paris
Pumain D, St Julien T 1984 Evolving structure of the French urban system. *Urban Geography* **5**: 308–25
Ramos A 1981 *Pasaporte Andaluz.* Planeta, Barcelona
Reece J E 1979 Internal colonialism: the case of Brittany. *Ethnic and Racial Studies* **2**: 275–92
Richez G 1983 Le parc naturel régional de la Corse onze ans après. *Méditerrannée* **1**: 35–44
Rickard P 1989 *A history of the French language* 2nd edn. Unwin Hyman, London
Risler J J, Vancon J P, Vigneron A 1986 Surveillance et gestion d'une nappe fortement solicitée, exemple de l'Alsace. *Hydrogéologie* **3**: 305–9
Rochefort C 1982 *Les petits-enfants du siècle.* Harrap, Bromley
Rollat A 1991 Challenge facing France's overseas départements. *Guardian Weekly* **144** (14 April): 15

Roose E J, Masson F 1985 Consequences of heavy mechanization and new rotation on runoff and on loessial soil degradation in northern France. In El-Swaify S A, Moldenhauer W C, Lo A (eds) *Soil erosion and conservation*. Soil Conservation Society of America, Ankery: 24–33

Safran W 1984 The French left and ethnic pluralism. *Ethnic and Racial Studies* **7**: 445–61

Safran W 1985 The Mitterrand regime and its policies of ethnocultural accommodation. *Comparative Politics* **18**: 41–63

Saint-Exupéry A de 1943 *The Little Prince* trans K Woods. Reynal and Hitchcock, New York

Sallnow J, Arlett S 1989 Green today gone tomorrow? *Geographical Magazine* **61**(11): 10–14

Salt J 1985 Europe's foreign labour migrants in transition. *Geography* **70**: 151–8

Sand P H 1987 Air pollution in Europe: international policy responses. *Environment* **29**(10): 16–20, 28

Saxenian A 1984 The urban contradictions of Silicon Valley: regional growth and the restructuring of the semiconductor industry. *International Journal of Urban and Regional Research* **7**: 237–62

Scargill D I 1990 French energy: the end of an era for coal. *Geography* **76**: 172–5

Scarth A 1983 The Villerest dam and the control of the Loire. *Geography* **68**: 335–8

Schain M A 1987 The National Front in France and the construction of political legitimacy. *West European Politics* **10**(2): 229–52

Sheridan G J 1979 Household and craft in an industrializing economy: the case of the silk weavers of Lyons. In Merriman J M (ed) *Consciousness and class experience in nineteenth-century Europe*. Holmes and Meier, New York: 107–28

Sheriff P 1980 Industry and commerce: problems of modernization. In Vaughan M, Kolinsky M, Sheriff P (eds) *Social change in France*. Martin Robertson, Oxford: 87–112

Singer D 1991 The resistible rise of Jean-Marie Le Pen. *Ethnic and Racial Studies* **14**: 368–81

Sjoberg G 1960 *The preindustrial city, past and present*. Free Press, New York

Smith A D 1991 *National identity*. Penguin, Harmondsworth

Smith F B 1984 The fate of airborne pollution. In The Watt Committee on Energy (eds) *Acid rain*. The Watt Committee on Energy Ltd, London (Report 14): 1–13

Smith K 1988 Avalanche hazards: the rising death toll. *Geography* **73**: 157–8

Soudan M 1975 Biological effects of industrial and domestic waste discharge in coastal waters. *Progress in Water Technology* **7**: 985–99

Sporton D, White P E 1989 Immigrants in social housing: integration or segregation in France? *The Planner* **75**: 28–31

Stevenson I 1980 The diffusion of disaster: the phylloxera outbreak in the département of Hérault, 1862–80. *Journal of Historical Geography* **6**: 47–63

Stevenson R L 1879 *Travels with a donkey in the Cevennes*. Kegan Paul, London

Story J 1990 Europe's future: western union or common home? In Crouch

C, Marquand D (eds) *The politics of 1992: beyond the European single market*. Blackwell, Oxford: 37–67
Stubbs M W 1977 Exceptional European weather in 1976. *Weather* **32**: 457–63
Sullerot E 1978 *La démographie de la France: bilan et perspectives*. La Documentation Française, Paris
Sutcliffe A 1981 *Towards the planned city*. Blackwell, Oxford
Szabolcs I 1974 *Salt affected soils in Europe*. Martinus Nijhoff, The Hague
Thody P M W 1982 Introduction. In Rochefort C *Les petits-enfants du siècle*. Harrap, Bromley: vii–xxiv
Thompson S A 1982 *Trends and developments in global natural disasters, 1947 to 1981*. Natural hazard research working papers **45**, Institute of Behavioural Science, University of Colorado, Colorado
Thumerelle P J, Momont J P 1988 Eléments pour une géographie des familles monoparentales en France. *Espace, Populations, Sociétés* **1988/1**: 128–32
Tilly C 1979 Did the cake of custom break? In Merriman J M (ed) *Consciousness and class experience in nineteenth-century Europe*. Holmes and Meier, New York: 17–44
Tipton F B, Aldrich R 1987 *An economic and social history of Europe from 1939 to the present*. Macmillan, London
Tufnell L 1984 *Glacier hazards*. Longman, London (Topics in applied geography)
Tuppen J 1983 *The economic geography of France*. Croom Helm, London
Tuppen J 1991 *Chirac's France*. Macmillan, London
Tuppen J, Mingret P 1986 Suburban malaise in French cities. *Town Planning Review* **57**: 187–201
Turnbull P 1976 *Corsica*. Batsford, London
UNHCR (United Nations High Commission for Refugees) 1983 Indochinese refugees. *Refugees Magazine* **3**: 9–16
Van de Walle E 1979 France. In Lee W R (ed) *European demography and economic growth*. Croom Helm, London: 123–43
Van Steijn H, De Ruig J, Hoozemans F 1988 Morphological and mechanical aspects of debris flows in parts of the French alps. *Zeitschrift für Geomorphologie* **32**: 143–61
Vesey F 1841 *Decline of the English language: the cause and probable consequences*. Saunders and Benning, London
Viers G 1984 Les inondations du 26 août 1983 en Pays Basque (Pyrénées-Atlantiques, France). *Annales de Géographie* **93**: 372–6
Vigouroux P, Vançon J P, Drogue C 1983 Conception d'un modèle de propagation de pollution en nappe aquifère – exemple d'application à la nappe du Rhin. *Journal of Hydrology* **64**: 267–79
Voltaire 1805 *Voltairiana* trans M. J. Young. Hughes, London
Vuddamalay V, White P E, Sporton D 1991 The evolution of the Goutte d'Or as an ethnic minority district of Paris. *New Community* **17**: 245–58
Wackermann G (ed) 1992 *La France dans le monde*. Nathan, Paris
Wallerstein I 1961 *Africa the politics of independence: an interpretation of modern African history*. Vintage Books, New York
Waltham A C 1978 *Catastrophe: the violent earth*. Macmillan, London

Ward R 1978 *Floods: a geographical perspective.* Macmillan, London (Focal problems in geography)
Weber E 1977 *Peasants into Frenchmen.* Chatto and Windus, London
White P E 1982 The structure and evolution of rural populations at the sub-parochial level: post-war evidence from Normandy, France. *Études Rurales* **86**: 56–75
White P E 1984 *The west European city.* Longman, London
White P E 1986 International migration in the 1970s: revolution or evolution? In Findlay A M, White P E (eds) *West European population change.* Croom Helm, London: 50–80
White P E 1991 Games and circuses: the state and the creation of spectacle in Paris. Paper presented at Institute of British Geographers Annual Conference, Sheffield
White P E, Winchester H P M 1991 The poor in the inner city: stability and change in two Parisian neighbourhoods. *Urban Geography* **12**: 35–54
White P E, Winchester H P M, Guillon M 1987 South-east Asian refugees in Paris: the evolution of a minority community. *Ethnic and Racial Studies* **10**: 48–61
White P E, Woods R I 1983 Migration and the formation of ethnic minorities. *Journal of Biosocial Sciences Supplement* **8**: 7–22
Whittlesey D 1939 *The earth and the state: a study of political geography.* Holt, Rinehart and Winston, London
Wilkes D, Donoghue P, Sutton C 1987 Marne-la-Vallée a new town or a dormitory suburb for Paris? *Built Environment* **9**: 255–65
Winchester H P M 1977 *Changing patterns of French internal migration 1891–1968.* Research Paper **17**, School of Geography, University of Oxford, Oxford
Winchester H P M 1986 Agricultural change and population movements in France 1892–1929. *Agricultural History Review* **34**: 60–78
Winchester H P M 1989 The structure and impact of the post war rural revival: Isère. In White P E, Ogden P E (eds) *Migrants in modern France: population mobility in the later 19th and 20th centuries.* Unwin Hyman, London: 142–59
Winchester H P M, Ilbery B W 1988 *Agricultural change: France and the EEC.* John Murray, London (Case studies in the developed world)
Winchester H P M, Ogden P E 1989 France. In Champion A G (ed) *Counterurbanization: the changing pace and nature of population deconcentration.* Edward Arnold, London: 162–86
Winchester H P M, White P E 1988 The location of marginalised groups in the inner city. *Environment and Planning D: Society and Space* **6**: 37–54
Wise M 1989 France and European unity. In Aldrich R, Connell J (eds) *France in world politics.* Routledge, London: 35–73
Wistrich E 1989 *After 1992: the united states of Europe.* Routledge, London
Woillard G M 1978 Grande Pile peat bog: a continuous pollen record for the last 140,000 years. *Quaternary Research* **9**: 1–21
Wolkowitsch M 1989 L'assurance en France. *Annales de Géographie* **550**: 646–67
Womack J R, Jones D J, Roos D 1990 *The machine that changed the world.* Rawson Associates, New York

Wrathall J E 1985a The hazard of forest fires in southern France. *Disasters* **9**: 104–14
Wrathall J E 1985b The Mistral and forest fires in Provence-Côte d'Azur, southern France. *Weather* **40**: 119–24
Wright L W, Wanstall P J 1977 *The vegetation of mediterranean France: a review*. Occasional Paper **9**, Department of Geography, Queen Mary College, University of London, London
Wrigley E A 1985 The fall of marital fertility in nineteenth-century France: exemplar or exception? *European Journal of Population* **1**: 31–60
Zeldin T 1973 *France, 1848–1945* (3 vols). Clarendon Press, Oxford
Zeldin T 1983 *The French*. Collins, London
Zukin S 1985 The regional challenge to French industrial policy. *International Journal of Urban and Regional Research* **9**: 352–67

Index